THE VOLGA

YALE UNIVERSITY PRESS
NEW HAVEN AND LONDON

THE VOLGA

JANET M. HARTLEY

A HISTORY OF RUSSIA'S GREATEST RIVER

For information about this and other Yale University Press publications, please contact:
U.S. Office: sales.press@yale.edu   yalebooks.com
Europe Office: sales@yaleup.co.uk   yalebooks.co.uk

Set in Adobe Garamond Pro by IDSUK (DataConnection) Ltd
Printed in Great Britain by TJ Books, Padstow, Cornwall

Library of Congress Control Number: 2020942541

ISBN 978-0-300-24564-6

A catalogue record for this book is available from the British Library.

10 9 8 7 6 5 4 3 2 1

This book is dedicated to my husband, Will Ryan

CONTENTS

CONTENTS

PART 3: THE VOLGA IN THE RUSSIAN EMPIRE: LIFE AND IDENTITY ON THE RIVER

PART 4: SOVIET AND POST-SOVIET VOLGA: CONFLICT, IDENTITY AND MANAGING THE RIVER

ILLUSTRATIONS

PLATES

ILLUSTRATIONS

MAPS

EDITORIAL NOTE

Transliteration of Russian in the notes and the bibliography is according to the Library of Congress system. The transliteration has been amended in the main text: 'soft signs' have been omitted; Russian surnames ending in 'ii' and 'yi' have been standardized to 'y'.

Russian names – forenames, patronymics and surnames – have for the most part been given in this system (for example, Nikolai Ivanovich Ilminsky), except in the case of tsars, where the anglicized form of the name has been used when it is more familiar (for example, Ivan III, but Peter I, Catherine II, Alexander I, Nicholas II), and where there is a commonly used English spelling of prominent Russians, such as literary figures (for example, Tolstoy and not Tolstoi, Alexander Pushkin and not Aleksandr).

The words 'Russia' and 'Russians' are controversial in the medieval period. I have used the word Rus in the medieval period for the east Slav peoples and for the names of the principalities which were located in what is present-day European Russia and part of present-day Ukraine. Ivan IV took the title 'tsar of all Russia' at his coronation in 1547; after this date, I refer to the principality of Moscow as 'Russia' and to the population of the principality as 'Russians'.

Notes referencing the archives in Russia use Russian descriptors: *fond, opis', tom, delo* and *listy* (abbreviated to l. or ll.).

The *Polnoe sobranie zakonov Rossiiskoi Imperii*, first and second series, is abbreviated in the text to *PSZ*, the *Sbornik Imperatorskogo Russkogo istoricheskogo obshchestva* is abbreviated to *SIRIO*.

Distances and weights are given in metric measures; Russian weights and measures are explained in the text.

From January 1700 to February 1918, Russia used the Julian calendar, which lagged behind the Gregorian calendar used in the West by 10 days in the seventeenth

EDITORIAL NOTE

century, 11 days in the eighteenth century, 12 days in the nineteenth century and 13 days in the twentieth century until February 1918. Dates in this book before February 1918 are given in the Julian form (that is, the February and October revolutions in 1917, not the March and November revolutions). Dates after February 1918 follow the Gregorian calendar.

ACKNOWLEDGEMENTS AND THANKS

I owe thanks to many kind colleagues and friends. The text has been much improved by suggestions for further reading from Liisa Byckling in Helsinki, Leah Bushkanets in Kazan, Andrei Stepanov in St Petersburg and Aleksandr Kiselev in Volgograd, by the anonymous readers for Yale (both of the proposal and of the text) and by the meticulous copy editor. I have been helped in the identification and reproduction of appropriate archival materials by Iaroslav Golubinov in Samara, Viktor Kulikov in Iaroslavl and Mikhail Pavlenko in Astrakhan. In particular, I should like to thank Chulpan Samatova in Kazan, whose enthusiasm and scholarly dedication resulted in far richer archival material in Kazan than I could possibly have found by myself. Kind friends read all or part of the text of the book and made excellent suggestions for improvements. I am very grateful to them all: Professors Simon Franklin, Tomila Lankina, Dominic Lieven, Margot Light, Peter Waldron. My research outside the United Kingdom was conducted in Moscow and St Petersburg, but also in Kazan, where I received enormous hospitality from members of the Federal University. In September 2018, Viktor Kulikov took me and my husband on a wonderful and unforgettable trip from Tver to the source of the Volga. My project also benefited from financial support from the Paulsen programme at the London School of Economics and Political Science. This book is dedicated to my husband, Will Ryan, not only for his editing precision, but also for his constant support and enthusiasm – not only for this project, but for all my work on Russian history.

A NOTE ON PLACE NAMES

A considerable number of the towns cited in the text, mostly on the Volga, changed names at various times (usually, but not always, in the Soviet period). The following places appear more than once in the text. They are listed below in alphabetical order by their current name, but the name used most commonly in the text is highlighted in **bold**.

DIMITROVGRAD - **MELEKESS** (1714–1972), DIMITROVGRAD (1972–)

A village in Simbirsk/Ulianovsk province, situated at the confluence of the river Melekesska and the Bolshoi Cheremshan, which is a tributary of the Volga. Melekess was named after the river; it was renamed in 1972 to honour Georgi Dimitrov, the Bulgarian revolutionary and first leader of the communist People's Republic of Bulgaria. It has retained this name.

ENGELS - POKROVSKAIA SLOBODA (1747–1914), ALSO KNOWN AS KOSAKENSTADT, **POKROVSK** (1914–31), ENGELS (1931–)

Engels is in Saratov province, on the left bank (eastern side) of the Volga, opposite the town of Saratov. The town was founded as Pokrovskaia Sloboda (*sloboda* means settlement in this context). It was given town status as Pokrovsk in 1914, and in 1918 became the capital of the short-lived German Autonomous Soviet Socialist Republic. The town was also commonly known as Kosakenstadt ('Cossack town' in German) from the eighteenth century. In 1931, it was renamed Engels in honour of Friedrich Engels, and that name has been retained.

A NOTE ON PLACE NAMES

IOSHKAR-OLA – **TSAREVOKOKSHAISK** (1584–1919), KRASNOKOKSHAISK (1919–27), IOSHKAR-OLA (1927–)

The town is situated on the river Malaia Kokshaga. It was founded in 1584 after the death of Ivan IV as the 'tsar's town on the river Kokshaga', on territory which had been within the khanate of Kazan. It was renamed Krasnokokshaisk, the 'town on the red Kokshaga river'. The old Mari name for the town was Charla. The town was renamed in 1927 and means 'red city' in Mari. It is now the capital of the Mari El Republic, a federal republic within the Russian Federation, which borders the river Volga in the west.

NIZHNII NOVGOROD – **NIZHNII NOVGOROD** (1221–1932), GORKII (1932–90), NIZHNII NOVGOROD (1990–)

The town was founded in 1221 by Prince Iurii of Vladimir at the confluence of the Volga and Oka rivers. Between 1932 and 1990 it was named after the writer Maksim Gorky (the pen name of Aleksei Peshkov), who was born in Nizhnii Novgorod in 1868 and located some of his stories there. The name reverted to Nizhnii Novgorod in 1990.

ORAL – **IAITSK** (1613–1775), URALSK (1775–1991), ORAL (1991–)

The town was founded in 1613 as a Cossack fort on the river Iaik. After the suppression of the Pugachev revolt, the town was renamed Uralsk and the river was renamed the Ural. The town is now in the independent state of Kazakhstan and was renamed Oral in 1991.

ORENBURG – **ORENBURG** (1734–1938), CHKALOV (1938–57), ORENBURG (1957–)

The town was founded in 1734, although its location changed over the next decade, at the confluence of the Ural (formerly Iaik) and Samara rivers. It was chosen as the location of the Muslim Spiritual Assembly from 1788 until 1917 and had jurisdiction over Muslims in Kazan and elsewhere on the Volga (and in Siberia). It was capital of the short-lived Kirgiz Autonomous Soviet Socialist Republic from 1920 to 1925 and then became part of the Kazakh Soviet Socialist Republic. The city was renamed Chkalov from 1938 to 1957 in honour of the test pilot Valerii Chkalov.

A NOTE ON PLACE NAMES

RYBINSK – UST-SHEKSNA (*C.* 1071–1504), RYBNAIA SLOBODA (1504–1777), RYBINSK (1777–1946), SHCHERBAKOV (1946–57), **RYBINSK** (1957–84), ANDROPOV (1984–89), RYBINSK (1989–)

Rybinsk wins the award for the most name changes! It was founded as the settlement of Ust-Sheksna (meaning 'at the mouth of the river Sheksna'). From 1504, it was known as Rybnaia Sloboda ('the fishing settlement') until 1777, when it was renamed Rybinsk and re-categorized as a town because of its important location for transporting goods from the Volga to the canal system which linked the Volga with St Petersburg. From 1946 to 1957, Rybinsk was named in honour of Aleksandr Shcherbakov, a Soviet politician, and then between 1984 and 1989 in honour of Iurii Andropov, general secretary of the Communist Party from 1982 until his death in 1984 (he was educated at the Rybinsk Water Transport Technical College and was at one time secretary of the Young Communist League at the College and then at Rybinsk shipyard). The town reverted to the name Rybinsk in 1989. I have never seen an account of the costs of such frequent name changes.

ST PETERSBURG – **ST PETERSBURG** (1703–1914), PETROGRAD (1914–24), LENINGRAD (1924–91), ST PETERSBURG (1991–)

St Petersburg (Sankt Peterburg in Russian; its first name was Sankt-Piter-Burkh, which is in origin a Dutch name) was founded by Peter I in 1703, on the site of the small Swedish fort of Nyenskans. In September 1914, after the outbreak of the First World War, the city was renamed Petrograd ('Peter's town') to remove the German-sounding words of 'Sankt' and 'Burg'. In 1924, it was renamed Leningrad in honour of Vladimir Lenin, five days after Lenin's death. It reverted to St Petersburg in 1991.

SAMARA – **SAMARA** (1586–1935), KUIBYSHEV (1935–91), SAMARA (1991–)

Samara was founded in 1586 on the left bank (eastern side) of the river Samara and named after it; the city is located at the confluence of the Samara and Volga rivers. It was named Kuibyshev from 1935 to 1991 in honour of Valerian Kuibyshev, Bolshevik revolutionary and Soviet politician, who was president of the Samara soviet in 1917 and chaired the revolutionary committee of Samara province during the Civil War. The name reverted to Samara in 1991.

A NOTE ON PLACE NAMES

TOLIATTI (OR TOGLIATTI) - **STAVROPOL** (1737–1964), TOLIATTI (1964–)

Stavropol ('the city of the cross') was founded in 1737 as a fortress for Buddhist Kalmyks who had converted to Orthodoxy. It is on the left bank (eastern side) of the river Volga in Samara province, and was sometimes referred to as Stavropol-on-Volga to distinguish it from the town of the same name in south-west Russia. The town had to be built on a new site after the construction of the Kuibyshev dam and hydroelectric station which created the Kuibyshev reservoir and which flooded the original town. The new town was called Toliatti (or Togliatti) in honour of Palmiro Togliatti, the secretary of the Italian Communist Party from 1927 to 1964, in part because it was home to an enormous factory producing Lada cars, which originally received technical assistance from Fiat. It has retained the name Toliatti.

TVER - **TVER** (C. 1135–1931), KALININ (1931–90), TVER (1990–)

Tver was founded at the confluence of the Volga and Tvertsa rivers in around 1135; it is located on both banks of the Volga. It was renamed Kalinin between 1931 and 1990 in honour of Mikhail Kalinin, Bolshevik revolutionary and Soviet politician, who was born in a village in Tver province. The name reverted to Tver in 1991.

ULIANOVSK - **SIMBIRSK** (1648–1924), ULIANOVSK (1924–)

The town was founded in 1648 on the right bank (western side) of the river Volga. In 1924, the town was renamed Ulianovsk in honour of Lenin (Vladimir Ilich Ulianov), who was born there in 1870 and lived there until he enrolled in Kazan University in 1887. The town has retained the name Ulianovsk.

VOLGOGRAD - **TSARITSYN** (C. 1589–1925), STALINGRAD (1925–61), VOLGOGRAD (1961–)

Tsaritsyn was founded at the confluence of the Volga and Tsaritsa rivers (hence the name Tsaritsyn, which has nothing to do with tsars – it means 'yellow water' or 'yellow river' in Tatar). It was renamed Stalingrad in 1925, in recognition of the role Joseph Stalin allegedly played in the defence of the town against the Whites in the Civil War. It was renamed Volgograd in 1961 as part of the de-Stalinization process inaugurated by Nikita Khrushchev after Stalin's death. The town has retained the name Volgograd, although there have been calls for it to revert to Stalingrad.

TIMELINE

TSARS OF RUSSIA

Ivan IV (1547–84)

Fedor I (1584–98)

Boris Godunov (1598–1605)

Time of Troubles pretenders (1605–12)

Michael (1613–45)

Alexis I (1645–76)

Fedor III (1676–82)

Sophia (regent) (1682–89)

Ivan V (ruled with Peter I) (1682–96)

Peter I (1682–1725)

Catherine I (1725–27)

Peter II (1727–30)

Anna (1730–40)

Ivan VI (1740–41)

Elizabeth (1741–61)

Peter III (1761–62)

Catherine II (1762–96)

Paul I (1796–1801)

Alexander I (1801–25)

Nicholas I (1825–55)

Alexander II (1855–81)

Alexander III (1881–94)

Nicholas II (1894–1917)

TIMELINE

LEADERS OF THE SOVIET UNION

Vladimir Lenin (1922–24)
Joseph Stalin (1924–53)
Georgii Malenkov (1953)
Nikita Khrushchev (1953–64)
Leonid Brezhnev (1964–82)
Iurii Andropov (1982–84)
Konstantin Chernenko (1984–85)
Mikhail Gorbachev (1985–91)

PRESIDENTS OF THE RUSSIAN FEDERATION

Boris Yeltsin [Eltsin] (1991–99)
Vladimir Putin (2000–08)
Dmitrii Medvedev (2008–12)
Vladimir Putin (2012–)

INTRODUCTION

The VOLGA! There is a mystery, a charm, in all mighty rivers, which has ever made us gaze upon them with an interest beyond that inspired by other great and glorious sights; but to look on the largest of the European rivers – the king of our fair tides and oft-sung streams – gave a thrill of joy surpassing all former pleasure of the kind. Those who know that the first glimpse of some great object which we have read, or dreamt of from earliest recollection, is ever a moment of intense enjoyment, will forgive the foolish transport that is felt while first standing on that commanding height, and devouring the majestic stream that rolls in such gloomy grandeur below.[1]

The comment is by Robert Bremner, surveying the river from the town of Nizhnii Novgorod in the 1830s. His awe at the size of the Volga, which he noted was four times the size of the river Thames at Blackfriars Bridge,[2] has been echoed by many others – Russians and non-Russians – over the centuries.

It is not hard to see where this fascination with the river comes from. The Volga is one of the great rivers of the world, along with the Amazon, the Nile and the Mississippi. At 3,530 kilometres (2,193 miles), it is the longest river in Europe, flowing from north-west of Moscow to the Caspian Sea, through the forest zone of northern European Russia to the steppe and then to arid semi-desert in the south of Russia.

The Volga river basin is the largest in Europe and, along with its major tributaries (including the river Oka at Nizhnii Novgorod and the Kama south of Kazan), provides water and drainage for the most densely populated region of Russia west of the Urals, including the Moscow region. The very simplicity of the name given to the river by the various peoples who lived on its shores possibly signifies that to them it is *the* river and not just *a* river. The word 'Volga' is derived from Slavic words for 'wetness' or 'moisture'. The river was known by Turkic peoples who lived

1

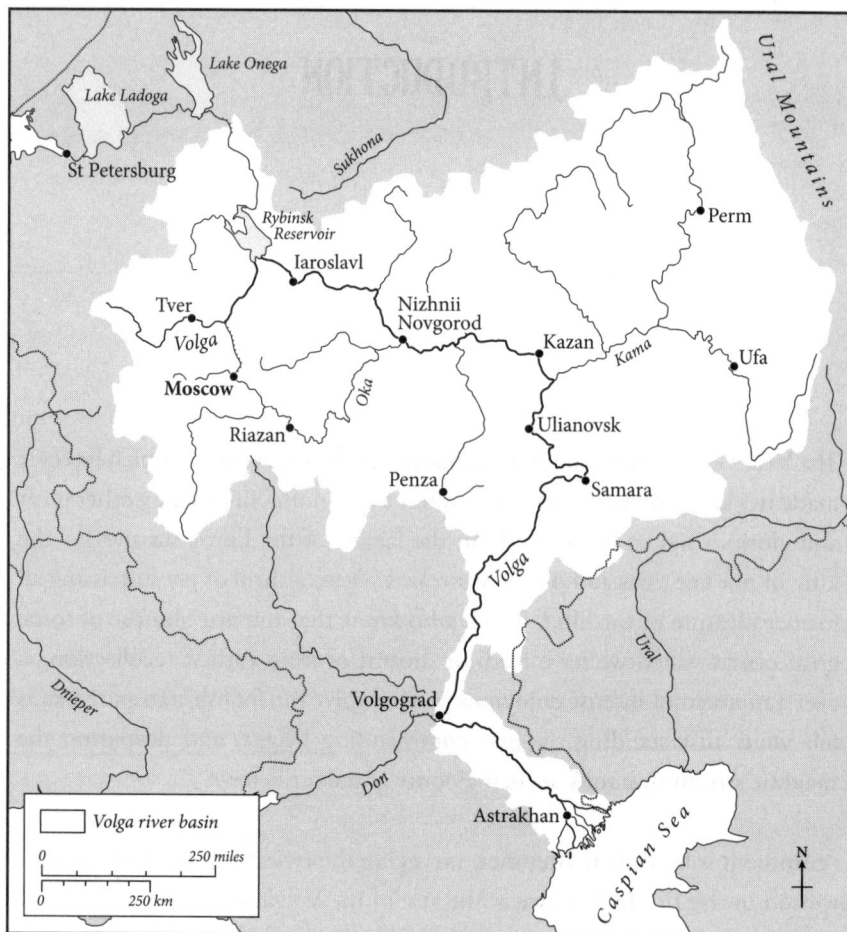

1. The River Volga

along it as Itil (or Itel or Atil; Idel in modern Tatar and Adӑl in modern Chuvash), which means 'big river'; the Mordvin name for the river, Rav, derives from the earlier Scythian word *Rā* which means 'wetness'.

The Volga flows through the heart of what is now European Russia, and significant towns were built on its banks. The source of the river is in the Valdai hills in Tver province, north-west of Moscow. It is simply a small pool on the edge of a wood leading to a stream. The site is of great charm, because a new wooden chapel has been built over the source in the style of the original chapel that was constructed in the late nineteenth century. Two plaques on the wall note that the waters were blessed by the patriarch of Moscow and of all Russia in 1995 and 2017. The whole site is peaceful and tranquil, with a few stalls selling souvenirs and dried fish. On a meadow sloping down to the source is a late-nineteenth-century church, which has been restored after neglect in the Soviet period and damage during the Second World War, when the region was occupied by German troops.

The river passes through a series of rapids and then broadens out at Rzhev, a town which became a trading centre and was also the site of a major battle in the Second World War. The river then becomes a major artery for the movement of goods and people, as it flows through the ancient town of Tver, and then north and east to Rybinsk, which became a key transit port for the Volga trade in the eighteenth century, after the construction of the canal system which transported goods from the Volga to St Petersburg, and which is now the location of a major hydroelectric power station constructed in the early Soviet period. The enormous statue of a woman, *Mother Volga*, at the Rybinsk hydroelectric power station stretches out her hand to the reservoir and has Lenin's slogan of 1920 engraved on its base: 'Communism is Soviet Power plus Electrification of the Whole Country'.

The river continues south-east through the important commercial towns of Iaroslavl and Kostroma. It then flows further east through Nizhnii Novgorod (just over 400km east of Moscow), which developed as the major trade fair in Imperial Russia in the second half of the nineteenth century, primarily because of its position at the confluence of the Volga and Oka rivers. When historians and geographers talk about the 'upper Volga', they mean the stretch from the source to at least Rybinsk, but normally also include Iaroslavl and Kostroma, down to Nizhnii Novgorod. However, the definition is never very precise, and Nizhnii Novgorod is regarded as being on the boundaries of the upper and middle Volga.

The Volga then flows from Nizhnii Novgorod through Cheboksary to Kazan. Cheboksary was a minor town in the eighteenth and nineteenth centuries (although home to a few major merchant families), but is now capital of the Chuvash Republic. The town of Kazan was the capital of the Kazan khanate until its conquest by Ivan IV in 1552. It remained an important trading, administrative and cultural centre throughout the imperial and Soviet periods, and became the capital of Tatarstan within first the Soviet Union, and now the Russian Federation. In April 2009, Kazan was officially allowed to brand itself by the Russian patent office as the 'third capital' of Russia (after Moscow and St Petersburg). Kazan by road is just over 800km east of Moscow. South of Kazan, on the eastern bank of the river Volga, are the archaeological remains of the city of Bolgar, the capital of the Volga-Bolgar state from the eighth to the thirteenth centuries.

Further south of Kazan is the port of Ulianovsk (formerly Simbirsk, but renamed in 1924 because it was the family name of Lenin and the place of his birth). The river then turns east via the 'great bend', which was the haunt of pirates until well into the nineteenth century. The route eastwards to Samara passes the major industrial town of Toliatti, known in the Soviet period for its massive Lada car production plant and hydroelectric power station; the reservoir for the power station was created by flooding the small town of Stavropol (literally the 'city of the cross'), which was founded in 1737 as a military base for Buddhist Kalmyks who had converted to Christianity. Samara lies at the confluence of the Samara and Volga rivers, over 1,000km by road from Moscow. To the south and east of Samara is the steppe, which stretches, almost treeless, in both directions. The town was founded as a Russian fort in 1586 and became an important commercial and trading centre in the eighteenth and nineteenth centuries. It played a significant part in the Russian Civil War (1918–22), when for a brief time it was a rival seat of government to the Bolsheviks; later, in 1941, it was earmarked as the alternative capital for the Soviet Union if Moscow fell to the German army. The stretch of the river from Kazan to Samara is normally designated as the 'middle Volga', although Samara is considered to be on the border between the middle and lower Volga.

The river then turns south and west to Saratov. Founded as a Russian fort in 1590, it not only developed as an important commercial and administrative centre in the eighteenth and nineteenth centuries, but was also the administrative centre of the German colonists, who settled mainly in Saratov province at the invitation of Catherine II in the 1760s. At least some of the land in Saratov province is rich black-earth soil and was the location of large noble estates. South of Saratov is the town of Volgograd (originally Tsaritsyn in the imperial period, and then Stalingrad from 1925 to 1961). Volgograd is at the confluence of the Volga and the Tsaritsa rivers (hence its original name). It was founded as a Russian fort in 1589, but became a 'boom town' in the late nineteenth century after the railways linked it with the river Don and the heartland of Russia. It was the site of the bloodiest battle of the Second World War (Stalingrad) – perhaps the bloodiest recorded battle of all time, now commemorated in a massive memorial complex.

The land on the lower Volga is poor, particularly on the eastern side of the river, and the region is subjected to hot, dry winds from the east which have been known to cause drought and destroy harvests. Not far south of Volgograd is the site of the town of Sarai, which was the capital of the Golden Horde in the thirteenth and fourteenth centuries, and was itself built on or near the site of the town of Itil, which was the capital of Khazaria from the mid-seventh century to the end of the tenth.

Further south is the port of Astrakhan, 1,393km by road from Moscow. It is sometimes included as part of the 'lower Volga' region and sometimes categorized as a separate delta region. For about a century, Astrakhan was the capital of the Astrakhan khanate, until it was conquered by Ivan IV in 1556. The Volga discharges into the Caspian Sea, an inland sea, through the largest river delta in Europe, estimated today at over 27,224 square kilometres, stretching some 160km and encompassing as many as 500 smaller rivers and channels. The delta was a rich source of fish, including the beluga sturgeon from which caviar is taken. It suffered serious ecological damage in the twentieth century.

❖

The river Volga has played a crucial role in the history of the people who are now part of the Russian Federation. The river was particularly important in the following areas:

1. For trade and commerce;
2. As a meeting place of different peoples, ethnicities, religions and cultures;
3. In conflict and empire/Soviet state building;
4. For the evolution of Russian and non-Russian culture and identity.

The Volga both united and divided the lands through which it flowed. Trade and commerce along this waterway united the country south to north, north to south and east to west, even though it was frozen for five or six months of the year (its importance was accentuated by the poor quality of the roads until recent times); but competition over trade also led to conflict between the early 'states' on the river. The location of key cities in the Khazar and Bolgar states was determined by the river and by commerce. Vikings sailed down the Volga from the Baltic Sea and from settlements near Lake Ladoga, attracted in particular by silver coins from the east. They brought furs and slaves to the markets of the Khazars and the Bolgars, who in return traded goods from the east and the south. The arrival of the Mongols in the thirteenth century devastated many towns, but the Golden Horde – and its successor khanates in Kazan and Astrakhan – also traded actively on the river Volga. Ivan IV established control over the length of the river after his conquest of Kazan and Astrakhan, and the river Volga then became the key artery for trade within the Russian Empire.

The river shaped the patterns of trade and exchange in Imperial Russia and the Soviet Union, and is still of great importance today. Major canal building from the eighteenth century onwards ensured that the waterway became the major route for moving vital products such as grain, fish, timber and salt from the lower and middle Volga region to the towns of the north and the Baltic ports, and from there to other ports in Europe. Goods were transported from Persia/Iran and India via the Caspian Sea and Astrakhan, and from China, Central Asia and Siberia via Volga towns such as Samara, Kazan and Nizhnii Novgorod. The construction of railways from the second half of the nineteenth century changed the trade routes, but to an extent created a new divide between the right and left banks of the river below Nizhnii Novgorod, until the construction of major rail (and then road) bridges across the river. The development of commerce and ports on the Volga – and indeed the river's role in the economy of the country – was determined in part by the initiatives taken by tsars, Soviet and post-Soviet leaders to stimulate modernization and industrialization from the eighteenth century to the present. Today the flow of the Volga is controlled by dams and reservoirs, from Rybinsk to just south of Volgograd. It was intended that this would develop the economy of the whole of European Russia, but it has also had severe environmental consequences.

Although the river Volga was never the geographical border between Asia and Europe, in many ways the middle and lower Volga does draw a line between the Christian, Russian, European West and the Islamic and Asiatic East. 'I am in Asia', declared Catherine II in a letter to Voltaire from Kazan in 1767.[3] The topography of the land on either side of the middle and lower Volga accentuates this feeling of a divide: the land on the western (right) bank is hillier, more cultivated and richer in vegetation; the land on the eastern (left) bank is low-lying, flat, mostly scrub land, stretching to the borders of Kazakhstan. Certainly, the German settlers who had been encouraged to come to the Volga region by Catherine II in the 1760s felt that those who had been granted land on the western 'mountain side' in Saratov province were fortunate – not only because the soil was better than on the eastern 'meadow' side, but also because they were not so exposed to dangerous raids by Kalmyk and Nogai horsemen. In their eyes, the river was a dividing line between European civilization and Asian barbarism.

However, the Volga was not simply an east–west, Europe–Asia divide. It was also a meeting place and (to an extent) a melting pot of many different peoples within, first, the multi-ethnic and multi-confessional Russian Empire, and then the Soviet Union. Astrakhan became home to Armenian, Persian and Indian traders, all of whom had their own quarters, religious and trading buildings and their own institutions. Kazan is today a Tatar and a Russian city; the countryside to the north, east and west of Kazan is inhabited by Russians, Turkic and Finno-Ugric peoples. In the Soviet period, autonomous republics were founded for non-Russian peoples in the 1920s within the Russian Soviet Federative Socialist Republic, and these still exist within the Russian Federation today (with the exception of the short-lived Volga German Autonomous Republic). All the republics, however, contain a significant ethnic Russian population and are important today for the forging of identities in a new Russia.

The religious composition on the Volga is complex. Finno-Ugric settlers originally followed shamanistic beliefs, although many converted, at least nominally, to Orthodoxy after they became subjects of the Russian Empire. The ruler and the elite in Khazaria probably converted to Judaism sometime in the early ninth century. Kalmyks in the south and south-east of the Volga were Buddhists (the only Buddhists in Europe). The Bolgar state, the Golden Horde and the khanates of Kazan and Astrakhan were, or became, Muslim. The Russian and Soviet states were conscious of the potential threat of Islam in the Volga region from the time of the conquest of Kazan in 1552. The history of the Volga is, in part, the history of (often forced) conversion to Orthodoxy by the Russian government and the reaction to this of the local inhabitants. In many cases, the conversion process was incomplete or, in the case of Islam, could be reversed. The remoteness of much of the Volga countryside attracted Old Believers – that is, schismatics from the Russian Orthodox Church who did not accept the changes in liturgy and practice in the middle of the seventeenth century. Eighteenth-century German settlers could be Catholic or Protestant. Settlements on the river Volga were a microcosm of the ethnic and cultural complexity of the Russian Empire and the Soviet state, and this study will examine the relationships between different groups of people on the Volga, and between non-Russians and the government.

The river Volga played a key role in the creation and evolution of early 'states', the Russian Empire and the Soviet Union. The importance of trade and strategic access to the Volga and the Caspian Sea led to rivalry and conflict between Khazaria and the Bolgars and Kievan Rus; and then between the Golden Horde (and its successor khanate in Kazan) and Moscow and other Rus principalities in what is now European Russia. After the conquest of Kazan and Astrakhan, the tsars of Russia took the title 'tsar of Kazan and Astrakhan' (and Siberia), as well as 'tsar of all Russia'. From this time on, Russia can be considered an 'empire', although the western title of 'emperor', rather than tsar, was only taken by Peter I in 1721.[4] The middle and lower Volga regions were the first significant non-Russian and non-Christian lands where the Russian Empire had to establish and exercise control. In many ways, they provided a testing ground, and then a model, for imperial (and to an extent Soviet) policies towards non-Russian peoples.

THE VOLGA

The major Cossack revolts of the late seventeenth and eighteenth centuries were in effect Volga revolts, as the rebel armies of Stenka Razin and Emelian Pugachev sailed up and down the river and sacked important Volga towns. Settlements on both banks of the Volga were ransacked by nomadic tribesmen from the east and the south. In the late nineteenth and early twentieth centuries, the region was beset by rural and urban resistance and revolt. The imperial government and the Soviet state responded by suppressing rebellious subjects and strengthening their administrative control, but also by intensifying their cultural and educational presence in the region. Imperial Russia and the Soviet Union were to an extent shaped by the experiences of ruling over lands acquired in the sixteenth century on the Volga.

The Volga became a crucial point of conflict in the twentieth century, and was central to the establishment and survival of the Soviet state. The river, and several key Volga towns, played a crucial role in determining the outcome of the Russian Civil War in 1918–22. Samara became, briefly, the centre of resistance to the new Soviet state, and towns on the middle and lower Volga were of vital strategic significance for both Whites and Reds. The failure of the White armies to join up on the river Volga largely determined the outcome of the Civil War. In the Second World War, the battle of Stalingrad was pivotal in the defeat of the German army and the survival of the Soviet Union. Both at the time and since it has been projected as the greatest of all 'patriotic' sacrifices made by the Soviet people. The river Volga at Stalingrad was regarded in 1942–43 as the key boundary, the Rubicon, which the Germans must not cross. Part of the enormous memorial complex of the battle shows German soldiers only crossing the river as defeated and demoralized prisoners of war.

Finally, the Volga became the subject of poetry, literature and art, and helped shape a sense of Russian identity through a shared experience of the river. Late-eighteenth-century odes to Catherine II both glorified the river and also 'tamed' it to honour the ruler. In the nineteenth century, writers, artists and tourists fully 'discovered' the Volga as something unique and special to Russia and all Russians. The Volga became 'Mother Volga' and the protector of the Russian people. The 'gloomy grandeur' of the Volga, as Bremner put it in the opening quotation of this chapter, was also, however, a common theme in Russian poetry, literature and art. The famous painting by Ilia Repin, *Barge Haulers on the Volga*, uses this image to depict the exploitation and suffering of ordinary people in late tsarist Russia. The battle of Stalingrad reinforced the special importance of the river as a barrier that protected the Soviet state and all its people, Russian and non-Russian, from those who wished to destroy them, and this was reflected in contemporary poems and popular songs.

INTRODUCTION

The change from regarding the river as the border with the 'other' (Asia) to adopting it as a symbol of Russianness within Russia is in part a reflection of the evolution of a *Russian* identity, in which the Volga played a significant part. However, it is also the river of many non-Russian peoples, and features in poetry and prose by non-Russians, as well as by Russians. The Volga today plays as important a part in the identification of Tatars in Tatarstan as 'Volga Tatars' and as descendants of the Bolgar state, as it does in Russian identity.

The river is of crucial importance to all those who live on its banks, but also to the Russians and non-ethnic Russians who have lived within the Russian Empire and the Soviet Union. The river, in the words of the popular 'Song of the Volga' in the 1938 film *Volga, Volga,* was:

Mighty with water like the sea,
And just as our motherland – free.

This book is an attempt to describe and explain the importance of and our fascination with the river Volga.

Part 1

Early History of the Volga

1
THE FIRST STATES ON THE VOLGA
Khazaria, Bolgar and Rus principalities

Flow the waters of the Itil [Volga] river
And crash on the cliffs.
There are many fishes and frogs there.
Marsh and reed banks are flooded as well.

Mahmud al-Kashgari[1]

None of the territories which are discussed in this chapter – Khazaria, Bolgar, Kievan Rus, the Rus principalities in what is now northern European Russia – are modern states in our understanding of the word. They can best be characterized as loosely defined geographical regions, which had a ruler (or joint rulers) who was supported by an elite military force and had enough authority to exact tribute from nomadic, semi-nomadic and settled peoples, and customs dues from local and non-local traders. These 'states' overlapped in terms of territory because there were no clearly defined land borders; of their subject peoples (who could be Turkic, Slav or Finno-Ugric) who paid them tribute; and of the authority which they could exercise over other territories. Both Bolgar and Kievan Rus, for example, continued to pay tribute to Khazaria even when both had begun to surpass it in wealth and military power.

The river Volga was a main, if not *the* main, artery for trade from south to north and from east to west for all these 'states'. The goods traded and transported up the river were immensely valuable: honey, wine, furs, spices, slaves, weapons, silver coins. Customs dues imposed on traders on the Volga and other rivers and trade routes were an important source of income for Khazaria and Bolgar, and a means of asserting their authority. The river Volga, and the commerce and wealth it generated, determined the location of the capitals of Khazaria and Bolgar and of other major towns, and contact and conflict between rulers were in part determined by competition over who controlled the river Volga trade. The towns and settlements on the Volga were also the places where diverse peoples met and interacted – people of different ethnicities and religions – which set the pattern for future contacts and conflicts.

Of course, the lands were not empty before these 'states' were established. The steppe lands of what is now southern Russia and Ukraine, east and west of the river Volga, had attracted migration from the east by nomads from roughly 1000 BC onwards. Until *c.* AD 370, it is thought, the nomads of this region were mainly Persian speaking; they were followed by Huns, who were Turkic speaking. The Pechenegs, a Turkic-speaking nomadic people, had migrated westwards through the Volga region to the lands east of the river Dnieper in the ninth century, and from there they attacked both Khazaria and Kievan Rus over the next century. The nomads were not literate, and so accounts of their activities come primarily from their neighbours and the victims of their raids. They have commonly been portrayed as 'uncivilized', aimlessly roaming the steppe and wantonly killing, raping and enslaving peaceful settlers on the lands. In reality, the nomads were highly organized in tribes, subdivided into clans, and followed a strict annual cycle so that their herds of sheep (and horses) had access to good pasture land in the right season. Raids were not random or constant, and they targeted newly harvested crops or valuables (including people) that could be traded or retained.[2] Furthermore, archaeological evidence shows that nomadic societies were capable of sophisticated craftsmanship.[3] Much of southern Russia continued to be home to nomadic and semi-nomadic peoples in the period covered by this chapter, and indeed they were not assimilated into the Russian Empire until the second half of the eighteenth century. In broad terms, this was a clash between settled and unsettled peoples, between the steppe and the forest, which has never fully been resolved in the lands of the Russian steppe.

Khazaria (sometimes known as the Khazar Khaganate) existed from *c.* 650 to the end of the tenth century. It stretched from the Aral Sea in the east to the river Dnieper in the west, and from the north Caucasus in the south to Bolgar in the north on the river Volga (some 190km by road south of present-day Kazan). At its greatest extent, Khazaria incorporated Kiev and the territory of what is currently most of eastern Ukraine and a swathe of the southern steppes of Russia. Khazaria was in effect a buffer zone between the Byzantine Empire in the west and Muslim states in the south (in the north Caucasus and Persia).

2. Khazaria

The river Volga was at the heart and core of Khazaria. It ran through the centre of the state, and the capital city from *c.* 750 to 965 was known as Itil (or Atil). This was located on both banks of the Volga, almost certainly just north of the delta (although the presumed archaeological site of Itil cannot be determined with precision, because remains have become submerged in the soft mud of the Volga delta). Itil means 'big river' and was the Turkic name given to the river Volga, as well as to the capital of Khazaria.

The ethnic origins of the Khazars are a matter of dispute, in particular because there are no written records in the Khazar language. The king and the ruling elite spoke a Turkic language; meanwhile the vast territory of Khazaria comprised nomadic and semi-nomadic peoples who almost certainly spoke several varieties of primarily Turkic languages. We do not know what the Khazars looked like, although

Arabic sources, on which we rely for much of our knowledge, claimed that there was a distinction between the so-called 'White Khazars', who were the ruling caste, and the 'Black Khazars', who were the subject peoples. The former were commonly described as 'strikingly handsome', while the subject people were dark skinned and swarthy, like 'a kind of Indians'.[4]

We also know only a little about the way in which Khazaria functioned. We know it had a dual kingship structure, which was typical of nomadic organization: one king, the senior or great king, commanded the army; and the other, lesser, king performed mainly religious and sacred duties. We know from the account by Ibn Fadlan, an Arab traveller who undertook a diplomatic mission as envoy of Muqtadir, the caliph of Baghdad, to the Bolgars in the late tenth century, that elaborate rituals took place on the death of a great Khazar king, in which the river Volga played a part in the ceremony, and:

> . . . it is customary to build him a house composed of twenty chambers and in each chamber to hollow out a tomb for him . . . Beneath this house is a river, a great river that flows rapidly, which they divert over the tomb.[5]

There was a landowning noble elite, drawn from nine tribes or clans, whose lands were in the southern steppes. At the peak of Khazaria's power, the king had an army of some 7,000–12,000 men – a figure that could be increased two or three times by using men from the retinues of this noble elite. There were also merchants, primarily located in Itil (discussed in more detail below), who traded along the Volga and on land routes from the south and east through present-day European Russia and eastern Europe.

An elaborate fiscal and judicial structure was established in Itil. Traders crossing Khazaria on the river Volga or on land routes paid significant dues in cash or kind (hides, honey, silver coin or livestock). Tribute was exacted from subject peoples, including from the Rus and the Bolgars (both of whom are discussed more below). In the middle of the ninth century, east Slavic tribes were paying Khazars one white squirrel skin per household in tribute.[6] The Bolgars paid a similar tribute to the Khazars up to the tenth century. The king of the Bolgars 'gives a sable skin for each household in his kingdom', reported Ibn Fadlan in the late tenth century.[7]

Khazaria was a multi-confessional state. The subject peoples followed traditional animistic and shamanistic beliefs, and at least some of them focused on the sky god Tengri (their beliefs are therefore known as Tengrism), which involved the worship of thunder and lightning and horse sacrifices. Khazaria was a buffer zone between two proselytizing religions: Christianity from Byzantium in the west and Islam in the south (there were also Buddhists in the east, but there is less evidence about the influence of Buddhism on the subject population of Khazaria). We know that there were Christian, pagan, Jewish and Muslim traders in Itil, and there is evidence that there were some converts to Islam among ordinary people.

Historians generally agree that sometime in the early ninth century the Khazar kings and the noble elite adopted Judaism.[8] In the late tenth century, the Persian traveller and geographer Ibn Rusta wrote that:

Their supreme authority [the major king] is Jewish and so is the [lesser king], and those commanding officers and important men who support him. The rest of them follow a religion like the religion of the Turks.[9]

The reason for the conversion is not known for certain, but it could be that this was a way to assert the independence of the king and the elite from Christian Byzantium in the west and Islam in the south, and to deflect the pressure to convert to either Christianity or Islam. It seems that Judaism was chosen because it was a religion of the book, but was not the faith of neighbouring rival regimes which had ambitions to take Khazar lands.[10] Jews had migrated to Khazaria from the Balkans and the Crimea and from Armenia in the south, and there were Jewish merchants in Itil. There is no evidence that this conversion led to any pressure on the subject people to convert to Judaism. According to Ibn Fadlan, in 922 the Khazar king destroyed the minaret of a mosque in revenge for the destruction of a synagogue with the words: 'If I did not fear that not a synagogue would be left standing throughout the lands of Islam I would have destroyed the mosque.'[11] Apart from this remark, we know little of the relations between the Jewish elite and the Muslim, Christian or pagan subjects or traders.

The economy of the majority of the population of Khazaria was pastoral, and mainly comprised grazing sheep, cattle and horses (as with the nomadic Pechenegs). This has led to Khazaria being depicted as building its wealth purely on trade (mainly on the Volga and on other river and land routes); this is summarized in a comment by a contemporary: 'The Khazar country produces nothing which can be exported to other lands except isinglass. As to the slaves, honey, wax, beaver and other skins, they are imported to Khazaria.'[12] In fact, the agricultural land around the capital, Itil, was productive enough to supply the large city with grain, mutton, honey, fruit and wine. Fishing, in the Volga and other smaller rivers, was also extensive, and fish were traded within and outside Khazaria. The Khazar army was a cavalry force, and horses were also traded within and outside Khazaria. Archaeological finds have included agricultural tools for ploughing, harvesting and grinding grain, as well as pottery and jewellery.

It was, however, the transit trade, via the river Volga and various land routes, which made Khazaria prosperous, not least because of the customs dues demanded from merchants who conducted the trade. These were levied on all vessels carrying goods.[13] Caravans came from China and Persia to Itil, laden with cloth, silks, cinnamon and other spices. Goods were traded up and down the Volga by the Rus, from the Baltic Sea to the Caspian. Furs were highly valued throughout the Islamic world. They came from the north of European Russia and from Siberia, were transported along the rivers Oka and Volga, and were then traded south and east from Itil. Fox pelts were especially prized by the Khazar kings, but the Khazar elite also purchased the pelts of sables and martens for kaftans and hats. The slave trade, too, was immensely valuable: mainly Slav slaves, and mainly pagans, were traded in Itil by the Rus and sold as household slaves, workers or conscripts for the army.[14] The trade along the Volga to Itil was the source of Khazar wealth; it was also the root of competition with other states, and ultimately caused Khazaria's decline and defeat, as we shall see below.

The river Volga was similarly the main route by which Islamic silver coins were traded with Rus in the north. Furs that were brought by Rus merchants from the north to sell at Itil were traded for silver coins.[15] An examination of the burial sites in northern European Russia and the Baltic shows that silver coins appeared in about 800. They originated mainly from Khazaria and Persia (that is, not from Central Asia), and had been transported from south to north, through the Caucasus and across the Caspian Sea, and then up the river Volga, through Khazaria to northern Rus settlements. Silver coins were also used by the Rus to pay tribute to the Khazars.[16] Analysis of trading patterns from hoards of coins in burial chambers is not straightforward; for example, it is not always clear how long coins were in the possession of the deceased before they were deposited in the burial chambers. Nevertheless, the sheer quantity of coins found in burial chambers shows that a substantial trade existed. It has been estimated that hundreds of thousands of silver coins were transported from the south to northern Europe via the Volga, through Khazaria and then via Bolgar.[17]

The capital of Khazaria, Itil, evolved into a large, multi-ethnic, multi-confessional seat of government and the main trading centre. According to Ibn Fadlan: 'The king of the Khazars has a great city on the River Itil [Volga], on both banks of the river.

The Muslims live on one bank and the king and his followers on the other.'[18] The western side of the city housed the kings, the noble elite and the administrative institutions – the courts and the treasury. The merchants resided in the eastern part: they were Muslim, Christian, Jewish and pagan, and came from Persia, Central Asia, the Caucasus, Byzantium, Bolgar and Kievan Rus, and the north of what is now European Russia. The main commercial court in Itil comprised seven judges: two who dealt with cases involving Muslims, two for Jews, two for Christians and one for pagans. Customs officials, court officials and servants were required to service the elite and the Khazar kings. It is impossible to know the total population of the city, but it has been stated that there were some 10,000 Muslims, and so the total population must have been many tens of thousands.[19] The city housed some 30 mosques, religious schools, bazaars and public baths.[20]

Native Khazar merchants also played a significant role in commerce, both in trading in Itil and by trading up the Volga as far as Bolgar in their boats. The Arab traveller and historian Al-Masudi recorded in 943:

> The Khazars have ships which they sail on a river, above their city [Itil], into the great river [the Volga] which traverses it . . . This river, which flows from the land of the Bulkārs [Bolgars], carries vessels from both kingdoms . . . The pelts of black and red foxes . . . are exported from their country . . . The black furs are worn by Arab and non-Arab kings, who esteem them more than they do sable, ermine and other similar furs . . . The upper reaches of the Khazar river [Volga] communicate by one of its branches with a gulf of the Sea of Pontus [Black Sea, but here probably the Sea of Azov], also called the 'Sea of the Rūs' because the Rūs, who are the only ones to sail it, live on its shores. They form a numerous pagan nation that doesn't recognize authority or revealed law. Many of their merchants trade with the Bulkārs.[21]

❖

The Bolgars formed a separate state (also known as Volga Bolgar or Volga Bulgaria or Volga Bulgharia), but were subordinate to Khazaria and paid tribute to it until the destruction of Itil by the Rus in the late tenth century. The Bolgars were almost certainly Turkic tribes that had originally settled north of the Black Sea. The first Bolgar state – Old Great Bulgaria – had been destroyed by the Khazars in 668. The Bolgars then moved north and east, and by the eighth century had settled in territory on the Volga, south of present-day Kazan.

The original population in the lands occupied by the Bolgars almost certainly predominantly comprised Finno-Ugric peoples (who would be described today as Mari and Udmurt people; see the more detailed descriptions of these groups in Chapter 3). Inscriptions found by archaeologists in Bolgar lands are in Arabic (Bolgar became a Muslim state), but include words in Turkic languages, including some that are related to the modern Chuvash language.[22] Today, both Chuvash people and Volga Tatars trace their origins back to the Bolgar state.

In 922, the Bolgar elite adopted Islam. The Bolgar ruler of the time, Almish, requested religious instruction from the Caliph Muqtadir (reigned 908–32) in Baghdad, and this is why Ibn Fadlan was sent as secretary to a mission to the king of the Bolgars. It is mainly from his account, and from those of later Arab and Persian travellers and scholars, that we know something of the Bolgar state. Ibn Fadlan had been sent to have audiences with the king, and he noted the latter's exalted position within Bolgar society. According to Ibn Fadlan, all Bolgars wore 'tall pointed hats', which they doffed in the presence of the ruler.[23] Despite the conversion to Islam of the king and the elite, many ordinary subjects retained shamanistic beliefs which were the same as those in Khazaria – that is, almost certainly the same as (or not dissimilar to) Tengrism. Archaeologists have found pagan symbols in burial sites on Bolgar territory. Becoming officially Muslim determined the culture, particularly the legal and urban culture, of the elite and the customs and way of life of the people – as well as making Bolgar part of the wider Muslim world.[24]

The Bolgars were described by Ibn Rusta at the end of the tenth century as follows:

> [They] . . . are camped on the bank of the river that flows into the Khazar Sea [the Caspian] and is called the Itil [the Volga] . . . They trade with the Khazars and the Rūsīya [Rus], who bring their merchandise to them. The Bulkārs [Bolgars], who live on both banks of the Itil, offer various products in exchange, such as the pelts of sable, ermine and grey squirrel and other furs . . . The majority are Muslim and there are mosques and Qur'ān schools in their inhabited places . . . They dress like Muslims and their tombs are constructed like those of the Muslims. Their wealth consists above all of marten pelts.[25]

It was trade that created the prosperity of the Bolgar state and ensured its continued existence, even when it was later challenged by the Rus (discussed below). At the centre of that trade was the river Volga. The Bolgar state became the great emporium for goods. As Khazaria had been the great market for goods from the east, south and north, so the Bolgars evolved as intermediaries between the Rus in the north and mainly Muslim traders from the east, including from Central Asia and China, and Khazar and Muslim traders from the south and from Persia and India. Caravan trains arrived in the capital of Bolgar and other Bolgar cities from Central Asia (archaeologists have found camel bones on the site): 'caravans constantly go from there' was a comment by a contemporary.[26] An Armenian community existed within Bolgar; fragments of Armenian cloth have been found, as well as inscriptions in the city of Bolgar.[27] Bolgar merchants also exported goods, and were found in the Rus principality of Vladimir-Suzdal in the eleventh century. In this trade, the river Volga was crucial, although a secondary route was the river Oka (which joins the Volga at present-day Nizhnii Novgorod).

The variety of goods which passed through Bolgar was greater even than the goods traded by Khazaria, and was such that a contemporary thought it important to list them in full:

Sables, miniver [a type of fur used on ceremonial occasions], ermines, and the fur of the steppe, martens, beavers, spotted hares, and goats [hides]; also wax, arrows, birch bark, high fur caps, fish glue, fish teeth [probably walrus tusks], castoreum [beaver secretion used in medicine and perfumes], prepared horse hides, honey, hazelnuts, falcons, swords, armour, wood, Slavonic slaves, sheep and cattle.[28]

The city of Bolgar and the Bolgar state were one great market, and the state grew immensely rich in the tenth and eleventh centuries on this trade. Of particular value were honey, wax and grain exported from Bolgar lands; precious furs, falcons and mammoth tusks from the north and from Siberia; and prisoners of war, who were traded as slaves by the Rus and others. The city of Bolgar became the largest fur market in Europe in the tenth century. Swords and ornamented pins have been found in burial chambers in Finland and Sweden, and are thought either to have originated in Bolgar or to have come up the Volga via Bolgar from Persia.[29] In the mid-twelfth century, the Persian scholar and traveller Abu Hamid noted:

From Islamic countries the [Bulgar] people import swords, which are made in Zendjan, Abkhar, Tabriz, and Isfahan. The swords are in the form of blades without handles or decorations; [they consist] only of iron as it comes out of the fire. And they temper these swords [so well] that if the sword is suspended on a thread and hit with a nail or anything made of iron or wood, the sound will be

21

heard for a long time . . . The swords that are imported from Islamic countries to Bolgar bring a large profit. The Bolgars then carry them to Visu [north-east of Novgorod, west of Lake Beloozero], where beaver are found, and then the inhabitants buy them for sable skins and male and female slaves.[30]

In the remains of the city of Bolgar, beads have been found that originated in Egypt and Syria;[31] ornaments originating in India and dating from the eleventh to the thirteenth centuries; and ceramics and pottery from the Caucasus.[32] The lands of Bolgar included fertile black-earth soil, and the Bolgars cultivated a number of crops, including wheat, rye and oats. During periods of poor harvest in the eleventh and twelfth centuries, the Bolgars also exported grain to the Rus.[33] Goods reached Novgorod (in the north-west of what is now European Russia), Kiev and the town of Suzdal, east of Moscow.[34] Russian goods – including icons and other church objects – also entered Bolgar from the north and were then traded in the south and the east.[35]

The trade in silver coins was of immense importance to the Rus, as we have seen above. The Bolgar state supplemented Khazaria in the exchange of silver coins for furs and slaves, and in time came to outstrip Khazaria in economic performance and power. By the tenth century, the Bolgar state was issuing its own silver coins, with the names of Bolgar rulers struck on one side, which it initially modelled on earlier Arab silver coins. It was thus no longer dependent on exporting coins from Central Asia and Persia.[36] This gave it far more importance in the trade with Rus, and also made the Bolgars the first people in what is now European Russia to issue their own coinage. Coins struck in Bolgar have been found in burial sites in present-day Russian towns, including Tver, on the river Volga north of Moscow, and in Pskov in north-west Russia, as well as in Ukraine, Belarus, Poland, Finland, Denmark, Sweden and Norway.[37] Bolgar came to dominate the export of silver coins to Rus, and its trade in coins was far larger in the tenth century than the coin trade with Rus from Khazaria had been in the ninth.[38] That shift in power was also, however, an indication of the growing importance of the Rus and the Rus principalities of the north. The trade in coins between Bolgar and Rus remained of great importance throughout the eleventh and twelfth centuries.

The Bolgar state developed a sophisticated urban culture, which included mosques, *madrassas*, merchant houses, bath-houses and systems of irrigation. Bolgar was the largest city, but there were other towns – Biliar (Bilär), Suvar (Suwar) and Oshel (Ashli or Aşli). In all, archaeologists have identified over 140 settlements in former Bolgar. By the early nineteenth century, accounts by travellers and scholars who visited the ruins in Bolgar had established the existence of the remains of a palace, mosques and a number of royal mausoleums and large towers (almost certainly minarets) in the city, attesting to its importance and wealth as the capital of the Bolgar state. Plans and drawings from that period show monuments which have now disappeared.[39] The city was positioned in an elevated place above the river Volga and protected by massive fortifications.[40] The area within the fortifications is extensive, and there is also evidence of the existence of several satellite settlements around the town. Major excavations of the site of the city of Bolgar were undertaken by Soviet archaeologists in the 1930s. Evidence suggests that the population of the city of Bolgar and its surrounding settlements could have been between 10,000 and 20,000.[41]

Archaeological evidence has been able to establish the layout of other Bolgar towns, as well as the types of mosques, merchant houses and other buildings that were constructed. Many of these settlements had their own fortifications and were on raised ground for ease of defence. In Biliar, excavations in the early 1970s revealed a large building with warehouses, kitchens, a dining hall, underground heating and a small mosque, all of which suggested that this was an inn, or caravanserai, for visiting, primarily Muslim, merchants.[42] The excavations of commercial scales and weights not only in the towns of Bolgar and Biliar, but also in smaller settlements, demonstrate that even the latter were active in trade.

The city of Bolgar developed as an artistic and cultural centre, and archaeological finds have shown that there were craftsmen undertaking, among other things, skilled leather work and elaborate bone carvings, and producing sophisticated ceramics and delicate and elegant necklaces, rings, brooches and amulets from silver and other metals. These are now displayed in the excellent museum on the Bolgar site.[43] Richly ornamented spearheads, swords and sword cases found in Scandinavia are believed to have originated in Biliar and other Bolgar towns. Wheel-made pottery finds, including amphora, in Biliar and other towns show that there was extensive craft knowledge.[44] Bolgar towns also imported luxury goods: excavations of the remains of a mosque in Biliar in 1974 uncovered beads, amber and silver coins, but also glazed pottery of Iranian, north Russian, Central Asian and Byzantine origin.[45]

❈

Who were these Rus, who traded so extensively with the Bolgars and the Khazars and who, as we have seen, according to Al-Masudi, were 'a numerous pagan nation that doesn't recognize authority or revealed law'? The answer is not straightforward, because the ethnic origins of the Rus have become part of the controversy over the foundation of the modern Russian and Ukrainian states. The so-called 'Normanist' view is that the Rus were Scandinavians (probably Swedes) who penetrated Russia from the Baltic down the river Volga and other trading routes. The *Russian Primary Chronicle*, which was originally compiled in about 1113 but which only survives in later versions, refers to the inhabitants of Rus as having come from the Baltic. They were therefore assumed to be Vikings, or Viking Rus or Russian Vikings and have also been known as Varangians. The name 'Rus' could be taken from Swedish or Finnish (the Finnish for Swedes is *Ruotsi* and is derived from the Old Norse, meaning 'the men who row'). The 'Normanist' view has been controversial among Russian and Soviet scholars because, as we shall see below, this group of people produced the first ruling dynasty of the principalities from which modern Russia and Ukraine emerged (and indeed is the source of the word for Russians and Russia). The theory was considered by Russian scholars to be provocative when it was first formulated in a scientific manner in the eighteenth century (all the more so as the first presentation on the subject to the Russian Academy of Sciences in 1749 was made by an ethnic German), because it suggested that the east Slavs (Russians, Ukrainians, Belarusians) had invited Scandinavians to rule over them and to create what were to become Slav states (contemporary Russia, Ukraine and Belarus); put bluntly, it suggested that the descendants of the Normans (Scandinavians) were suited to rule, but the Slavs were not.

Archaeological and linguistic evidence suggests that the high-status Rus who encountered the Khazars and Bolgars on the Volga were Scandinavian in origin, but that they very quickly assimilated with the native east Slav population (as indeed the Vikings had done in western Europe, including in Britain and Ireland). One historian has expressed this as follows: 'In 839 the Rus were Swedes; in 1043 the Rus were Slavs.'[46] We know that Swedish (Viking) settlers appeared on the shores of Lake Ladoga, east of present-day St Petersburg, probably from the mid-eighth century. Excavations of the site at what is now called Staraia Ladoga (Old Ladoga), on the river Volkhov near Lake Ladoga, have demonstrated that this was an important trading post from the mid-eighth century to about 950. Objects which have been found include silver coins, beads, ornaments, combs, tools for working wood and metal, and shoes.[47] It has also been possible to reconstruct the layout of the wooden houses in this settlement, which is similar to Viking settlements elsewhere in Scandinavia. Objects of Swedish origin from the eighth to the early eleventh centuries have been found at other sites in north Russia, on the Baltic and on the Volga.

What encouraged the Vikings, the Rus, to move so far, and across often difficult terrain, from their original homes in Sweden and Norway? It is generally agreed that it was the immensely lucrative trade route down the Volga that brought these settlers to the forested and marshy lands of north Russia, and that river networks determined the places of settlement, as indeed was the case with Vikings in other parts of Europe and in Iceland.[48] Trade led to contacts and conflicts with other peoples; trade was the reason for the Rus to establish relations with the Khazars and the Bolgars. As we have seen above, the Rus traded fur and slaves for silver coins, honey and weapons.[49]

As with the Khazars and Bolgars, the earliest descriptions of the Rus on the Volga come from Ibn Fadlan, who came across them on the river Volga during his mission to Bolgar:

> I saw the Rūs, who had come for trade and had camped by the river Itil [Volga]. I have never seen bodies more perfect than theirs. They were like palm trees. They are fair and ruddy. They wear neither coats nor caftans, but a garment which covers one side of the body and leaves one hand free. Each of them carries an axe, a sword and a knife and is never parted from any of the arms we have mentioned.

Ibn Fadlan's admiration of the physique of the Rus was tempered by what were almost certainly Muslim prejudices against pagans, whom he regarded as unclean and primitive:

> They are the filthiest of God's creatures. They do not clean themselves after urinating or defecating, nor do they wash after having sex. They do not wash their hands after meals. They are like wandering asses . . . Each of the men has sex with his slave, while his companions look on.[50]

He also noted that offerings were made to wooden idols; that animal sacrifices were made; and that on the death of a great man, slave girls were buried alongside him.

Nevertheless, Ibn Fadlan's description of the Rus did show that they had skilled craftsmen who could make jewellery and weapons; that they were active traders; that they were fierce warriors; and, above all, that they had the capacity to build ships which could sail down the Volga and other rivers (indeed, we know that Vikings also crossed the Atlantic) and establish, and defend, settlements on the river banks. The river Volga, as we shall see, was not easy to navigate, even in the eighteenth and nineteenth centuries (see Chapter 10), nor is the climate of the region benign; thus, it is a tribute to the ship-building and navigational skills of the Rus that they were able to penetrate down the whole length of the Volga, from the north to the Caspian Sea, and consolidate their position on the river.

Ibn Fadlan also notes that the Rus had a king, protected by 400 chosen men, and it is significant that he uses the same word for king – *khāqān* – as he uses for the kings of the Khazars. In fact, some sort of political order, including a ruler and military social elite, had existed since the first third of the ninth century, when embassies had been sent from the Rus to Constantinople; Ibn Fadlan's later comments simply confirmed this.

The depiction of the Rus was echoed by Ibn Rusta in the early tenth century:

> They have a ruler . . . The Rūs raid the [Slavs], sailing in their ships until they come upon them. They take them captive and sell them in Khazarān [includes the slave market at Itil] and Bulkār [Bolgar]. They have no cultivated fields and live by pillaging the land of the Saqāliba [Slavs] . . . They have great stamina and endurance. They never quit the battlefield without having slaughtered their enemy. They take the women and enslave them. They are remarkable for their size, their physique and their bravery.[51]

Ibn Rusta also noted that 'The Rūs live on an island in a lake' and this island 'is pestilential and the soil is so damp that when a man steps on it, it quivers underfoot'. This was clearly information that was passed to him, but in fact there are several locations of Scandinavian settlements in north Russia which could have been the 'capital' of the Rus, all of which had access to the Volga and/or other rivers. Gorodishche, sometimes known as 'Rurik's fortress', is a likely candidate, positioned as it is on an elevated position above three rivers – that is, like an island – of which the most important is the river Volkhov, which flows out of Lake Ilmen. Archaeological excavations have shown that the settlement was heavily fortified (unlike Staraia Ladoga) and was an important trading centre. Another candidate is Sarskii fort on the river Kotorosl, which flows into the Volga at the present-day town of Iaroslavl on the upper Volga.[52] Both those settlements had access to furs from the north which could be traded for silver coins. What is clear is that the Rus had progressed beyond simply scattered or temporary settlements on the rivers and had established more permanent bases in what is now European Russia. It seems they were imposing tribute on Slav and Finnic tribes by the mid-ninth century. By this time, they were also attacking Khazaria and Bolgar and asserting their rights over their valuable trade.

The Volga was not the only river artery of crucial importance to the Rus. They also travelled down the rivers Don and Dnieper, and were said to have reached

Kiev by about 860, from where they mounted raids against the Khazars (and the nomadic Pechenegs – and against Constantinople itself). The presence of the Rus in the Dnieper area inevitably brought them into conflict with the Khazars. Khazaria had already fought a series of wars in the seventh and eighth centuries against Muslim Arab forces in the north Caucasus. During this earlier period, Khazaria had in effect protected the southern frontiers of the Rus settlements in the Dnieper region by controlling, at least to an extent, the raids by nomads, including the Pechenegs. By the late ninth century, the Rurik dynasty, which was of Swedish and Viking origin, established what has become known as Kievan Rus, centred on the city of Kiev. This city had been an outpost of Khazaria, to which it paid tribute, but it was taken by the Rus, under Prince Oleg of Novgorod, a member of the Rurik family, at the end of the ninth century. At its peak in the mid-eleventh century, Kievan Rus stretched from the Baltic to the Black Sea.

It has already been noted that none of the regions discussed in this chapter can be seen as a modern state in our understanding of the word. Kievan Rus is best characterized as a loose federation of Slavic and Finnic peoples, tied together by the Rurik dynasty. The succession as grand prince of Kiev did not pass from father to eldest son, but could be claimed by anyone who was a senior member of the house of Rurik. Instead of a modern concept of succession and legitimacy, power was distributed and redistributed through a kinship group.[53] The name 'Kievan Rus' is itself a later designation. At the time, the lands which comprised Kievan Rus were known more commonly as 'land of the Rus'; it was only in the nineteenth century that the term Kievan Rus was used by Russian historians to distinguish the period when Kiev was the centre of power (as opposed to the later dominance of Moscow and Muscovy, which is discussed in the next chapter).

What is clear is that Kievan Rus grew wealthy, and that trade was the basis of this wealth. The city of Kiev was crucial for controlling the Dnieper river route, through which access to the Caspian Sea (via Khazaria) and Byzantium was possible. Kievan Rus also incorporated the lands of Novgorod in north-west European Russia, which gave it control over the route from the river Volga through the river Volkhov to the Baltic Sea. Kievan Rus, like Khazaria and Bolgar before it, became an emporium for trade, north and south and east and west. The cities of Kiev and Novgorod grew in wealth and size: at the peak of their prosperity, Novgorod is estimated to have had a population of 25,000–30,000 and Kiev of perhaps 30,000–40,000.[54]

In 988, the grand prince of Kiev, Vladimir, adopted Greek Orthodox Christianity, which tied Kievan Rus culturally to Byzantium, and at the same time distanced it from both Muslim Bolgar and Jewish/Muslim Khazaria. Christianity spread from the top down throughout the Slavs in Kievan Rus (although paganism continued to exist in the outer reaches of the realm). By the middle of the eleventh century, Kiev was a well-fortified city which boasted a splendid royal palace, the impressive cathedral of St Sophia, monasteries, churches and merchant houses. The city had established itself as a centre of skilled craftsmen, including glass producers and potters, and as a cultural centre of eastern Christianity.

❋

By the late ninth century, Khazaria was being challenged from the north by Kievan Rus and from the south by states in the north Caucasus. By the tenth century, as we have seen, Bolgar had also emerged as more powerful than Khazaria, despite still paying it tribute. Khazaria was in decline and its power was being eroded from all directions.[55] The conflict between the Khazars and the Rus largely concerned control over the lucrative trade on the Volga.[56] In 912, the Rus attacked Itil from the river, but were repelled and massacred by the Khazars with the loss, it is claimed, of some 30,000 men.[57] The following year, the Khazars closed the Volga trade to the Rus, sparking Rus raids on the river down to the Caspian Sea, which continued over the next few decades. In the early 960s, the Khazar ruler, Joseph, wrote about the threat from the Rus:

> I protect the mouth of the river [Itil-Volga] and prevent the Rūs arriving in their ships from setting off by sea against . . . their enemies . . . I wage war against them. If I left them [in peace] for a single hour they would crush the whole land of the Ishmaelites [Arabs] up to Baghdad.[58]

His claim that he could protect his lands was, however, over-optimistic. In the 960s, the Rus attacked and sacked a number of Khazar towns and fortresses; Itil itself was sacked in *c.* 968 or 969. The town was razed to the ground, and it was said 'not a grape or a raisin' remained in the ravaged land. The elite fled and the subject population was subsumed into new states (or into successor nomadic hordes; see the next chapter). The formerly prosperous cities continued to exist for a while, but in much reduced circumstances. Khazaria and Itil were described at the end of the tenth century by the Arab geographer Al-Maqdisi as follows: 'Khazar, a grim, forbidding place, full of herd animals, honey and Jews . . . Itil, the large capital city, is situated on a river of the same name [the Volga] which flows into the lake [the Caspian].'[59] In time, all that was left of a once powerful state were a few place names, artefacts and some archaeological remains.

By the eleventh century, Kievan Rus had become more powerful than Khazaria. The dynastic structure of Kievan Rus led to a series of struggles for power – mainly between brothers and their armed followers – between the grand prince of Kiev and Rus princes in other (in principle) subordinate principalities ('principality' itself being a modern term – at the time, the territories were referred to as 'lands', e.g. 'Suzdal land'),[60] including Novgorod in the north and Vladimir-Suzdal in the north-east.

By the twelfth century, the principality of Vladimir-Suzdal was asserting its power vis-à-vis Kiev. It was perfectly located to control the river Volga, because it included the towns of Tver, Kostroma and Iaroslavl, all of which are located on the upper Volga. Vladimir-Suzdal also controlled the key tributaries of the Volga, including the Kliazma, which linked the river Oka (which in turn flowed into the Volga) with Moscow, and the Sheksna, which linked the Volga with Lake Beloozero to the north-west.[61] Vladimir-Suzdal was also remote from Kiev and very difficult, if not impossible, to control from there. The principality of Vladimir-Suzdal could probably have taken over the role of the trading emporium from Bolgar, had not the attentions of the ruling princes been diverted by challenging the grand prince of Kiev. Regular conflicts took place between the forces of Bolgar and Vladimir-Suzdal in the twelfth century. In 1152, the Bolgars attacked Iaroslavl, some 250km north-east of Moscow. By this time, the subjects of Vladimir-Suzdal were systematically pillaging Bolgar cities from the north and west. The Bolgars had to move their capital inland from Bolgar to Biliar, away from the river, in order to protect themselves from these raids along the Volga. In 1184, the prince of Vladimir-Suzdal attacked Biliar, but did not succeed in taking the city.

The conflict between Vladimir-Suzdal and the Bolgars was about who would control the Volga, and ultimately about who controlled the immensely valuable trade in furs and pelts which were transported down both those rivers. The princes showed their determination to push the Bolgar state back from the river. The fort of Gorodets on the river Volga (some 50km north-west of present-day Nizhnii Novgorod) was founded probably in 1152 on the Volga, precisely to assert control over the river from the north. In 1221, the prince of Vladimir-Suzdal founded the town of Nizhnii Novgorod, at the confluence of the rivers Volga and Oka, and by so doing asserted his control over the northern part of the river Volga and, in effect, over the fur trade.[62]

Both the Bolgars and the Rus, however, faced a far greater enemy than each other. That threat came from the east, and its power only became apparent in the early thirteenth century, when the armies of Batu, the grandson of Chinggis Khan, swept into the lands of the east Slavs and the Bolgars. The conflict which then ensued between the Golden Horde (and its successor states, the Kazan and Astrakhan khanates) and the Rus principalities is the subject of the next chapter.

2
THE VOLGA AND CONQUEST
Hordes, khanates and Moscow

This river [the Volga], he says, is the size of the Nile taken three times and even more; great vessels ply upon it and travel to the Russians and the Slavs. The source of the river is also in the land of the Slavs.[1]

Egyptian geographer Al-Umari, fourteenth century

The river Volga was at the centre of dramatic events from the thirteenth to the mid-sixteenth centuries, during which lands were seized and new regimes established. Peoples from the east and then the north arrived as warriors, administrators and settlers. Towns were sacked and villages destroyed or abandoned; but at the same time, other cities rose in prominence, trading routes continued to be developed and craftsmanship flourished in the new urban centres. Throughout this period, the river Volga was of crucial importance. It continued to be the major trading artery of what is now European Russia, and was used to transport goods from south to north and from north to south; it was also a significant route for bringing wares from the east to Europe. Administrative authority was linked to that trade, and the most significant cities of the Golden Horde and the khanates of Kazan and Astrakhan were located on the middle and lower Volga. To the north, the struggle for dominance took place between the principalities of Moscow and Tver; the latter is on the upper Volga, but both were well positioned within the river network of European Russia to dominate Volga trade. By the sixteenth century, Moscow was dominant and controlled the upper Volga as far as Nizhnii Novgorod. Ivan IV was determined to bring the whole of the river Volga under his sway – from its source to the Caspian Sea – and this was achieved by the conquest of Kazan and Astrakhan. The river Volga was thus wholly within what can now be termed the Russian Empire: a new era had dawned.

❖

In 1223, the forces of Kievan Rus, supported by nomadic Cuman (a Turkic people related to the Pechenegs who had settled north of the Black Sea and on the Volga and had displaced the original Pechenegs), fought a Mongol army at the river Kalka, in the region of present-day Donetsk in eastern Ukraine. The combined armies were routed by the Mongols and many of the Rus princes were slaughtered. The Mongols, however, did not press further and returned to Central Asia. In 1232, they returned and attacked the city of Bolgar, 'laid waste and conquered much of the lower land of the Bolgars and destroyed their cities'.[2] Four years later, they invaded again, with an army whose size has been estimated at anything from 25,000 to 140,000 men. Led by Batu, grandson of Chinggis (Genghis) Khan, this time they conquered and stayed. The city of Bolgar was taken in 1236, and the towns of Biliar and Suvar were ransacked. The armies now turned to towns in the principalities of what is today European Russia and Ukraine: Riazan was sacked in 1237; Vladimir (the capital of the principality of Vladimir-Suzdal) was burnt to the ground in 1238; Kiev was stormed and destroyed in 1240. Towns on the upper Volga, including Gorodets, Kostroma, Nizhnii Novgorod, Iaroslavl, Uglich and Tver, were all ransacked. Moscow, at the time a small town in the principality of Vladimir-Suzdal, was sacked in 1238.

By 1241, the Mongols were in present-day Hungary. They then retreated eastwards, but still controlled – directly or indirectly – swathes of land in eastern Europe and European Russia. The Mongols' invasion not only had a profound impact on the lands and people they controlled, but also played a significant role in the evolution of the principality of Moscow, which became the dominant principality in the Rus lands. By the middle of the sixteenth century, Moscow had taken over key lands in European Russia and had gained control of the river Volga from its source to the Caspian Sea.

The Golden Horde was originally the name given to the western section of the Mongol Empire. As that empire fragmented after 1259, it became a separate khanate. The term 'Golden Horde' was used by Russian chroniclers in the sixteenth century, but it is also known as the Kipchak khanate or the Ulus (realm) of Jochi. The origins of the name 'Golden Horde' are unclear: it could have referred to the golden colour of the tents used by the Mongols or by their leader, or khan, Batu; it could be a reference to the wealth of the khan; it could – far more prosaically – simply mean the central camp. It occupied a large part of the present-day Russian Federation – from western Siberia, across the Urals, to encompass much of what is nowadays Ukraine and southern Russia. The southern borders of the Golden Horde reached the shores of the Black Sea (including the Crimea), the north Caucasus and the Caspian Sea.

3. The Golden Horde and the Rus Principalities

The river Volga was crucial to the commercial life of all these regions. The capitals of the Golden Horde and of the khanates of Kazan and Astrakhan were all located on the river.

The Mongol hordes were loose structures under a ruler, a khan (known as a 'tsar' in the Russian chronicles), who was a descendant of Chinggis Khan (died 1227). He was supported by a military elite of Mongol and also Turkic ethnicities, and society was organized around tribes, which were subdivided into clans. The hordes were able to exact tribute from the subject peoples – Turkic, Slav or Finno-Ugric (these people are described more fully in the next chapter). For the subject people, this meant that the obligations owed to the Golden Horde (and the other Mongol hordes) were not unlike those owed to its predecessors, Khazaria and Bolgar. As in Khazaria, the subject people of the horde were of different ethnic background from the khan and the ruling elite, and were often of different religions. Originally the Mongols were shamanistic, but Berke, the khan of the Golden Horde, adopted Islam at some point during his reign (1257–66). There was, however, no attempt to impose Islam on the east Slavs or on the Finno-Ugric peoples of central and north Russia (or, for that matter, on Buddhists in the east). It could therefore be said that the lives of ordinary people – farmers, hunters, fishermen – changed little as a result of the Mongol invasion; but that would be to downplay the significance of over-lordship. Most rural people in the remote, heavily forested parts of present-day European Russia and the more open southern steppes had relatively little contact with government regimes – either then or in later centuries – until obligations were required of them. These obligations were primarily fiscal and military, and were not insignificant.

The horde was able to demand tribute from the princes of Moscow, Vladimir-Suzdal and Tver, but Khazaria had also been able to take tribute from the Bolgar state, and quite probably from Kievan Rus as well. Princes of the Rus principalities in what is today northern European Russia now had to travel to Sarai, or to other Golden Horde cities, to be formally granted the right, or patent, to retain their principalities – something that Khazaria had never attempted to impose on Kievan Rus or on the Bolgars. The following description, from the second half of the thir-teenth century, shows that the khan claimed the sole right to install a Russian prince and had established an appropriate ceremony to impose his authority:

> When Yaroslav [of Vladimir; son of Alexander Nevsky] arrived at the Horde, the khan received him with honour, gave him his armour and ordered that he be instructed in the ceremony [of appointing him] to the grand principality. He ordered Vladimir of Ryazan' and Ivan of Starodub, who were in the Horde at the same time, to lead his horse. And in August [1264] he let him go and sent him with his envoy Jani Beg and with the patent for the grand principality.[3]

It is clear from this account that the Mongols ruled Russia, albeit indirectly.

The Mongols have had a 'bad press'. They have been portrayed, both in contem-porary and later Russian chronicles on which historians have relied heavily, and also popularly, as bloodthirsty monsters who destroyed everything in their path, razed whole cities to the ground and mercilessly slaughtered all the inhabitants, regardless of sex or age. It is claimed that they demanded exorbitant tribute from the Rus principalities which crippled the finances and economy of the north and west of European Russia. The Mongols have been accused of causing depopulation and the economic collapse of towns, and of ruining trade and artisan skills. Indirectly, the Mongols have been blamed for what have sometimes been perceived as the more negative aspects of the way in which the principality of Moscow evolved. The charge was that, in order to defeat the khanates, Moscow had to adopt facets of the social and political organization of the Mongols, and in particular inherited attitudes and the practice of 'oriental despotism' from them.[4]

It is true that the military success of the Mongols led to the sacking of key cities in what had been Bolgar, Kievan Rus and Rus principalities. Kiev was said to have been reduced from a great city to 'almost nothing'.[5] In Kolomna (a town south-east of Moscow), according to a Russian source: 'No eye remained open to weep for the dead.'[6] After the first wave of attacks, the Mongols could impose a reign of terror and carry out brutal reprisals against anyone who dared to rebel against them. Destruction and plunder were often inflicted in reprisal, but sometimes attacks were seemingly random – Nizhnii Novgorod was sacked in both 1377 and 1378. As late as 1382, some 24,000 people were said to have been killed when the forces of the Golden Horde entered Moscow in a reprisal. Nor should the Mongol invasion be seen purely in terms of the destruction of Rus towns in the west and the north. They also sacked the city of Bolgar on the Volga and absorbed what had been the Bolgar state into the territory of the Golden Horde.

Some places, however, were left untouched by the Mongol invasion; and towns such as Novgorod, Tver and Moscow itself seemed able to recover quickly as cultural as well as economic centres.[7] Furthermore, the Rus princes often colluded with the Mongols in order to assert their position vis-à-vis other princes: the sacking of Moscow in 1382 came when other, rival, princes who were serving with the Golden Horde forces persuaded the Muscovites to open their gates, on the false promise that they would not be harmed. This was not the only time towns on the Volga and elsewhere in Russia were sacked, and they were able to recover from these attacks (Stenka Razin and Emelian Pugachev could be as destructive as the Mongols, and nothing could equal the appalling destructiveness of twentieth-century warfare).

The impact of the Mongols on the economy is hard to determine. The city of Kiev never recovered its dominant position in trade, and that was in part because the new land trading routes bypassed it. However, trade and commerce continued and expanded within the lands of the Golden Horde, as we shall see below. The town of Bolgar continued to be an important trading centre on the Volga. The collection of tribute was almost certainly a large burden on towns and villages in what is now European Russia, and indeed the violent revolts against the tribute indicate the resentment or desperation caused by the levy. In 1262, the upper Volga town of Iaroslavl was one of several to revolt against tribute collectors and to kill anyone who looked 'Muslim'. It seems, however, that the particular Golden Horde agent, who had converted to Islam from Christianity, had 'caused great offence to Christians' and was described as 'mocking the cross and the holy churches',[8] and this incident may have been an exception occasioned by his provocative behaviour.

Nor can the organization and functioning of the Golden Horde be character-ized as primitive or 'uncivilized'. At the top, a bureaucratic structure evolved around a central advisory council, with a senior official in charge of the treasury.[9] The very size of the territory required an administrative structure based on, in effect, regional overlords drawn from the leaders of clans who were officials of the khan and also served in his army. Mechanisms, including a census, were established to collect and organize the tribute and to settle legal disputes. The Golden Horde was capable of exercising indirect control over the Rus principalities – by appointing officials who collected tribute and conscripted troops – and of interfering in their internal poli-tics. The adoption of Islam brought with it religious institutions, including mosques and *madrassas*, and a religious hierarchy. The Mongols are also credited with estab-lishing the first postal system, in order to communicate across vast distances.[10] They established diplomatic relations not only with Rus princes, but also with Byzantium, the Caucasus and Egypt.

Furthermore, the Golden Horde established trading links and commercial cities.[11] The capital of the Golden Horde was the city of Sarai, on the lower Volga, built near to – or possibly on the same site as – the Khazar capital of Itil (and then rebuilt upstream as New Sarai in the 1340s). It was a great city, with a population estimated to have been up to 75,000 at its peak: that is, a city larger than Itil in Khazaria or Bolgar when it was capital of the Bolgar state. Archaeological excava-tions have found evidence that at least the wealthy inhabitants of Sarai dressed in silk and other rich fabrics, and that the women wore elaborate jewellery.[12] Sarai had all the sophisticated buildings and infrastructure of a great Muslim city, including aqueducts, water channels, palaces, administrative buildings, mosques, *madrassas*, shops and caravanserais. The city was impressive, with broad streets and rich villas constructed of brick. Craftsmanship flourished in Sarai – glass, leather, ceramics, gold and other metals – and coins were minted in the city.[13] The khan was said to live in luxury, eating from utensils made of gold and silver.[14] It was here that diplomatic missions from Rus princes and foreign rulers were received.

Contemporaries were in awe of the wealth and appearance of the city on the Volga. The Moroccan traveller Ibn Battuta visited it in the early fourteenth century and wrote that:

> Sarai is among the most beautiful of cities; and its grandeur is very considerable; it is located on a plain and is full of inhabitants; it has beautiful markets and wide streets . . . It is necessary to note that the houses there stand next to each other and there are neither ruins nor gardens. There are thirteen main mosques for Friday prayers . . . Sarai is inhabited by people of many nations . . . Each nation lives in a separate quarter, where it has its own markets.[15]

The Egyptian geographer Al-Umari did not visit Sarai personally, but described it from other accounts as follows:

> . . . the city of Sarai was built by Berke Khan on the banks of the Turonian river [the Itil; that is, the Volga]. It lies upon salt earth, without walls of any sort. The king's palace of sojourn there is a great palace, upon the highest point of which there is a golden crescent [weighing] two Egyptian quintals. The palace is surrounded by walls, towers, and houses, in which his emirs live. In this palace they have their winter quarters.[16]

Trade continued to flourish from east to west – bringing silks and spices from China and Central Asia to Europe via Sarai, and from there up the river Volga to the north – and from south to north, from Persia and India across the Caspian Sea. Sarai became an important centre for the slave trade, including prisoners of war. The principalities of Novgorod and Moscow benefited from the Golden Horde trade up the river Volga, and from there along the rivers Oka and Moskva. Furs came to Sarai from the north and the north-east of present-day European Russia (and tribute could be taken in furs, rather than cash). Indeed, the Bolgar merchants increased their fur trade with the north in response to the demands for tribute from their new Mongol overlords.[17] Other goods transported from Rus lands included fish, caviar and salt. Silver coins continued to reach the town of Novgorod from the south in the fourteenth century.

The vast territory controlled, directly or indirectly, by the Golden Horde meant that the trade routes stretched to the north of European Russia and the Baltic Sea. Merchants travelled from Sarai up the Volga, through Rus lands to the Baltic ports. It is said that, in 1474, some 3,200 merchants and 600 envoys journeyed from

Sarai to Moscow, where they sold 40,000 horses to the local inhabitants.[18] Sarai was home to many merchants from numerous cities and countries – from Nizhnii Novgorod and other towns on the river Volga, but also from places much further afield, such as China, Hungary and Genoa.[19] Ibn Battuta, quoted above, noted that the merchants included Muslims (Ossetians, Kipchaks, Cherkassians) and Christians (Rus and Greeks).

The Mongol Empire was, however, constantly disrupted by disputes over succession between claimants to the khanship. By the second half of the fourteenth century, the Golden Horde was in decline, weakened by almost constant internecine war and by the ravages of disease. The Black Death was devastating in the 1340s as it passed through Astrakhan and Sarai on the way to the Crimea (in later centuries, too, the river Volga would play a significant part in the spread of disease, see Chapter 7). In 1364, the plague struck Sarai for a second time, and then travelled up the Volga as far as Nizhnii Novgorod.[20]

In 1380, the horde was defeated by Prince Dmitrii of Moscow at the battle of Kulikovo field on the river Don (in present-day Tula province, south of Moscow). This was the first major victory against the horde by a Rus prince, but in itself it did not change relations dramatically between the horde and Moscow – not least because the Muscovites had lost so many men in battle that they could not follow up the victory or prevent further incursions by the horde (as we saw above, Moscow was sacked in 1382).[21] However, the khan did have to abandon a planned follow-up campaign against Moscow, because he was attacked from the east by the forces of Tamerlane (Timur). Tamerlane had taken control of the western (Chagatai) khanate, from which he campaigned in Central Asia, the north Caucasus and the southern Volga area to found the powerful, albeit short-lived, Timurid Empire.

As with the earlier conflicts between the Rus and the Khazars and Bolgars, much of the strife between the Golden Horde and the forces of Tamerlane centred on control of the lucrative trade on the river Volga. Tamerlane attacked towns on the Volga, sacked both Old and New Sarai in the 1390s, and then re-routed the spice trade to the south of the Caspian Sea, so bypassing the lower Volga.[22] The trade and customs dues on which the wealth of Sarai had been built were disastrously diminished. The Golden Horde underwent a brief revival in the late fourteenth century and the first quarter of the fifteenth, but by the mid-fifteenth century it had broken up and formed smaller successor khanates, including those of Kazan and Astrakhan, both on the river Volga.

✤

Kazan had originally been a border post in the Bolgar state. Archaeological evidence suggests that the town was probably founded at the very beginning of the eleventh century. The rise of Kazan to prominence was due to the decline of the Golden Horde; but the town also has a very advantageous geographical position at the confluence of the Kazanka and the Volga rivers, just north of where the Kama joins the Volga. By the beginning of the fifteenth century, Kazan seems to have become the administrative centre of the khanate and had eclipsed the Volga town of Bolgar; indeed, it was sometimes known as the 'New Bolgar'.[23] As the successor khanate to the Golden Horde, the Kazan khanate shared many of its features. The khan of Kazan and his immediate entourage were probably ethnically Mongols, but most of the landowning class were probably ethnic Bolgars and of primarily Turkic origin. The Islamic elite in Kazan was dominated by families claiming descent from the Prophet Mohammed; they also acted as diplomats, particularly with Central Asian states. The subject people, from whom tribute was exacted, were Finno-Ugric, east Slav or Turkic. The elite and some of the Turkic population were Muslim; the Finno-Ugric tribes were primarily shamanistic.

Kazan established itself as a major trading and commercial centre on the middle Volga (although there were other towns in the khanate). It took goods from the north – mainly furs and slaves – as well as from China and Central Asia from the east, and the Caucasus from the south. In this way it traded with the Rus, but was also a competitor with the Rus for control of the fur trade on the river Volga. Those merchants who resided in Kazan or who visited its fairs included Armenians, Slavs and Persians. Skilled craftsmen flourished in the town, working in leather, wood, ceramics, gold and other metals. Some 5,000 craftsmen were said to have been working in Kazan by 1552.[24]

Astrakhan was first mentioned in thirteenth-century Arab accounts, by the name of Xacitarxan. The herds of the Mongols were pastured near present-day Saratov from June to August, and then were moved down the river banks to near Astrakhan from August to December.[25] Ibn Battuta noted that the khan moved to Astrakhan in the winter, when it was too cold to stay further north. The town was destroyed by Tamerlane in 1395 when he was fighting the Golden Horde (see above), but was rebuilt.

Astrakhan's position on the delta made it a natural trading point between the Caspian Sea and the river Volga and the delta, and later it became a truly multi-ethnic and multi-confessional merchant town (see Chapter 9). It is also surrounded by the shallow waters of the delta, which are immensely rich in fish, including the beluga sturgeon, from which caviar is taken. We know very little about commercial

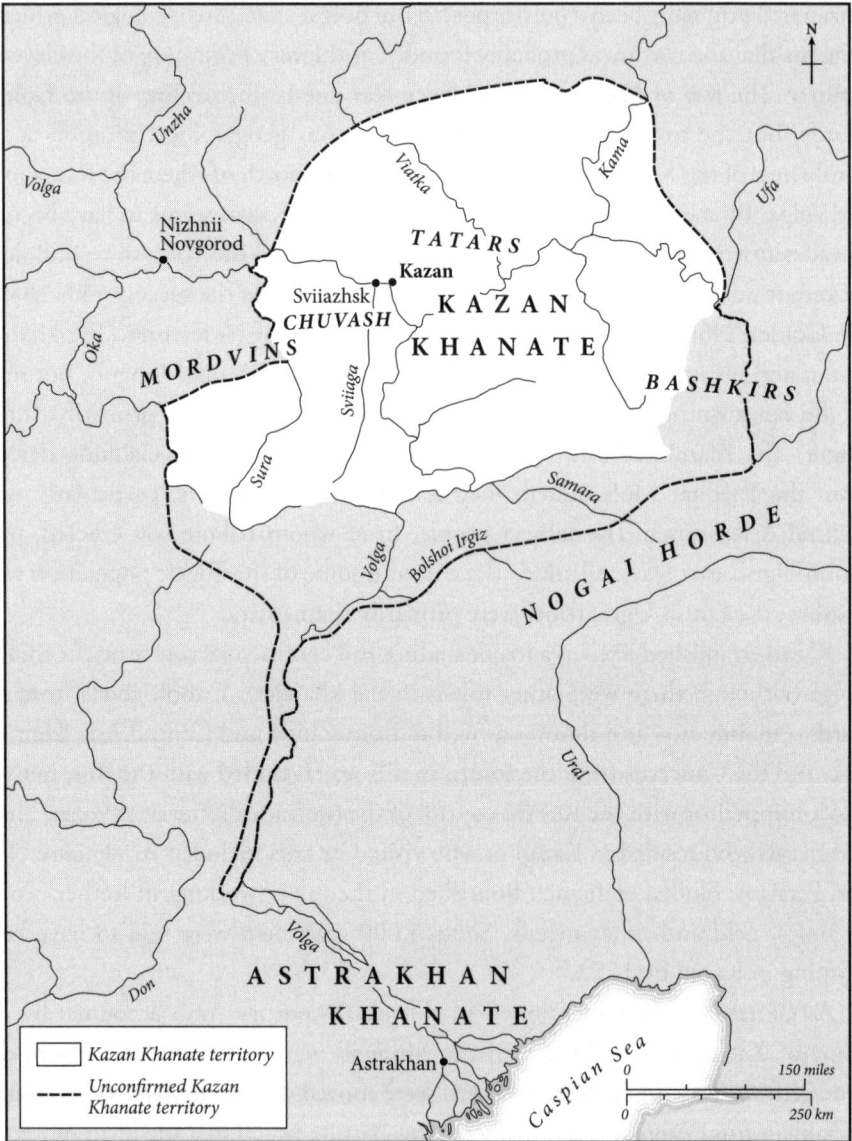

4. The Kazan Khanate

and everyday life in the khanate of Astrakhan in the century of its existence from the mid-fifteenth century until 1556, when it was conquered by Ivan IV (except the list of khans who ruled it).[26] One of the few comments on trade we have is by Sigismund von Herberstein, who was sent to Russia as the ambassador of the Holy Roman Empire and wrote extensively on the history and geography of the country.

He noted that when there was a conflict between the Russians and the Kazan khanate in the 1520s, the Russians attempted to sell their furs in Nizhnii Novgorod rather than in Kazan, but that this had a serious effect on their trade from Astrakhan:

> . . . by the removal of a fair of this sort, the Russians suffered as much inconvenience as the people of Kazan; for it produced a scarcity and dearness in many articles, which had been the custom to import through the Caspian Sea from Persia and Armenia by the Volga from the emporium of Astrachan, and especially of the finer kinds of fish, amongst which was the beluga [sturgeon], which is taken in the Volga, both on this side and the other of Kazan.[27]

Astrakhan sent ambassadors to Moscow with silk and horses from China and Persia to exchange for furs, swords and saddles.

At first, the Kazan khanate continued the policy of the Golden Horde and repeatedly asserted itself against the Rus princes. Kazan forces raided Russian lands several times and forced the princes to pay tribute. In 1439, the Kazan khan occupied Nizhnii Novgorod and then attacked Moscow. In 1444 and 1445, the khan defeated the forces of Vasilii II, the grand prince of Vladimir, at the town of Suzdal and forced him to pay a heavy tribute. By the 1470s, however, the Kazan khanate was in decline, torn apart by warring members of the khan's family. By this time, it was being challenged by a new Rus state – the principality of Moscow (sometimes known as the Grand Duchy of Moscow, or simply Muscovy).[28]

❖

It falls beyond the remit of this book to explain the rise of the principality of Moscow at the expense of other Rus principalities.[29] It is worth noting, however, that there was nothing inevitable about the dominance of Moscow. Indeed, in the mid-thirteenth century, Rus leadership was provided rather by the princes of Vladimir, and in particular by Alexander Nevsky, who is credited with repelling invasion by the Teutonic knights and Sweden from the west. Novgorod was the largest and most prosperous city in the Rus lands at the beginning of the fourteenth century. Principalities did, however, have a habit of subdividing between sons and other rivals to the throne on the death of a prince. In the second half of the thirteenth century, there were 14 principalities in the north-east of what is today European Russia. One of them – Kostroma, on the river Volga – could even have been a rival to Moscow. However, the principality of Kostroma was short-lived: it emerged as a separate principality in the 1250s, but ceased to exist in 1277, when its prince, Vasilii, died.[30]

Moscow's rise, from the beginning of the fourteenth to the sixteenth centuries, was based on a number of factors: the personal qualities of key rulers in Moscow; the success of the Moscow princes in centralizing government and fiscal institutions; their subsequent ability to raise money through taxes to pay for the armed forces; the improvement and modernization of Moscow's armed forces; the cultural and ideological advantage which ensued after the establishment of Moscow as the residence of the metropolitan of the Orthodox Church; the evolution of a system by which nobles were rewarded with lands in return for service to the prince, thus (usually) ensuring their loyalty; the skill of Moscow princes in negotiating with the Golden Horde and the Kazan khanate, even while paying them tribute and being formally subservient to them; the judicious alliances formed between Moscow and other enemies of the horde and the khanate, in particular the Crimean khanate and the nomadic Nogai horde (discussed in the next chapter); the luring of many Tatar nobles into the service of Moscow; and the in-fighting within other principalities which weakened them vis-à-vis Moscow.

The outcome of the contest between Moscow and its rival Tver to the north was crucial. Tver was located at the confluence of the rivers Volga and Tvertsa (which links the river Volga with Novgorod and later with St Petersburg) and had become a separate principality in probably only 1247. The principality comprised land either side of the banks of the Volga, from the town of Kashin in the east to Zubtsov in the west. Both Moscow and Tver were well situated within the river network to develop as trading centres; but had Tver triumphed, then there would have been a dominant principality directly on the river Volga.

In the fourteenth century, the principalities of Moscow and Tver competed with each other for control over the principality of Vladimir, and, in effect, for political dominance over north-east European Russia and its rivers. At first, the advantage lay with the principality of Tver, under Prince Mikhail, who managed to get the backing of the Golden Horde khan to award him the patent for the principality of Vladimir. Mikhail, however, made a series of blunders at home, including alienating the city state of Novgorod and the metropolitan of the Russian Orthodox Church. Iurii, prince of Moscow, took full advantage of this. In 1312, a new khan ascended the throne of the Golden Horde and favoured Iurii, who had tactfully visited him on the Volga at Sarai. Armed conflict ensued between Tver and Moscow: Mikhail triumphed, but his victory only made the khan suspicious of him as a potential military threat. Both Iurii and Mikhail were brought to

Sarai for the khan's judgement in 1318, whereupon Mikhail was accused of withholding payment of tribute and executed (he was later, in 1549, declared a saint in the Orthodox Church). The principality of Vladimir passed to Moscow.

Sarai was a dangerous place for Rus princes to visit. Seven years later, Iurii was murdered there by Mikhail's son, Dmitrii, while the khan was still trying to determine finally who should control the principality of Vladimir. It briefly passed back to Tver, but in 1327 disaster struck: the city was occupied by a Tatar force under the khan's cousin, who, it was claimed, intended to rule the principality of Tver himself. The citizens of Tver rose up in revenge and lynched any Tatars or Muslims they could find; the khan, not surprisingly, turned again to support the claims of Moscow. The new prince of Moscow, Ivan I (ruled 1331–40 and nicknamed 'moneybag'), promptly sacked the town of Tver. In effect, it was ruined and Moscow had regained the support of the khan. This was an important victory for Moscow, albeit it shows how dependent the Rus princes still were at this time on the support of the khans to further their ambitions. The struggle for supremacy with Tver continued over the next few decades, and the principalities fought each other inconclusively. By 1375, however, Moscow was able to assert its political superiority and the prince of Tver was forced to accept a junior status.

The victory at Kulikovo field (see above) in 1380 was significant in terms of raising the prestige of the prince of Moscow, Dmitrii, who by this time was also grand prince of Vladimir; but in itself, this triumph did not signify that Moscow had any independence from the Golden Horde. In 1392, coins issued under Dmitrii had 'Grand Prince Dmitrii Ivanovich' on one side, and on the other 'Sultan Tokhtamysh: Long may he live!'[31] In the following century, Moscow moved to control the city state of Novgorod, as well as Tver. Novgorod was forced to accept the overlordship of Moscow in the middle of the fifteenth century, and it was formally annexed in 1478. As a result, the lands of the Moscow princes now surrounded the principality of Tver. Many prominent Tver nobles recognized where power now lay and joined the service of the Muscovite Grand Prince Ivan III (ruled 1462–1505). At this point, Mikhail III, the last prince of Tver, sought help from Lithuania, which simply gave the grand prince of Moscow an excuse to invade his territory. In 1485, the city surrendered to Moscow forces and the principality of Tver met its abrupt demise. The town remained important in trade on the Volga, but was never able to assume such a prominent political role again. The principality of Pskov, in the north-west of European Russia, soon followed suit and became annexed to Moscow.

The principality of Moscow was now the undisputed leader of northern and north-eastern European Russia. It was, as one historian has stated, 'in every sense, a major power'.[32] The ambition of the Moscow princes was demonstrated by the titles taken by Ivan III. As well as grand prince, he began to use the title 'tsar' – an imperial title (although not recognized as such in other European countries at the time) that could even be seen to make him the Orthodox equivalent of the Holy Roman Emperor. Ivan also adopted the two-headed eagle on coinage, possibly in imitation of the Habsburgs. Moreover, he added the words *of all Russia* (*vseia Rusi* in Russian) to his titles, thus stressing his claim of sovereignty over all the Rus principalities in what is now European Russia and his right to conquer the lands of Kievan Rus, which were now part of Lithuania.[33]

By the late fifteenth century, Moscow was in a position to challenge the Kazan khanate. In the reign of Ivan III, it placed direct pressure on the khanate by asserting a share in the fur trade of the north and by giving military assistance to the enemies of the khanate. In 1469, Muscovite troops entered the territory of the Kazan khanate, but were beaten back. In retaliation, Kazan forces attacked towns in the Moscow principality. In 1480, the armies of the khan and of the grand prince of Moscow (with support from Crimean troops) faced each other in a standoff on the Ugra river (now in the province of Kaluga in western Russia and west of the river Oka), whereupon the khan retreated, rather than risk invading Muscovite territory. This was portrayed at the time, and later in Russian accounts, as the moment when Moscow, and in effect the Rus, threw off the 'Tatar yoke'. In itself, the incident did not lead to any dramatic change in the relationship between Moscow and the khanate, but it has lived on as one of the myths surrounding the ultimate defeat of the khans. A coin minted in honour of the occasion depicted on one side a mounted Christian warrior, probably St George, and on the other side Ivan III's name written in Arabic, rather than the name of the Mongol khan, symbolically implying that his dynasty was succeeding the khans.[34]

By this time, Moscow had almost certainly stopped paying tribute to the Kazan khanate (although at points of vulnerability it continued to send 'presents' to the khans) and was successfully interfering in the affairs of the khanate when, as frequently happened, there was a change of regime. In 1487, Ivan III occupied Kazan, deposed the khan, and put his own puppet leader on the throne. The khanate was certainly weak by this stage, although it is probably too much to state, as some historians have, that the Kazan khanate was already in a position of 'dependency' on Moscow.[35] One of the reasons for Moscow's failure to control the khanate indirectly (and in effect make it a client state) was that the very rapid changes of khan meant it was almost impossible for Moscow to keep its own candidates on the throne.

Ivan IV (ruled 1547–84) was the first tsar to be crowned 'Grand Prince of Vladimir, Novgorod and Moscow and *Tsar of All Russia*', consciously asserting himself during this ceremony as the only independent Orthodox ruler, who therefore could claim to be 'emperor' of the Eastern Church.[36] And it is from this point on that I will refer to the inhabitants of the principality of Moscow as 'Russians' rather than Rus or Muscovites, and to the principality of Moscow as 'Russia'.

Ivan initially continued the policy of trying to control the khanate, by installing a puppet khan, Shah-Ali, in 1546. It was only when his chosen candidate was overthrown within a month that Ivan decided the situation had to be resolved militarily. His first two campaigns failed, but by 1551 he was in a far stronger position, having not only once again installed his puppet, Shah-Ali, but having also brought over to his side nomadic and semi-nomadic Nogais from the Nogai horde and a significant element of the non-Muslim elite on the western bank of the Volga, by promising to recognize their privileges. To take Kazan, however, Ivan had to get control of the river from the north and the west to cut off reinforcements. The tsar founded the fortress at Sviiazhsk on a promontory at the confluence of the Volga and the river Sviiaga, which is west of Kazan (or indeed, 'on the shoulders of Kazan', as a Russian chronicle put it).[37] This gave Ivan control of the western bank of the Volga and, in effect, divided the territory of the khanate east from west.

The fortress at Sviiazhsk allowed the Russian forces to maintain a military base on the Volga from which they could supply their troops besieging Kazan from the north. In the first instance, Ivan tried to impose conditions on Kazan which would have made it totally dependent on Moscow, including the acceptance of a permanent Russian garrison in Kazan itself. When this was rejected in the usual way – that is, by deposing the khan and slaughtering the Russian garrison – Ivan raised another enormous army of some 150,000 men and attacked, laying siege to the town and blocking any relief not only on the Volga, but also on rivers to the north and south of Kazan. The Russians had skilled sappers and superior artillery, as well as a 12-metre-high siege tower, on which they could mount cannon and concentrate their fire against the walls and defences of the citadel. Completely cut off, the far smaller Tatar forces in the citadel put up fierce resistance, but could not match the Russian forces or resist their siege tactics. On 2 October 1552, the Russians and their allied troops stormed the Kazan citadel. With the exception of the khan, all the survivors were slaughtered and the buildings in the citadel destroyed.

The control of the western bank of the Volga was crucial for the success of Ivan's campaign. In popular folktales, the river itself had to be 'tamed' to the will of the tsar, but then went on to ensure his victory. The tales collected by Nikolai Aristov in the late nineteenth century included an account of Ivan forcing the raging river to bend to his will, so that his troops could cross successfully and approach Kazan:

'Send the executioner here!' yelled the tsar, 'I'll teach you a lesson!' The executioner arrived, a mighty man – and the tsar ordered him to whip the river with his knout, to teach it not to rebel against the tsar and his army. The executioner took his knout, rolled up the sleeves of his red shirt, took a run, and as he whipped the Volga, the blood sprayed upwards a yard in height, and a bloody wound appeared in the water, thick as a finger. The waves in the river went calmer, but the tsar cried, 'Show no mercy, strike harder!' The executioner took a longer run than before and struck harder: the blood sprayed higher and the wound was wider; the Volga calmed down more than before. After the third blow, which the executioner struck with all his might, the blood splashed three yards high and the wound was three fingers wide – then the Volga became completely still. 'Enough!' said the Terrible Tsar, 'That's taught you a lesson!' After that, all the troops crossed the river, and not one soldier-lad drowned, although they were very frightened.[38]

Over a thousand kilometres to the south of the khanate of Kazan was the khanate of Astrakhan, whose territory encompassed the lower Volga and the Volga delta. In 1554, Ivan installed a puppet ruler in the khanate of Astrakhan; but when that khan proved disloyal, an army of some 30,000 men was sent down the river Volga to meet up with the Nogais and lay siege to Astrakhan. The town fell with little resistance in 1556.

In the short term, the effects of the military campaigns on the local population were devastating. The Nogais, who had supported Ivan IV in the attacks on Kazan and Astrakhan, now found themselves in a subordinate position to the Russians and had to pay them tribute. Their leader, Ismail, appealed to Ivan IV for support, stating that the Volga (and the Iaik) had been Nogai rivers, but that 'now the Volga is yours, but it is the same as ours because we are brothers. And there are your people in the Volga estuary and there are our people there too, whom we left there poor and hungry.'[39]

Ivan IV not only destroyed Kazan, but also forced the Tatars away from the city. Even today, the villages immediately surrounding Kazan are predominantly Russian, and Tatar villages tend to be located some 30km away from the town. The Englishman Anthony Jenkinson travelled through south Russia in 1558 on behalf of the English Muscovy Company, looking for trade routes down the Volga and across the Caspian Sea into Persia. He noted the conquest of Kazan, but also the devastating impact of the military campaigns on the Tatar population who had fled and relocated to Astrakhan:

> There was a great famine and plague among the people, and especially among the Tartars who came thither in great numbers to render themselves to the Russians, to seek succour at their hands, their country being destroyed: they were but ill entertained or relieved, for there dyed a great number of them for hunger, which lay all the island through in heapes dead, and like to beasts unburied, very pitifull to behold: many of them were also sold by the Russes, and the rest were banished from the Island [Astrakhan]. At the time it had been an easy thing to have converted that wicked nation to the Christian faith, if the Russes themselves had been good Christians: but how should they show compassion unto other nations, when they are not merciful unto their own? At my being there I could have bought many goodly *Tartars'* children, of their own fathers and mothers, a boy or a wench for a Loafe of bread worth six pence in England, but we had more need of victuals at that time than of any such merchandise. This *Astracan* is the furthest hold that the Emperour of *Russia* hath conquered of the *Tartars* towards the Caspian Sea. There is a certain trade of merchandise there used, but as yet so small and beggarly, that it is not worth the making mention.[40]

The town of Bolgar, which had survived the Mongol invasion, also fell into complete ruin after the conquest of Kazan.[41] 'Bolgar, the holy refuge of Islam,/ Now lies in ruins; nothing else is there' are two lines from a poem by Gali Chokri on the 'Sacking of Bolgar'.[42] The town became a ruin, and it was ironic that it should be Tsar Peter I in the early eighteenth century who demanded that the remains of a once great city should be preserved – ironic, because at just that time he was pursuing an active policy of conversion, sometimes forced, of Muslims on the middle and lower Volga. This poem by Gali Chokri, however, was written in the mid-nineteenth century, as part of the Volga Tatar intellectual revival. The Bolgar state and the site of Bolgar were (and still are) of enormous significance in terms of the evolution of the identity of the Volga Tatars in the nineteenth century and today.

The conquest of Kazan (and then also Astrakhan) was seen, at the time and thereafter, as a triumph of Christianity over Islam, and was justified and legitimized in those terms – something which set the scene for the assertion of Orthodoxy over the whole region. A Russian chronicle recorded that:

With the aid of our Almighty Lord Jesus Christ and the prayers of the Mother of God . . . our pious Tsar and Grand Prince Ivan Vasilevich, crowned by God, Autocrat of all Russias, fought against the infidels, defeated them finally and captured the Tsar of Kazan . . . And the pious Tsar and Grand Prince ordered his regiment to sing an anthem under his banner, to give thanks to God for the victory; and at the same time ordered a life-giving cross to be placed and a church to be built, with the uncreated image of our Lord Jesus Christ, where the Tsar's colours had stood during the battle.[43]

The day after the citadel was taken, the foundations were laid for an Orthodox cathedral, which was completed as early as 1557. The foundations of an Orthodox cathedral in Sviiazhsk had been laid even before Kazan was taken. In January 1553, an official diplomatic note was written on the orders of Ivan IV to the Lithuanian nobles; in it the name of Kazan appeared among the tsar's titles. The note stated that:

. . . the Mussulman nation of Kazan which had shed Christian blood for many years and which had caused our Sovereign much annoyance before he reached his mature age, by God's grace this Mussulman nation of Kazan died by the sword of our Sovereign [Ivan IV]; and our Sovereign appointed his viceroys and governors in Kazan, and he enlightened this Mussulman abode with the Orthodox Christian faith and he destroyed the mosques and built churches [in their place], and God's name is being glorified now in this city by the Christian faith, and this land obeys our Sovereign, and it has been united with the state of our ruler . . . And we praise God for this, and may God also grant in the future that Christian blood be avenged against other Mussulman nations.[44]

The fall of Kazan and Astrakhan gave the Russians control over the whole of the river Volga, from its source to the Caspian Sea, and meant the replacement of Mongol/Tatar power with Russian power. From this point on, Russia was a multi-ethnic and multi-confessional empire, with a ruler who claimed to rule over all Russia and to be the secular head of the Christian Church in the east, but also to be the ruler of predominantly Muslim Kazan and Astrakhan. How the Russian tsars settled and governed their newly acquired territory on the middle and lower Volga is the subject of Part 2.

Part 2

The Volga in the Russian Empire

Violence and Control on the River

3
NON-RUSSIANS AND RUSSIANS ON THE NEW VOLGA FRONTIER

The conquest of Kazan and Astrakhan made Russia a multi-ethnic and multi-confessional empire. The borders of the empire had moved significantly to the south, and the middle and lower Volga were now frontier territory. This was the first significant ethnically non-Russian and non-Christian territory to be incorporated into the Russian Empire, and in many ways the policies pursued by the tsars in the second half of the sixteenth and the seventeenth centuries set the pattern for control of both the borderlands and non-Russian peoples for the whole imperial period – and, to an extent, for the Soviet and post-Soviet periods as well. The middle and southern Volga region demonstrates how 'empire' worked in the early modern period.

The government had to assert control over the region, and did so in four main ways:

- through force, and particularly a military presence;
- through government and the establishment of administrative organs of the Russian state;
- through ideology, by the extension of Orthodoxy, with the construction of churches and monasteries and with conversion policies towards Muslims and pagans;
- through assimilation, by incorporating non-Russian elites into the Russian state through service to the state and land grants.

This last element was particularly important in a non-ethnically Russian and non-Christian region, and shows that the Russian Empire could be both determined and flexible. Historians have sometimes characterized this as an 'empire of difference' in which some diversity could be tolerated on the peripheries in order to achieve social stability, provided rule from the centre was accepted.[1]

❧

The first groups of non-Russians that the Russians encountered on their route down the river Volga were Finno-Ugric peoples. One such were the Mordvins, who speak a Finno-Ugric language. Some of them had settled in the Nizhnii Novgorod area and lower down the Volga. It is hard to establish reliable figures for the number and ethnicity of the population of Russia before the first censuses were taken in the eighteenth century (and even then, figures have to be treated with caution and can only be taken as a rough estimate). In the second half of the eighteenth century, some 114,000 to 176,000 Mordvins were listed in the censuses in the middle Volga region, with a further 32,200 to 59,000 on the lower Volga.[2] Today, some 700,000 people in the Russian Federation identify as Mordvins. Udmurts (sometimes also known as Votiaks) are another Finno-Ugric people; they settled north of Kazan province, in what is today the Udmurt Republic, or Udmurtia. There were also Udmurts in Nizhnii Novgorod province, although they were always a small minority compared with Russians.

Mari (known earlier as Cheremiss by the Russians) are another Finno-Ugric people who settled in the northern region of the middle Volga, between the Volga and the Kama. These people are subdivided into so-called 'mountain' Mari, who had settled on the western bank of the Volga, and the 'meadow' Mari, on the eastern bank. Their language is also Finno-Ugric, and is divided into four mutually intelligible regional dialects. In the second half of the eighteenth century, some 48,000 to 54,000 Mari were listed in censuses on the middle Volga. In Russia today, there are some 500,000 people who identify as Mari (the majority of whom live in the present-day Mari El Republic). Most of the Finno-Ugric peoples practised shamanistic or animist beliefs at the time of the conquest of Kazan (although some Mordvins who lived in the northern Nizhnii Novgorod region had already converted to Orthodox Christianity before their lands were incorporated into Russia).

Foreign travellers provide contemporary ethnographic accounts of the peoples on the Volga. The descriptions of non-Russians are, however, almost always disparaging by comparison with accounts of the appearance and physique of western or central European people, including Russians. Johann Gottlieb Georgi, a German botanist employed in the Russian Academy of Sciences in St Petersburg in the late eighteenth century, was disparaging about both Mari and Mordvins: the women of the former, according to him, were 'much inferior to those of Russia, as well in regard to comeliness, as in gaiety of temper, and vanity of dress'; the latter 'have commonly brown, harsh hair, a thin beard, and lean face; it is very rare to find a pretty woman amongst them'.[3] The main characteristic commented on by foreign travellers, however, was that the Mari were known for sorcery and witchcraft (as indeed were other Finno-Ugric people within the Russian Empire at the time and later): 'they are a disloyal, thieving, and superstitious people'; they are 'barbarous, treacherous, and cruel, and much given to sorcery and robbery'.[4]

The Chuvash had settled on the Volga, mainly to the south and west of the Mari on the western bank, in what is today the Chuvash Republic. Both the people and their language are Turkic, although there is some dispute about the ethnic origins of the Chuvash people, and in particular about whether they are descended from the Bolgars. However, they have been in close contact over a long period with Finno-Ugric peoples, such as the Mari, with whom Georgi compared them (even more) unfavourably. They had, according to him, 'indeed a paler complexion, are more lazy, and are not so sharp-witted as the [Mari]; besides this, they are not so cleanly, and are less nice in the choice of food and necessaries'.[5] The Chuvash were more numerous than the Mari, the Udmurts or the Mordvins. In the second half of the eighteenth century, there were between 240,000 and 300,000 Chuvash in the middle Volga region, and a further smaller number on the lower Volga. Some Chuvash had become Muslim by the eighteenth century, but many still followed animist beliefs. Today there are some 1.5 million people within the Russian Federation who identify as Chuvash, and they are heavily concentrated around Cheboksary, the capital of the Chuvash Republic.

The largest ethnic groups that the Russians encountered in the Volga region were, and still are, Tatars. They formed the largest non-Slav ethnic group in the Russian Empire – and still are the largest in the Russian Federation today. It was estimated that there were some 800,000 Tatars in the Russian Empire in 1795 (out of a total population of some 40 million).[6] There are over 5 million Tatars in the Russian Federation today, living in Tatarstan, Bashkortostan and western Siberia (and in the Crimea, annexed by Russia from the Ottoman Empire in 1783), as well as significant numbers in Uzbekistan and Kazakhstan. The word 'Tatar' in English conjures up an image of savage tribesmen from the east, partly because the word is often synonymous with 'Mongol', and not least because the old spelling of 'Tartar' designated someone from Tartarus, which in the Greek-Roman world was the part of the infernal regions reserved for the punishment of the wicked. Today, the designation 'Tatar' is used more narrowly to designate Turkic-speaking people. The ethnic origins of the Tatars in the Volga region have been (and still are) a matter of some dispute, and we shall see that in the nineteenth century the Volga Tatars traced their origins to the Bolgar state, rather than to the Mongols.

The largest group of Tatars in the Russian Empire were in the middle Volga region, around Kazan. In the early eighteenth century, ethnic Russians were not in a majority in Kazan or Sviiazhsk, although by the end of the century they were, with Tatars accounting for only some 12–14 per cent of the population of Kazan province. By this stage, there were over 300,000 Tatars in the middle Volga region, and another 43,000 on the lower Volga.[7] Some Tatars, unlike the majority of Finno-Ugric peoples and the Chuvash, had settled in towns as well as in villages, and there were Tatar merchants of some significance. Tatars also had an aristocracy, with princely families at the top, and a landowning class. Some princely Tatars had joined the army of the tsar even before the conquest of Kazan. They were the first Muslim peoples encountered by the Russians.

The more sophisticated social and economic position of Tatars in the Volga region was recognized by foreign travellers. Possibly this also accounts for the fact that the physical descriptions of Tatars are, on the whole, more favourable than the descriptions of the other ethnic groups in this region. Georgi stated that:

> A tall man is rarely met with among them: they are commonly lean, have a small face, a fresh complexion, a little nose, small mouth and eyes, which last are generally black; a sharp look, deep chestnut-coloured hair, which is lank, and turns grey long before old age. For the most part they are well-made; their sprightly manner, straight shape, and modest or timid mien, give them a certain very agreeable air. They are haughty and jealous of their honour, but of a very middling capacity, negligent, without being lazy; apt at every sort of handicraft; fond of neatness; given to sobriety, frugality, and compassion. These virtues they acquire by education, and the precepts of their religion, to which they are zealously attached.[8]

Baron August von Haxthausen, a German scientist, economist and folklorist, travelled to Russia in the mid-nineteenth century and commented on Tatars as follows:

> The face is oval, the eyes black and animated, and the nose noble and curved, the mouth of a refined expression, the teeth excellent, and the complexion that of the Caucasian race . . . all their movements are active, graceful, and often dignified: the women are small, and disfigured by paint . . . The character of the

Tartars is amiable; they are sociable, honourable, friendly, confiding, orderly, and cleanly. The old antipathy to, and mistrust of, the Russians still prevail, but to the Government they are loyal and obedient.[9]

The non-Russians commented on so far were settled peoples, in villages or towns; but as the Russians moved south after the conquest of Kazan, towards Astrakhan, they came into contact with nomadic peoples, whose territories touched the river Volga but spread far beyond: Nogais, Kazakhs, Kirgiz (also written as Kyrgyz or Kirghiz) and Kalmyks. The origins of the Nogais can be traced back both to a Mongol tribe called the Mangits and to Turkic tribes; their language is Turkic. By the turn of the sixteenth century, the Nogai horde comprised 18 Turkic and Mongol tribes and had its own distinct dynasty. Nogais spread over the north Caucasus, the lower Volga region and the river Iaik (renamed the river Ural in the late eighteenth century) further east of the Volga; by the mid-sixteenth century the Nogai had split into the lesser and greater hordes (it was the greater horde that was located on the river Volga and the river Iaik).[10] By the time Ivan IV conquered Kazan and Astrakhan, the Nogais had converted to Islam. Today there are some 100,000 Nogais in the whole Russian Federation, but fewer than 5,000 in the Astrakhan region. Nogais frequently attacked settlements and were regarded with suspicion or outright hostility. An English traveller, Mary Holderness, commented in the nineteenth century on the Nogais: 'The moral character of the Nogays is of the worst description, and there is hardly any kind of mischief which they will not perpetrate.'[11]

Kazakhs in the sixteenth century were a confederation of Turkic and Turkified Mongol tribes, organized in three hordes. Most of the lands of the Kazakhs were to the east of the river Volga (incorporating present-day Kazakhstan); but the lands of the lesser horde included territory in the vicinity of Astrakhan. The Kirgiz are of mixed ethnic Mongolian, Chinese and Turkic origin, but probably came from the western parts of present-day Mongolia. Their language is Turkic. European travellers in the eighteenth and nineteenth centuries often saw the Kazakhs and Kirgiz as the same people, and sometimes referred to them as Kazakh-Kirgiz.

Encounters with the above groups meant encounters with Muslims (Tatars, Nogais, Kazakhs, Kirgiz and some Chuvash), people practising animist religions (Mari, Chuvash, Udmurts, Mordvins) or converts to Orthodoxy (Mordvins). But in the south of the Volga, the Russians also came into contact with Buddhists. Kalmyks are from a western branch of the Mongols known as Oirats, and their language is Mongol. In the late sixteenth century, Kalmyks converted to Tibetan Buddhism. They constituted an anomaly: as one historian has commented, they were 'an outpost of Buddhism on the south-eastern fringe of Europe, and a foreign body in the middle of the Turkic-speaking and Islamic world of the steppe'.[12] Most Kalmyks in the sixteenth century were subject to the Kazakhs, but a group of Kalmyks from the Torgut tribe settled in the southern Volga region, at least until 1771, when some 150,000 Kalmyks moved east from the Volga back towards China (see Chapter 5). It is perhaps not surprising that the word 'Kalmyk' comes from the Turkic 'to leave behind' – perhaps indicating that they were regarded as the perpetual 'outsiders' who were not fully incorporated into the Mongol Empire or its successor states.[13]

As far as the Russian settlers and foreign travellers were concerned, nomadic peoples were dangerous, untamed people of the steppe, and in the main contact with them was violent, with nomadic raids from the east and south-east on the settlers' villages and towns, and on individual travellers. The river Volga was a border in this respect, with the dangerous eastern steppe far wilder and more dangerous than the more settled and cultivated lands on the western side of the river.

❖

The conquest of the towns of Kazan and Astrakhan was only the first step in establishing the Russian presence over the lands of the middle and lower Volga. The state needed to control key strategic points on the river and to stabilize the countryside. It did so by establishing Russian administrative control over the region, and by having the lands settled by Russian landowners and Russian peasants. At the same time, a strong Christian, Orthodox, presence was established on the new frontier. These were deliberate acts of internal colonization, and demonstrated the intent of Ivan IV and later tsars to establish their authority over their new lands and new empire.

In the immediate aftermath of the taking of Kazan, all the Tatar soldiers defending the citadel were slaughtered, and all the mosques were destroyed, in a clear statement that the town was now in Russian – Christian – hands. Tatars were forcibly moved away from the town. In the short term, military repression had to be used to subdue revolts by Mari and Chuvash peasants on the middle Volga, and led to the temporary establishment of military outposts in Mari and Chuvash territory.

A more permanent and visible sign of Russian authority was then ensured by the construction, at considerable cost, of forts which gave the Russian state a military and urban presence along the Volga. A fort – as well as a church – was established at Sviiazhsk in 1551: that is, before and in anticipation of the conquest of Kazan. This was soon followed by the construction of the forts of Cheboksary in the early 1550s on the western bank of the Volga, and then Tetiushii further down the river, south of Kazan and Samara (on the eastern bank). In the mid-seventeenth century, a new series of forts (called the Simbirsk line) was constructed to the south of this line; these ran west to east, via the towns of Kozlov and Tambov, to reach the fort of Simbirsk (founded 1648) on the Volga, via Saransk and Sursk (on the river Sura, which runs parallel to, but to the west of, the Volga through the province of Penza). Another fortified line was constructed which branched off from this line through the town of Syzran, on the Volga, west of Samara and south of Simbirsk.[14] Further south, Saratov, Tsaritsyn and Chernyi Iar were established as forts on the river to create a Russian military and administrative presence lower down the Volga,[15] something that was made necessary by the continuing raids from the east and south from the Nogais, Kirgiz and Kalmyks.

Forts were usually a collection of wooden buildings in a stockade, surrounded by earthen ramparts. The first stone buildings to be constructed were a clear demonstration of the presence of the Russian state: a *kremlin* and an Orthodox church. Kremlins were fortresses constructed either on the high bank of a river or on raised ground, and were defended by wooden and then stone walls; within the kremlin were the main administrative buildings and barracks, but also an Orthodox church. Kremlins were built to impress, as well as for defence: they were the symbols of Russian military power and of legitimacy. The uniformity of the architecture of both kremlins and churches was also significant and deliberate: this was a distinctively Russian style and was very different from local, Tatar, architecture. It was a clear statement that the territory was now part of the Russian state.

Russian control was exercised from these forts at various points on the river Volga. Indeed, control of the river was synonymous with control of the new territory. These settlements were military outposts in the first instance, and only later developed commercial and economic significance. This can be seen from the forced relocation of military personnel to these new towns on the river. Ivan IV had founded special military units, normally called musketeers (*streltsy* in Russian), which had participated in the assault on Kazan in 1552. These were infantrymen, originally recruited from free tradesmen and peasants. In the second half of the sixteenth century and the early seventeenth century, this force was relatively effective and became the backbone of the garrisons in the newly founded river Volga forts. By the 1560s, there were already some 600 musketeers in the garrison at Kazan.[16] In the early seventeenth century, Adam Olearius, the secretary to the ambassador of the

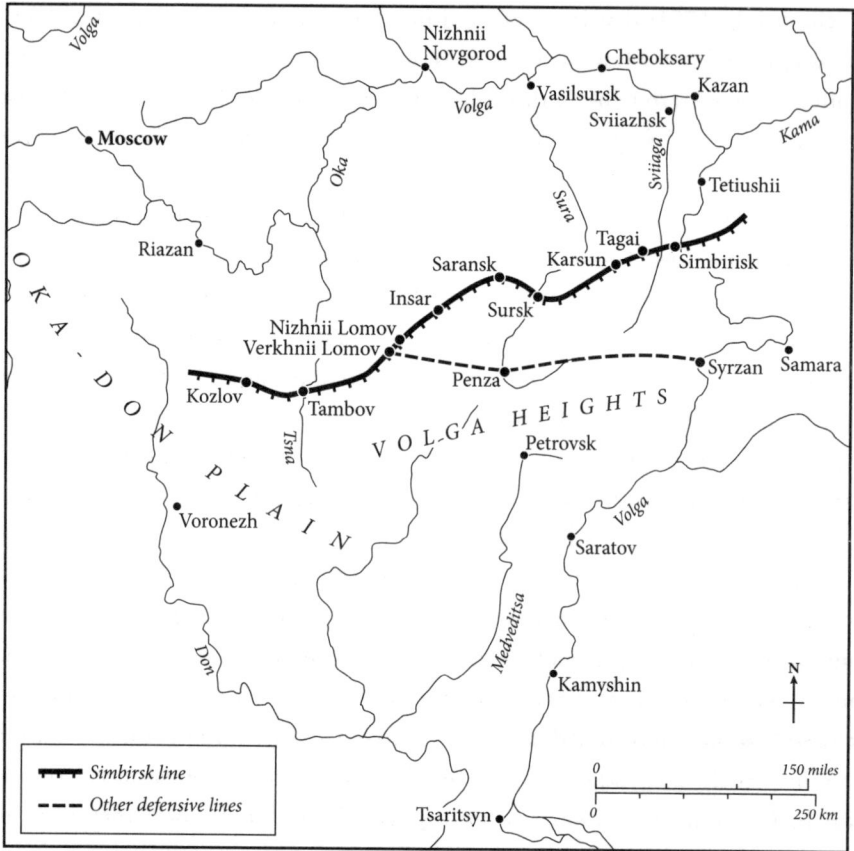

5. The Establishment of Russian Control of the Volga

duke of Holstein-Gottorp, remarked of the town of Saratov that 'only streltsi live here' and their purpose was to protect the town from Kalmyks. Further down the Volga, he noted that the fort of Chernyi Iar had been constructed with eight towers for defence and was 'inhabited exclusively by the Streltsi, who are supposed to deal with the many marauding Cossacks and Tartars hereabouts'.[17] By the middle of the seventeenth century, 1,776 musketeers were manning the garrison in Simbirsk, some 6,000–8,000 were stationed in Astrakhan, and over 2,000 in Tsaritsyn.[18] The presence of musketeers was supplemented by Cossacks (discussed further in the next chapter), who could be moved from what had been the previous border of the Russian Empire (in what is now Ukraine and Russia's Belgorod region) to the Volga. Cossacks could also be moved down the chain of forts on the Volga as the northern part became more settled and secure, to defend the southern and eastern towns from nomadic raiders.

The musketeers and Cossacks were supplemented by Tatars who had served in the Russian army, at both the level of ordinary soldiers and as officers, including as commanders of forces. This had been the case before the conquest of Kazan and was irrespective of the Muslim faith of the serving soldiers and officers (or, to put it another way, it was not necessary at this time to convert to Orthodoxy to serve the tsar at any level). Tatar commanders fought for the tsar during the Lithuanian Wars in the first half of the sixteenth century. The number of Tatars in the army could be considerable: in 1581, it is said, over 5,000 Muslim Tatars (out of a force of only 7,100) fought for Ivan IV at a battle at Rzhev (the uppermost town on the river Volga).[19] In some locations, the proportion of Tatars in Russian service was very high. In 1669, in Sviiazhsk, it was estimated that the 149 men in military service included 82 Muslim Tatars.[20] Tatars fought in the Russian army during the wars against Poland in the early seventeenth century, in the Russo-Polish War of 1654–67 and in the Great Northern War of 1700–21.[21] Tatars who served the Russian state in this way received special trading privileges;[22] but this was also a way of controlling the newly conquered Tatar territory, by assimilating Tatars within Russian service and, in particular, by providing the opportunity for the Tatar elite to become loyal subjects of the tsar. A significant number of Tatars rose to prominence in the Russian Empire and became members of the aristocracy.

The Russian government also created a bureaucratic structure to govern the new lands by establishing a special office, the Kazan Office (*Kazanskii prikaz*). This was not simply a body to administer the town: it had authority over the whole Volga basin south of Kazan to Astrakhan – that is, the new territory acquired by the tsar. In the second half of the sixteenth century, the Kazan Office became one of the most important offices of state, with responsibility for collecting taxes, overseeing the military establishment, constructing defences, supervising trade caravans in Kazan and elsewhere and settling local disputes. The key function of the office was, however, to collect money from both taxes and tolls from goods passing up and down the Volga. The office headquarters were in Moscow, at the heart of state power, and the office acted in the name of the tsar – that is, it acted as a central office, and in effect as a colonial office, rather than as a local organ.[23] In this respect, the Russian state kept a central control over its borders – in effect, its 'colonies': there was no equivalent of the independent authority exercised by the East India Company in India, for example. The office continued to be a key body for formulating colonial policy (along with the Siberian Office, which performed the same function for new lands acquired east of the Urals). It was only in the reign of Peter I, in the early eighteenth century, when a major restructuring of central administration took place, that the work of the Kazan Office became absorbed into other central institutions. At the same time, the Russian state increased the number of *Russian* officials in the Volga region (and in Siberia), although the figure was still tiny: in the 1690s, there were only 2,739 officials in Moscow and 1,918 local officials to govern the whole empire.[24]

To a large extent, Russian authority in the second half of the sixteenth and in the seventeenth centuries was exclusively *urban* and was asserted from towns on the river Volga, as outposts of Russian power which housed not only soldiers, but also local officials appointed from Moscow. In order to take full control, the Russian state needed to move population into the countryside as well. This was a long process, and in fact was never fully achieved, because different ethnic and confessional groups tended to live apart from each other, in separate villages. In the seventeenth century, settlement patterns on the Volga changed for three reasons: the granting of lands to military servitors, mainly but not exclusively ethnically Russian; the settlement, planned and unplanned, of Russian peasants; and the granting of lands and peasants to Orthodox monasteries.

The Russian state operated through a system whereby land grants were made, on condition that the noble landowner performed military service. The lands were still the property of the tsar and were granted temporarily (at least in principle), on condition that the noble landowner continued to serve the state either in active military campaigns or on frontier guard duty. These lands were usually inhabited by peasants, and the income which the landowner derived from the land was in the form of dues paid by the peasants. These could be in cash, but in the early period were more often in kind – such as grain, chickens, eggs, honey, etc. This was therefore a primitive reward system (not dissimilar to other states in the period, for example the Ottoman Empire). The military servitor could be given pieces of land all over the country, and only after the major Law Code of 1649 (known in Russian as the *Ulozhenie*) were they allowed to exchange lands to try to consolidate them. Even then they needed the state's permission (and the same code specified the punishments – knout and confiscation – if the landowner ceased to serve the state or pay a special fee for non-service).[25] The result was that the noble landowner was often not present on the land, and in practice could neither control the work undertaken by the peasants nor stop them fleeing his estate.

These noble landowners provided an unreliable military force: it was difficult and slow to raise an army this way, and men often arrived ill-equipped and untrained. It was for this reason that Ivan IV set up the musketeers as a corps of more professional troops; but the bulk of the army remained rooted in the land-allocation system. The fundamental problem was the inability of the state to raise – and to pay for – an army in any other way in this period. The position only changed in the early eighteenth century, when Peter I professionalized the army officer corps and established a regular recruitment system; but lands continued to be granted for service, and only became fully hereditary in 1762.

The price paid for this landowning system was serfdom, which was institutionalized in the same Law Code of 1649. Serfs were peasants who lived on the estates granted by the state to noble servitors. Until the mid-seventeenth century, peasants had some ability to move away from these estates, although this right became more and more circumscribed in the century before 1649. After this date, peasants were bound to their noble landowners: they could not move away without his permission; they could be moved, with or without their families, to other estates or noble-owned factories in the Urals; they had to work a certain number of days (normally three a week) on the landowner's fields, or else had to pay dues in lieu of labour; they could be obliged to work as house serfs and taken from the fields; and they were subject to the jurisdiction of the landowner for all but the most serious crimes. In effect, they were 'owned' by their noble landowners and had no separate legal existence outside his estate. In 1649, these restrictions on peasant movement were forced on the Russian tsar (Alexis I) by the poorer nobles – that is, those who served in the army and received land in return for service, but who were not aristocrats, so that it was more difficult for them to retain their peasants on their land. Serfdom remained in place until 1861, in the first instance because land grants with peasants continued to be the reward for service to the state, and then because serf labour began to be perceived as the only way to cultivate land in the Russian countryside.

For all the weaknesses of this military land system, and of serfdom, this policy did mean that lands could be granted to servitors in the new areas on the Volga; and in turn, the beneficiaries could be expected to be loyal to the Russian government. The recipients could be minor nobles, given land inhabited by 10 or so peasants; or great magnates, who already had massive estates elsewhere in the country. In either case, the landowner took over the villages that were on the land, and the peasants – be they Mari, Chuvash, Udmurt, Mordvin, Tatar – became the property of the new owner, his serfs. By 1568, 34 nobles had been given 30 villages and 485 peasant homes in Sviiazhsk district.[26] Some of this land in Sviiazhsk, however, was empty, because this was not a heavily populated region; in fact, it is estimated that at first only roughly half of the land granted on the middle Volga had villages on it.[27]

Where the land was empty, the new landowners transported their *Russian* serfs from their estates elsewhere in Russia to the new properties. The size of the estate depended very much on the quality of the land. North of Kazan, the land was heavily forested, and many of the Mari practised a slash-and-burn economy. Where the land could be ploughed, then large estates could be allocated by the state. Prince G.F. Dolgoruky, one of the greatest magnates in Russia, was granted land with over 400 Mari peasants; and M.G. Sobakin was granted no fewer than 47 Mari villages, inhabited by over 2,000 peasants.[28] Not all Russian servitors welcomed lands in what they regarded as remote and dangerous territory, particularly on the eastern bank of the Volga, where lands, as we have seen, were subject to raids from nomadic Nogais and Kazakhs. The new landowner was obliged to defend these lands for the tsar, as well as to serve in the army. Some ethnic Russians in military service managed to refuse to accept these lands, or at least managed to negotiate for what they saw as a better deal.

As far as the tsars were concerned, the *service* to the state was more important than the ethnicity of the servitors, or even their religion. Indeed, this was a way of assimilating non-Russian elites into the nobility and ensuring their loyalty to the regime. It was quite common for Tatars in Russian service to be granted land on the middle and lower Volga in return for their military service. In general, Tatars who had converted from Islam to Christianity were more handsomely rewarded with land; but land could also be granted to Muslim Tatars. For example, in Kazan province lands were granted in return for service to Iakov Asanov, a converted Tatar, and to members of the Khozaishev family, who remained Muslim.[29] Muslim Tatars from Astrakhan were given land in the 1570s.[30]

It is not surprising that non-Russian peasants should often have protested against the intrusion into their lands of landowners, who then demanded dues from them. Mari and Chuvash peasants had paid 'tribute' to the Golden Horde and now found that they had to continue to pay Russian landowners, usually in kind until the tax reforms of Peter I in the early eighteenth century. Chuvash peasants in Iadrin district complained that the musketeers came to their villages to force them to pay dues and punished them 'without mercy' when they could not pay, with the result that some of them had fallen into 'debt slavery' and others had fled from the villages.[31] They also complained about forced conversion to Christianity, something that will be discussed further in Chapter 6.

Many Russian peasants were forcibly settled on the middle and lower Volga. Some of these were the serfs of great magnates (like the Dolgoruky family mentioned above), who had the right to move their serfs from estate to estate. There was especially intense settlement of Russian peasants by landowners on the rich black-earth agricultural lands in the Saratov region. It was estimated that by 1678, over 200,000 ethnic Russian peasants were settled in that region; and by 1719, the number had risen to almost 500,000.[32] By the end of the eighteenth century, the majority of serfs in the Volga region were ethnically Russian, comprising 64 per cent of the population of the middle Volga, 71 per cent of the lower Volga and 41 per cent of the very southern delta Volga region. That proportion remained much the same through to the mid-nineteenth century.[33]

Some Russian peasants came on their own initiative, and were fugitives from other landed estates. These fugitives included landless peasants from central European Russia; poverty-stricken peasants fleeing from poor harvests or in debt to their landlord; and schismatics (Old Believers; see Chapter 6) who were being persecuted after the rupture in the Russian Orthodox Church in the mid-seventeenth century. In 1662, officials in Saransk region noted that, along with servitors, there were in this region 'many fugitive peasants and other people'. Other peasants, the officials noted, had fled south from Kostroma and Nizhnii Novgorod (both towns on the river Volga) because of 'hunger', 'scarcity' or 'harvest failure'; meanwhile one group comprised 'Vaska, Ermolka and Demka, in 1671, with their wives and children, because of the Church schism'.[34] The reality was that the government could not control the movement of peasants (or for that matter, other social groups or nomads) into the Volga lands – or at least not until the late eighteenth century, when it had greater means to check movement and recover fugitive serfs.

The role of Russian Orthodox monasteries was also crucial – not only as the physical symbol of the Russian state, but also for the control of the newly settled and existing peasants and for the economic development of the Volga region in the second half of the sixteenth and the seventeenth centuries. Indeed, it can be said that the state and the Church acted together in this period to assert *Russian* domination through a Christian, Orthodox presence, as well as a military and administrative presence in the newly acquired and non-Christian areas. Monasteries were granted lands and peasants on the middle and lower Volga, and particularly along the banks of the river, along with fishing and other economic concessions. The state granted the lands, and often paid for construction. Abbots were regarded almost as officials of the state, in that they asserted the Christian ideology of the state and loyalty to the tsar as protector of the Orthodox faith. In return, the monasteries cultivated the land and encouraged trade, which was often conducted at fairs in their own grounds, in part because it was a safe place within the monastery walls.

Monasteries on the river Volga performed, in effect, state obligations. On the one hand, their strong walls were another line of defence against enemies; on the

THE VOLGA

other, at least in the later period they were supposed to attempt to convert the pagan peasantry to Christianity. That would in turn, it was assumed, tie the peasants more firmly to the ideology of the Christian, Russian state (although in practice, monasteries had no objection to having non-Christian peasants on their lands, because they could exact valuable tribute from them). Peasants on monastic land were either the original inhabitants, of whatever ethnicity, or Russian peasants transferred there from other monasteries in central European Russia. They were categorized as 'monastic' or 'church' peasants, a classification that differed very little from that of serfs. Only in the second half of the eighteenth century, monastic and church land was secularized and these peasants became in effect state peasants, owing fiscal and other obligations to the state, was the financial power of the Church weakened, on the Volga and elsewhere.

The first monastic settlement in the new lands was in Sviiazhsk, on the Volga, which had been founded by Ivan IV in 1551, as a place to gather his troops before attacking Kazan, and where the troops had been blessed before going into battle. By 1568, the monastery owned 14 villages and 83 peasant households.[35] Monasteries were quickly established in Kazan, and were then set up as outposts of Russian power all the way down the river, from Kazan to Tsaritsyn. Some of these monastic complexes became so large that they were like small towns. The Spaso-Preobrazhenskii (Transfiguration of the Saviour) monastery near Samara on the middle Volga had massive lands and owned over 1,000 peasants by the 1670s, including Chuvash peasants and fugitives.[36] The Makarev Zheltovodskii monastery was founded in the village of Makarevo on the river Volga and owned almost 250 peasant houses by 1650. As well as growing the normal grain crops, it had two granaries, gardens and a stud farm, and was involved in the fishing trade. Its annual income by the beginning of the eighteenth century was the not inconsiderable sum of over 2,000 roubles (to put this into some sort of context, a vice governor in the late eighteenth century earned between 1,000 and 2,000 roubles a year, and a governor between 3,000 and 4,000 roubles).[37] The monastery was the location of the main fair on the river Volga, until this was transferred up river to Nizhnii Novgorod in the second half of the nineteenth century.

In the short term, the monasteries gained great wealth not only by farming, but in particular by being given lucrative rights to fish, and to control the fishing industry, on the Volga. Until the 1750s, tolls were paid for transporting goods up and down the river; the monasteries were often situated on the waterway, so that they could exact tolls, as well as transport their own fish and other goods. Many monasteries grew enormously wealthy on this trade. For example, in Kazan the

Zilantov-Uspenskii (Assumption) convent was permitted to charge tolls on all goods that passed through its lands on the Volga.[38] The fishing trade of the Savvo-Storozhevskii monastery in Simbirsk was estimated to be worth a staggering 10 million roubles by 1700.[39]

The Volga was becoming a key trade route within Russia, and the control exercised by the monasteries was another way of asserting Russian state control over the passage of goods on the river. The salt trade in Russia was very important, not least because salt was a royal monopoly. It came from Lake Elton, the largest salt lake in Europe, near the border with present-day Kazakhstan. The trade was described by Olearius in the 1630s as a 'thriving business', with Russians 'transporting [the salt] up the Volga . . . and shipping it all over Russia'.[40] The Bogoroditsa (Mother of God) monastery in Sviiazhsk established an important role in this lucrative trade on the Volga, as it was permitted to import over 135 tonnes of salt from Astrakhan.[41] Monasteries were able to collect the tolls as salt was transported up the river Volga. Even when internal tolls were abolished in the mid-eighteenth century, the monopolies on goods such as salt, vodka and medicinal rhubarb remained. A special wharf was opened in Samara in 1600 to handle the salt exports from Lake Elton and Astrakhan.

The Russian government had shown its intent to absorb and control the new lands in the south. It had done so through a combination of a show of force – soldiers in garrisons – and of settlement, by granting land to military servitors and monasteries and by either settling peasants on those lands or by enserfing the inhabitants who lived there. The adoption of the title 'tsar of Kazan and Astrakhan', the creation of a Kazan Office to govern the new territory and the encouragement of the establishment of monasteries made it clear that these lands were to be fully incorporated into the Russian Empire.

The lands of the middle and lower Volga were, however, still a wild frontier. Towns and monasteries were built with defensive walls, but these could not prevent attacks by nomadic Nogais, Kirgiz, Kazakhs and Kalmyks. The military defences were intended to defend the towns, but in practice they did not prove strong enough to deter such assaults: Nogais attacked Samara in 1615 and 1622;[42] and Kalmyks stormed the town in 1639 and 1670[43] and set it alight.[44] In the early eighteenth century, John Bell, a Scottish doctor and traveller, noted that the fortifications of Samara were designed to resist incursions by the Tatars.[45] Samara was particularly vulnerable, because it was located on the eastern bank of the Volga, and so was exposed to raids from nomadic peoples from the east; but Saratov, on the western bank, was also attacked in 1612 and raided in the 1620s and 1630s by

Nogais.[46] Each onslaught led to strengthened fortifications and an increased military presence, but this did not eliminate the incursions. And if the towns were vulnerable, the villages were totally defenceless.

The middle and lower Volga represented the border between the more settled west and the open steppe of the east which was vulnerable to attack from that area's nomadic peoples. But Tsaritsyn was also attacked from the south: Kalmyks descended on the town in 1681 and 1682. And in 1705, Tatars from the Kuban attacked Tsaritsyn from the south, stole livestock, took Russian prisoners as slaves, and in general were said to have 'ruined' the town.[47] It continued to be threatened by Tatars from the regions of the Black Sea and the north Caucasus. Nor was the 'enemy' purely external. Serfs – Russian and non-Russian – could rise up in revolt against landowners and government tax collectors; meanwhile the soldiers stationed on the Volga – the musketeers and the Cossacks – could be unreliable and were known to cause disturbances, as well as to quell them. It is to the Volga as a location for violence and revolt that we now turn.

4
VIOLENCE ON THE VOLGA
Pirates, raiders and Cossacks

In 1743, the British merchant Jonas Hanway travelled down the Volga to Persia, and witnessed the horrific punishment of thieves designed to be a warning to others. They were floated down the Volga on rafts, on which gallows had been erected with:

> . . . a sufficient number of iron hooks, on which they are hung alive by the ribs. The float is launched into the stream, with labels over their heads signifying their crimes; and orders are given to the inhabitants of all towns and villages on the borders of the river, upon pain of death, not only to afford no relief to any of these wretches, but to push off the float, should it land near them . . . these malefactors sometimes hang thus three, four, and some five days alive.[1]

The agonizing death was supposed to be a deterrent, but was also a reflection of the serious threat posed by robbers on the Volga.

Violence was an everyday occurrence in a country which lacked any rural or urban police forces until the very end of the eighteenth century, and where the state relied on small garrisons of often unreliable troops to put down any disturbances or to track down fugitives and robbers. Peasants could turn on their noble or monastic owners; powerful nobles attacked the neighbouring estates of poorer landowners; deadly riots broke out in Moscow and other towns, particularly at times of economic distress or when plague struck; soldiers billeted in towns or villages were often in conflict with their reluctant hosts; forays into the countryside to collect taxes or round up conscripts often ended in violence on both sides; gangs of fugitive peasants, army deserters and escaped convicts preyed on the unwary and attacked remote villages. Religious festivals led almost invariably to drunkenness and disorder: Easter was the most dangerous time, according to one diplomat, 'when all the Rabble are drunk and mad'.[2] The fluid borders of Russia were open to raids from hostile neighbours, but towns in central Russia could also be surrounded by robbers. In 1728, Nizhnii Novgorod, on the middle Volga, had to establish a special foot regiment to rout robbers in and around the town.[3] Wooden towns were susceptible to fires, and this could also lead to opportunities for disorder. In 1750, peasants were publicly whipped in Nizhnii Novgorod for looting after a fire in the town.[4] Violence could break out between inhabitants at any time, on the Volga and elsewhere, and the poor and less powerful normally came off worst. The Russians are in general 'quarrelsome people who assault each other like dogs', remarked Adam Olearius, secretary to the ambassador of the duke of Holstein-Gottorp, in the 1630s.[5] In 1767, the merchants of Syzran, on the middle Volga, claimed that a wealthy merchant had 'held one merchant, Fedor Zabirzin, for a long time, whipped him painfully, and no one knows for what'.[6]

Any journey was a hazardous undertaking for travellers. Indeed, a number of magic rituals were normally practised before setting out. Roads were poor and deep ruts caused carriages to overturn. Spring and autumn rains turned roads into morasses of mud. Sledge travel in the winter was often swifter, but sudden storms and blizzards could be fatal. Human intervention could be as dangerous as the weather: gangs of robbers and deserters roamed the Russian countryside, attacking unprotected travellers. In Siberia, bands of escaped convicts attacked the few roads which crossed the vast territory; they were known as 'General Cuckoo's Army' because they appeared in the spring. All the trees along one of the roads leading to St Petersburg were chopped down, so that bandits could not shelter in them. Travel along the river Volga between Nizhnii Novgorod and Astrakhan was, however, notoriously dangerous, not least because there was a dearth of suitable alternative routes or roads that could be used. Gangs comprised Cossacks, deserters from the army, fugitive peasants and escaped factory workers from the Urals.

In 1466, Afanasii Nikitin, a merchant from Tver, travelled down the Volga on a journey to India. His party passed Kazan without difficulty, but was attacked and robbed by Tatars at Astrakhan. One of his party was shot, and four Russians were taken prisoner, although two Tatars were shot in return.[7] Adam Olearius travelled down the Volga in 1636, in a ship 'that was equipped with all sorts of powder and missiles, metal- and stone-firing cannons, grenades, and other weapons' because of the fear of brigands; he was warned about Cossack pirates who were 'a barbarous and inhumane people, and more cruel than lions', and local Mari people who were 'a cruel unhuman lot, who loved pillage more than God'.[8] In 1734, a gang of fugitive peasants located on an island in the Volga near Cheboksary was said to 'beat and rob' many traders on the river.[9] An account written in 1739 tried to reassure the prospective traveller, but in doing so revealed the danger: 'Going down the *Volga* from *Saratoff* is not the least dangerous, provided one has a good Boat and a well-armed Company: Then he has nothing to fear from *Calmucks* [Kalmyks] or *Russ* Pirates.'[10] Hanway, quoted above and travelling in the 1740s, noted that robbers plagued the river in gangs of 30, 40 or even 80 men, so that boats had to travel in convoys for protection.

It was not only the unfortunate traveller who could be relieved of his money and valuables. Bandit gangs were drawn to the Volga by the immense value of the goods being transported upstream from Persia and the east (in particular, fish, salt and silks) and downstream from north Russia (furs), and by the portability (and saleability) of these goods. Booty seized from ships could then be traded at towns like Astrakhan with few questions asked. Gangs had to take advantage of the short sailing season (usually from March/April to November), before retreating inland or focusing their attacks on border territory. Robbers used oared flat-bottomed longboats (called *strugi*) which were perfectly designed for hit-and-run raids on shipping. The boats could move swiftly and manoeuvre easily among the many sandbanks, islands and small tributaries of the middle and lower Volga (see Chapter 10), making it easy to attack lone ships without warning and then to escape. North of Saratov, the 'great bend' in the river east to Samara was a particularly good hiding place for such boats.

Bandits were able to operate on the river almost with impunity. Ships travelling upstream against the current were the most heavily laden with goods and were slow. Even when boats travelled in convoys, it was easy for one or two ships to become separated, especially when navigation was difficult because of sandbanks and low water levels. Arming a ship sufficiently to repel large gangs was almost impossible. Even if men prepared to be armed guards were available, the cost of hiring and feeding them would have been prohibitive and would have made the journey uneconomic. Furthermore, there was the danger that armed men could go over to the bandits when attacked, and that they could be joined by disgruntled bargees and other boat workers. Traders were resigned to the loss of a number of cargoes, whether from robber gangs or natural disasters. In 1744, the Senate instructed authorities to 'eradicate robbers and thieves' on the rivers, but the instruction was meaningless without armed support from the government.[11] It was only when robbery and raiding evolved into something that could be regarded as a general rebellion that regular military forces were deployed, as we shall see below.

The river was the greatest source of booty, but settlements on the middle and lower Volga were also subjected to robbers and raiders. On the eastern bank, nomadic horsemen would descend on villages, terrifying the inhabitants and leaving a trail of destruction. The nomadic Nogais were particularly feared. In part, Nogai hostility was a reaction against the encroachment onto their grazing lands of settlers, who ploughed the land and restricted the movement of the Nogais' herds; but the villages also provided them with precious booty. Horses were particularly valuable, but people (of both sexes) could also be seized and either sold at high prices as slaves or held for ransom. German colonists who were settled on the eastern side of Volga were subjected to devastating raids by Nogais and others. In spring 1774, several colonies on the eastern side were attacked by Kirgiz horsemen; one village was razed to the ground, its inhabitants tortured and slaughtered, and some 300 survivors were taken into captivity.[12]

❖

We saw in the last chapter that Russian forts and towns were subjected to raids by Nogais and Kalmyks. On the Volga, however, the main threat to law and order came from Cossacks. Cossack raiders were active from the middle of the sixteenth century, around the time of the conquest of Kazan and Astrakhan. It was reported in 1554 that some 300 Cossacks had 'committed theft on the Volga'.[13] In 1600, some 500 Cossacks sacked Tsaritsyn: having committed robbery 'on the Volga and on the sea' against 'trading people', it was reported that the 'thieving Cossacks . . . entered Tsaritsyn and killed people'.[14] In 1631, over a thousand Cossacks descended on the Volga from the river Don and carried out raids in the Volga delta, in Astrakhan itself and on the Caspian and the Persian coast.[15] Cossacks could clash with nomadic tribes and other non-Russians from the east and the south (in 1655, Cossacks were said to have captured 400 Tatars),[16] but they were at their most dangerous when they were able to attract nomadic peoples and other resentful rural and urban inhabitants to their side: at that point, raiding became a much greater potential threat to the stability of the state.

Who were the Cossacks? The word 'Cossack' in Russian is of Turkic origin and originally referred to freebooters, fighters and brigands on the steppes in the fifteenth century, who could be Slavs or Turkic people. By the seventeenth century, most Cossack communities on the frontiers of the empire were ethnically Russian or Ukrainian, of which many were runaway former serfs or state peasants who had fled south and established communities in the borderlands and on the great rivers there (in particular on the Don, but also on the Dnieper, the Iaik/Ural, the Terek and the Volga). Cossacks are not, therefore, a separate ethnic group, although they were categorized as such in the Soviet Union; but they certainly developed a separate identity and distinctive way of life within the Russian Empire, which lasted throughout the imperial period and has resurfaced in post-Soviet Russia.

As the Russian Empire expanded further south, it established 'lines' of forts on the river Volga and other main routes, as we saw in the last chapter. Not only did these forts have to be manned by soldiers, but the newly acquired land had to be occupied by settlers. To this end, a number of Cossack communities, called Cossack 'hosts', were either formally established by the Russian state or evolved in the borderlands and on the great rivers in the south. Cossacks were required to serve the state by providing a protective barrier in the south, and had also to serve in garrisons or military campaigns. In return, they were given lands, lucrative fishing rights on the rivers, and a salary from the government for military service; moreover, they were also allowed to exercise considerable autonomy in their affairs: they chose their own leaders (*atamans*) and held their own mass Cossack assemblies, which practised a rough sort of democracy and meted out their own judicial punishments.

Although, in practice, Cossack communities were dominated by rich and powerful elders, and relied on the Russian government for payments for military service, they certainly regarded themselves, and were regarded by others, as genuinely 'free' in spirit and as operating by their own rules. Skilled horsemanship and bravery were valued most highly; agricultural work was eschewed, at least in the early seventeenth century. For the Russian government, the Cossacks were a mixed blessing: valuable when they repelled attacks by Tatars, Nogais and Kalmyks, and as irregular troops in battle, but always a potential source of disturbance, and a threat to law and order. It was only in the late eighteenth and early nineteenth centuries that the Cossacks were finally tamed (discussed in the next chapter), so that they became the bastion of tsardom and were associated with repression of all those who opposed the state.

Cossacks were settled on the river Volga in forts in the wake of the conquest of Kazan and Astrakhan. An early famous Volga Cossack was Vasilii Timofeevich, better known by his nickname 'Ermak', who entered the service of the wealthy and influential Stroganov family in the early 1580s, led a group of Cossacks in the service of the Stroganovs across the Ural mountains, and defeated the Siberian khan at a great battle outside his capital of Isker. Ermak subsequently claimed 'Siberia' for Tsar Ivan IV, demonstrating the region's value to the tsar by a gift of immensely valuable furs. Little is known of Ermak's origins, or even what he looked like (later chronicles described him as 'flat-faced, black of beard and with curly hair, of medium stature and thickset and broad shouldered',[17] but this was speculation).

This has not prevented the town of Novocherkassk on the Don claiming Ermak as its own and erecting a statue of him – or, for that matter, the post-Soviet Russia issuing a ten-rouble postage stamp with his image in 2009. We know he had worked on the Stroganov fleets on the Kama and Volga rivers, and had operated as a pirate on the Volga – and was probably from the Volga. Ermak was just the sort of daring, fearless, ne'er-do-well who could lead a small group of mainly Cossacks to a series of stunning victories over tribesmen who greatly outnumbered them – albeit armed with much more primitive weapons. After his death (probably in 1585), Ermak acquired mythical status, including miraculous power to cure diseases, which merged with the folklore and popular songs surrounding Stenka Razin and Emelian Pugachev, whose exploits on the Volga are described below.

A formal Volga Cossack host was established in the early eighteenth century, and a thousand Cossack families were moved to headquarters in the town (the Cossack *stanitsa*) of Dubovka, on the right bank of the Volga, north of Tsaritsyn. The Volga Cossacks were always small in number and, although they participated in revolts on the Volga in the eighteenth century, they did not instigate those revolts. That honour fell to the Cossacks of the Don.

The Don Cossack host had grown by the late seventeenth century to be the most numerous and significant Cossack community within the Russian Empire. Don Cossacks lived by a mixture of fishing and herding, supplemented by raids on the Black Sea and the Caspian Sea. Don Cossacks were proud of their autonomy: their assembly met in Cherkassk, their own capital, and they insisted on dealing with the Russian government through its foreign office (*posolskii prikaz*). Nevertheless, the Don Cossacks in practice depended on the Russian government for payment for military services, and by the late seventeenth century were experiencing considerable economic and social strains. These were caused partly by wartime taxation and poor harvests; partly by the influx into the Don region of new fugitive peasants who put pressure on the limited resources, which led to increasing divisions within the Cossack community between 'old' and 'new' settlers, and between wealthy and poor; and partly because the Ottoman Empire had fortified its bases on the northern coast of the Black Sea against Cossack raids. It was this combination of factors that pushed the increasingly desperate Cossacks eastwards towards the river Volga and the Caspian, in search of booty. In fact, the Don Cossacks were fighting a losing battle against a Russian state that was increasingly asserting its authority over the borderlands. But this was not obvious at the time, nor was it of any consolation to the victims of the two major Cossack revolts on the Volga in this period: Stenka Razin in 1667–71 and Emelian Pugachev in 1773–75.

Stenka Razin was from an established and relatively wealthy Don Cossack background, but at some point – perhaps following the execution of his brother for desertion in the Russian–Polish War of 1654–67 – he acquired a loathing for anyone in a position of authority. He was a charismatic leader – brave, striking in appearance, a bold strategist – who seemed instinctively to understand the mindsets of ordinary Cossacks, peasants and townspeople. Supporters included poor Cossacks desperate for plunder; impoverished townspeople who shared his hatred of wealthy merchants and tsarist officials; soldiers in undermanned garrisons who changed sides to save themselves; exploited river workers (Razin's uncle had worked as a barge hauler on the Volga for three years);[18] peasants (Russians and non-Russians) who sought revenge on their noble or monastic landowners; Kalmyk tribesmen who took advantage of the disorder to carry out raids on the left bank of the river; Tatars who had been forcibly converted to Christianity; and Old Believers (schismatics who had rejected changes in Orthodox rituals in the mid-seventeenth century; discussed more fully in Chapter 6) who had left central European Russia and settled in remote villages on the Volga.

For four years, Razin used his considerable skills, and his ability to recruit followers, to wreak havoc on the lower Volga. His band of about a thousand men left the Don in April 1667 and almost immediately attacked and overran a large convoy of ships laden with rich cargo; the survivors were invited to join his gang, and all those who resisted were thrown into the river. His band sailed past Tsaritsyn, on the lower Volga (where the gunners failed to fire on him from the fortress), and took the fort of Iaitsk (renamed Uralsk after the Pugachev revolt, and now called Oral in Kazakhstan) on the river Iaik. Some soldiers defected to Razin; those loyal to the garrison commander were slaughtered. The following March, Razin and his gang left Iaitsk and sailed into the Caspian, raiding the coast of what is now Dagestan as far south as Baku. His Cossacks held off the Persian army and the fleet over the next 18 months, but at a very heavy cost; and in August 1669 Razin headed back to the Volga. Here a deal was struck with the Astrakhan governor: the Cossacks would be pardoned by the tsar if they surrendered their heavy guns and ships and restored the goods and men they had taken from the Persians. Razin accepted and entered Astrakhan in triumph. He then promptly refused to give up his ships and weapons.

Razin's legendary status was already in the making, but was further embellished by the tale (probably fictitious) that in order to ensure harmony among his Cossack companions, he cast his beautiful Persian (or Tatar) princess into the river Volga as a sacrifice. An eye witness, Ludvig Fabritius, a Dutchman in Russian service, gave the following explanation, which is far more colourful:

Stenka had not kept the promise which he had made to the god Gorinovich at the beginning when he first came with his boats to the river Iaik – 'if I attain happiness through your help, then you can expect me to give you in return the most cherished of my acquisitions'. And so he took the poor woman and threw her in her full attire into the river, saying: 'Take her, my patron Gorinovich, I have nothing better to give or to sacrifice to you than this image of beauty.'[19]

In September 1669, Razin and his band left Astrakhan, weighed down with treasure, and returned to the Don, pausing only to enter Tsaritsyn and release prisoners from jail.

So far, Razin had carried out a traditional, if very extensive, series of raids for plunder on the lower Volga; but his next foray, in 1670, was far more dangerous, coloured as it was by a direct attack on the social elite and officials, ostensibly to protect the tsar from their evil influence. He proclaimed that he would lead his men, now some 7,000 strong, 'from the Don to the Volga and from the Volga into Rus against the Sovereign's enemies and betrayers, and to remove from the Muscovite state the traitor boyars [nobles] and Duma men [advisers] and the *voevodas* [governors] and officials in the towns . . . and to give freedom to the common people'.[20] His revenge on the elite was brutal. The first town to fall was Tsaritsyn, where the terrified townspeople opened the gates in the hope they would be spared. The governor and a few loyal soldiers were slaughtered, and the one survivor, Timofei Turgenev, was tortured by being pierced with lances and finally speared and thrown into the river Volga.[21] A relief force was obliged to surrender to Razin: all the officers were drowned in the Volga and the ordinary soldiers were compelled to become oarsmen for the rebels.

The insurgents then turned south, taking Chernyi Iar, a fort on the Volga, where soldiers mutinied against their officers; and they acquired more booty by capturing goods and money from traders. Razin's forces then took Astrakhan, his surprise tactics helped by the support of mutinous and panic-stricken garrison soldiers, who turned on their officers. Razin enacted a reign of terror, described in a government report as 'like a wolf falling on a Christian flock',[22] torturing and slaughtering officials and their children, but also targeting wealthy merchants (irrespective of whether or not they were Russian) and clergy. Blood flowed in the streets. Fabritius had already recorded that Razin's gang would 'hang people up by their feet, stick holes through their ribs, and hang them up on iron hooks'. This was the fate of the secretary to the governor in Astrakhan; the governor himself was tortured and thrown off the tower. No one was spared if they were associated with the rich or privileged:

> Both the sons of Boyar [noble] Prozorovsky were hanged by the legs, but after they had been hanging a whole day, a Cossack officer named Luzar took pity on them and cut them both down. The younger [aged 8] was still alive except for his feet, which were quite dead, and he was taken to his mother. The elder [aged 16] was thrown down from the tower half-dead, and the good youth went the way of his father.[23]

In the summer of 1670, Razin, now with some 6,000 loyal Cossacks, sailed up the Volga from Astrakhan, acquiring more recruits as he passed by Chernyi Iar and Tsaritsyn. Saratov was protected by only a small garrison and surrendered without a fight. Along the route, peasants – Russian, Chuvash, Mari, Mordvin – were influenced by Razin's simple message against the 'evil boyars', and took revenge on the landowners, and sometimes on members of the clergy who failed to support their revolt. Razin even installed on one of his barges a fake 'Patriarch Nikon' and invited peasants to kiss his cross. In fact, some of Razin's support came from Old Believers, who virulently rejected the real Patriarch Nikon's reforms of the Orthodox Church; but Nikon was born on the Volga, and Razin may have thought that the presence of the impostor gave his army the appearance of real authority. By September, the rebels had gone further up the river to Simbirsk, hoping to move on from there to Kazan and Nizhnii Novgorod. But despite their superiority in numbers, even after a month-long siege they proved unable to breach the well-defended citadel.

In October, a relief force reached Simbirsk and routed Razin's army. He and many of his Cossacks escaped by boat down the Volga, leaving those behind to be cut down. Abandoned by many of his followers, Razin found that the gates of the Volga towns of Saratov and Samara were closed to him, and he was forced to withdraw to the Don, where he was captured by Don Cossacks loyal to the tsar and handed over. He was taken to Moscow and brought into the city on a cart, chained and in rags, with his brother, Frolka, tethered to the cart like a dog. Razin was cruelly tortured, and was finally quartered on Red Square in June 1671. His head and limbs were mounted on stakes, and his torso thrown to the dogs. The revolt was over.

The rebels had committed atrocities in every town they had taken. Government reprisals were, if anything, even more brutal and designed not only to punish but to terrorize the local populations. Tens of thousands were slaughtered, either in battle, in retreat or in reprisals. Rebels were tortured, impaled or flogged to death. Public executions and torture took place in all towns, and the mutilated bodies were exposed in market squares or the centres of villages, or floated down the Volga on barges. Whole Cossack and peasant villages were razed to the ground, and it has been estimated that over 11,000 rebels were executed.[24] Non-Russians were equally suppressed: some 400 Mari peasants were knouted and over a hundred had their limbs severed.[25] The extent and brutality of the reprisals allowed psychopaths to act without restraint: when Prince Iakov Odoevsky retook Astrakhan, many inhabitants were tortured and executed by being buried alive:

> Finally, when there were only a few people left, he had the whole town destroyed and ordered the houses to be rebuilt outside the town. When these were half finished they had to be taken down again and transferred to the castle. Now this had to be done by the men, their wives and children, who had to pull the carts themselves, as there were no horses. It even happened that many pregnant women fell as a result of the heavy work and died in child-birth like animals. After this long tyranny there remained only old women and small children.[26]

The brutal suppression was supposed to ensure that there would never be another rebellion. In the short term, that was successful; but in the longer term, the suppression only led to simmering resentment and further disorders on the Volga. Astrakhan had been brutally sacked by Razin; but no lessons seemed to be learned – either about the causes of the revolt or about the military defence of the town. In 1705, a revolt broke out among the soldiers in the garrison and spread to the townspeople. The soldiers had heard rumours that their grain ration was going to be reduced, while the townspeople hated the corrupt governor, Timofei Rzhevsky, who, it was claimed, had taken bribes and stolen horses from the inhabitants for eight years, thus bringing many to 'great ruin'. An underlying cause of the revolt was reaction against the modernizing changes implemented by Peter I, including his 'German' (in effect, foreign) dress and the pressure to shave off beards (which offended peasants and merchants alike). As in the Razin revolt, the tsar was not challenged directly, but rumours spread that Peter I was a 'false tsar', who was setting out to 'change the faith, bring in Latin, and force the wearing of German dress and the shaving off of beards and moustaches'.[27] Rzhevsky was executed after a mock trial, and some 300 nobles and officials were slaughtered. The rebels moved towards Tsaritsyn, but were defeated by regular troops; those captured were tortured and executed. In 1708, Tsaritsyn was again sacked, the governor killed and the whole town 'ruined' by Cossacks under the leadership of Kondratii Bulavin.[28]

❖

The last great Cossack revolt in Russia was led by Emelian Pugachev in 1773–75. Many of the social groups involved in the Razin revolt also participated in this revolt, including Cossacks, peasants, non-Russian peoples and Old Believers. The Pugachev revolt, however, spread further east from the Volga and also became a protest among the Bashkirs against the encroachment of settlers onto their traditional grazing lands and against the imposition of taxes by the state. Recruits were also drawn from workers in the factories of the Urals, who were often serfs forcibly transported by their landowners from the countryside and obliged to work in appalling conditions in the mines. The river Volga played a significant role in the second stage of this revolt, when Pugachev's forces moved down the river and attacked major Volga towns, and the unrest spread on either side of the river as Pugachev incited serfs to rise against their masters.

Pugachev followed the tradition of Stenka Razin in his hatred of the noble and bureaucratic elites and his desire for revenge on them, but he went one step further and presented himself as the true tsar, Peter III, the husband of Catherine II, who had been deposed and then murdered in 1762 by Catherine's supporters. Tsarist impostors were not unusual in Russia, particularly in remote areas and in the borderlands, because they gave a kind of spurious legitimacy to the revolts and made it easier to convince simple people that they were not doing anything wrong.[29] Pugachev maintained the fiction by establishing an elaborate and grotesque 'court' in the town of Berda (near Orenburg), where he dressed in finery, sat below a portrait of the son of Peter III (the future Tsar Paul) and referred to him as his own son, and was addressed obsequiously by a group of fellow Cossacks on whom he bestowed the names of Catherine's main advisers. That he bore no physical resemblance to Peter III was not an obstacle – at least not until he was defeated and betrayed by his fellow Cossacks, who knew full well that he was simply one of them.

Pugachev was, like Razin, a Don Cossack (and came from the same settlement). By the 1770s, the Don had become less volatile, mainly because it had become a more settled and socially stratified community, and the state had established greater control over the Cossack host. Pugachev therefore appealed to the more primitive Cossack settlements on the river Iaik, to the east of the river Volga, where sympathy with Old Belief was rife and where government bureaucracy was only just making

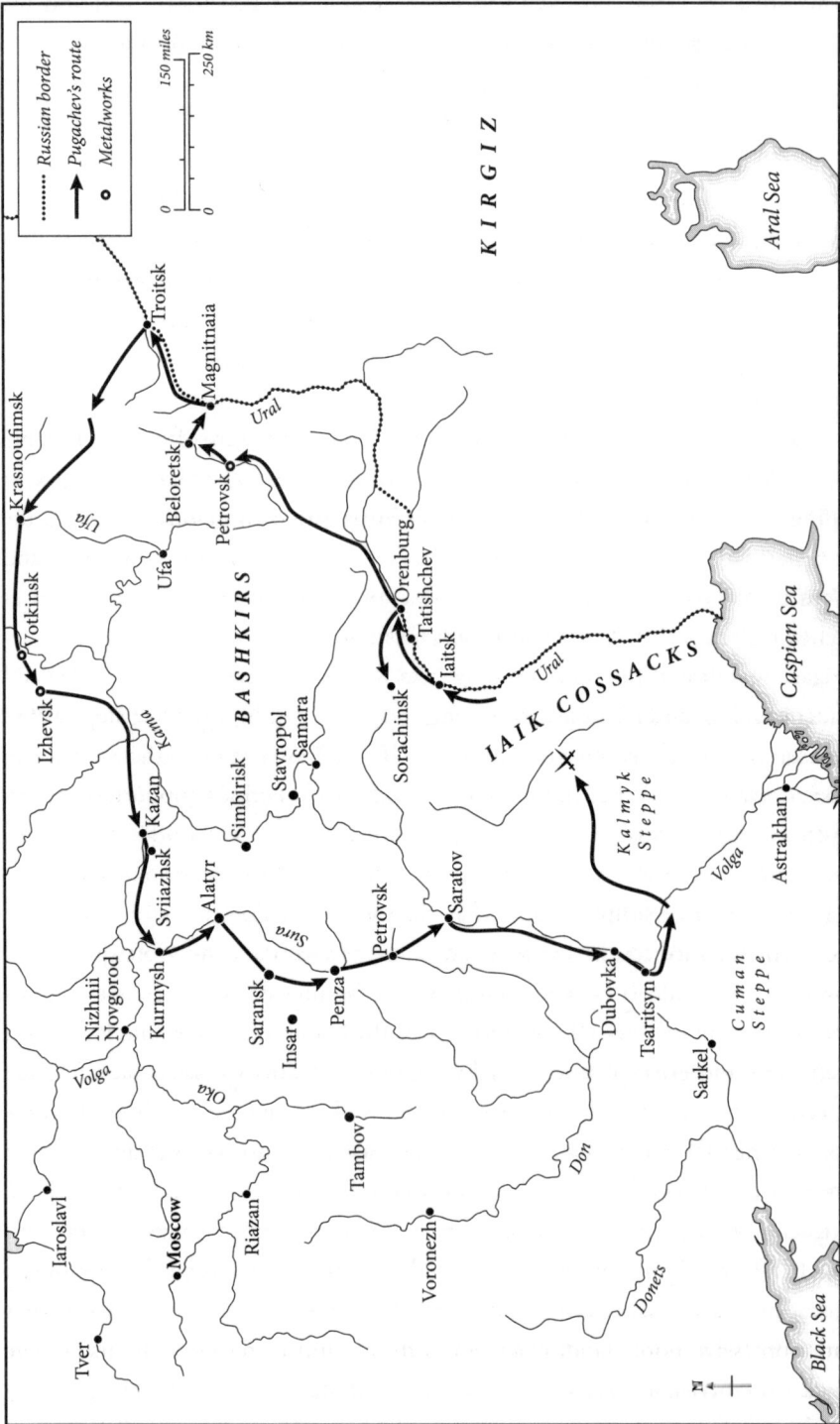

6. The Pugachev Revolt on the Volga

inroads into the Cossack way of life. The year before Pugachev's arrival, a mutiny by Iaik Cossacks – sparked by rumours that they would be fully incorporated into regular Russian forces, which would have meant shaving off their beards – was brutally suppressed. The area was therefore ripe for revolt, and Pugachev's message that he was the 'true tsar' and would restore Cossack freedoms and punish the mighty was enthusiastically received. Forts and settlements up the river Iaik greeted Pugachev's forces with bread and salt (he sensibly bypassed the well-fortified garrison fort of Iaitsk itself). Numerous recruits were drawn in from factory workers and Bashkirs, who at this point comprised almost half of Pugachev's army. There followed a six-month siege of Orenburg, while at the same time Bashkir tribesmen besieged the town of Ufa. Both had been founded as administrative centres to control the new lands and people acquired by the Russian state, and were therefore seen as symbols of state control populated by hated tax collectors and other officials. The garrisons in both towns resisted, conscious that surrender would lead to bloody reprisals. In March, both towns were relieved: regular forces, supplied with artillery, inflicted a crushing defeat on the Bashkirs outside the walls of Ufa and on Pugachev's forces at the fort of Tatishchev. Thousands of rebels were killed, although Pugachev himself escaped.

The revolt looked to have been snuffed out, but it was at this point that Pugachev turned towards the river Volga, focusing on Kazan as the route to Moscow itself. In July 1774, along with some 7,000 men, he stormed the outskirts of the town and unleashed a mass orgy of looting and murder on an almost Razin scale. In one day, 162 people were killed, 129 were wounded, 468 went 'missing' and 2,063 of the 2,873 houses were destroyed by fire. The town was reduced to ashes. The rebels targeted state institutions, burning down the courthouse and the treasury (most of the money had already been removed) and opening the prison. Many of the wealthier citizens had already escaped from the town, and many of Pugachev's victims were poor artisans, lowly clerks and women.[30] The soldiers in the citadel of Kazan, however, did not surrender, and the arrival of regular troops the following day forced the rebels to retreat from the town and escape across the Volga.

Remarkably, Pugachev was able to regroup, and he turned south along the river, sparking outbursts of peasant violence involving up to 3 million people over a vast swathe of the countryside on both banks of the Volga, from Nizhnii Novgorod down to Saratov. Pugachev played on peasant resentments by undertaking to exterminate oppressive noble landowners and officials and to relieve peasants of state obligations, including conscription and taxes; he promised them 'the old cross [with three bars] and prayers, heads and beards, liberty and freedom'.[31] In short, he offered them complete freedom and restoration of what simple people believed to be the old ways and beliefs – all of which was a powerful rallying cry. Peasants – Russian and non-Russian – rose up and formed bands loyal to Pugachev which roamed the countryside, setting fire to barns, granaries and manor houses, and murdering landowners, officials and clergy who resisted them.

The Volga countryside was gripped with fear. The nobleman Andrei Bolotov wrote that after the sacking of Kazan, the rumour was that Pugachev had murdered 'all nobles and lords' and was on his way to Moscow.[32] Ivan Kozyrev, an estate steward, commented to his landlord from Simbirsk:

> They have murdered and are murdering all the lords, landowners, stewards, and authorities of any rank who fall into their hands. They ravage and plunder to the foundation manor houses and all belongings without exception, and on the spot of the killings (without leaving one alive) they take dolts into their crowd, thereby multiplying their brutality over them.[33]

It was estimated by General Pavel Panin that in all 1,572 nobles were killed (796 men, 474 women, 302 children), of whom 665 were hanged, 635 were bludgeoned or tortured to death, 121 were shot, 72 were stabbed, 64 were beheaded and 15 were drowned.[34] Small towns panicked and opened their gates to the rebels, although this did not prevent the sacking of courts and treasuries, or the torture and murder of urban officials and rich merchants. The German colony south of Sarepta was plundered and the fields 'ruined'.[35]

In August 1774, the gates of Saratov were opened to Pugachev and there followed a three-day orgy of drunkenness, looting and destruction. Many of the wealthy merchants had already left, but the rebels took their revenge on the 'enemies' who remained, and executed 22 nobles, 11 officials and 10 merchants.[36] Government offices were ransacked and official papers thrown into the Volga.[37] The Cossack town of Dubovka followed suit. With the support of barge haulers in Saratov, Pugachev was able to float some 1,000 men down the Volga towards Tsaritsyn. His army meanwhile attacked Tsaritsyn, but the garrison stood firm and the bombardment forced Pugachev's troops to withdraw. His flotilla of boats only arrived after the battle, and had to withdraw in the face of intense bombardment. Pugachev fled further south on the river, towards the fort of Chernyi Iar where his, by now primarily peasant, army met a regular force; in a decisive battle, Pugachev's forces were completely routed. He himself escaped again, but was soon betrayed to the authorities by disillusioned Cossacks.

Pugachev was dispatched first to Iaitsk, and from there to Moscow in a specially constructed cage in which he had to crouch like an animal. Catherine II, not only sensitive to the danger of creating another martyr like Razin (who had suffered an agonizing execution), but also conscious of the impression that such savagery would create abroad, refused to let Pugachev be tortured. He was tried in a secret court, albeit with an inevitable verdict of 'guilty', and sentenced to be hanged, drawn and quartered; however, to the great disappointment of the crowd, Catherine instructed that he first be beheaded. His followers were subjected to more savage punishments: the hangings, floggings, torture and mutilations in Volga villages and towns went on for several months.

❖

The two revolts, although crushed, lived on in memory and folklore. Magical powers were attributed to Razin, as they had been to Ermak almost a century earlier. Razin, it was claimed, could fly over the Volga and back on a special raft.[38] In folklore and legends, Pugachev was presented as a Christ-like and saintly tsar, who had meekly accepted the treachery of his wife (Catherine II) and her courtiers in dethroning him; had returned to Russia only after pilgrimages to Jerusalem and Constantinople; and had accepted leadership of the revolt to rescue his people. Pugachev had claimed to have special signs on his chest and head which proved that he was the real tsar.[39] Magical powers of escape attributed to Razin were also attributed to Pugachev; the latter could allegedly escape from prison by drawing a horse on the wall of the cell. Both Razin and Pugachev became part of the legend of a second coming, and both are referred to in poems and ballads as 'a resplendent sun' – an epithet associated with earlier messianic figures, going all the way back to Grand Prince Vladimir of Kiev, who brought Christianity to Russia.

Legends about Razin and Pugachev merged. Some believed Razin had not been executed in Moscow; other stories were that Razin never died, but was shut up in a mountain, ready to return on judgement day. A hundred years later, many simple people saw in Pugachev the second coming of Razin – a new messiah. Pugachev was not forgotten by the ordinary people in the Volga region: for many years after his death, peasants in Saratov are said to have reckoned the date according to the birth and death of Pugachev, rather than of Christ.[40] Sixty years after the revolt had been crushed, the poet Alexander Pushkin was told by a peasant in Berda (where Pugachev had set up his 'court') that 'It may be Pugachev for some people, your honour, but for me he is our Father, Tsar Peter Fedorovich.'[41] In turn, Pugachev's execution did nothing to prevent the appearance of other 'Peter IIIs' in the Volga region and elsewhere.

The river Volga was central to the folklore of the Razin revolt in particular. The story of Razin sacrificing his beloved in the river at Astrakhan, described above, lived on in popular tales and songs. In Pushkin's unpublished 'Onegin's Journey', Onegin meets Volga boat haulers, who sing of Razin's exploits:

The Volga swells. The haulers leaning against boat hooks of steel
In plangorous voices sing
About that robbers' den,
About those daredevil incursions
When in the old times Stenka Razin
Begored the Volga wave.
They sing of those unbidden guests
Who burned and butchered.[42]

The lyrics to the popular song 'Stenka Razin', or 'Volga, Volga, Mother Volga', were written in 1883, but the words and music were based on a folksong that records the sacrifice:

Volga, Volga, Mother Volga
Volga, Russian river
You have never seen such a present
From the Cossacks of the Don!
So that peace may reign forever
In this band so free and brave,
Volga, Volga, Mother Volga
Make this lovely girl a grave!

The river Volga itself was seen as symbolic of the freedom that the Cossacks craved and promised others.[43] The river was their home, their ally and their defence: a means to attack enemies, but equally an escape route from superior forces; a passage south to the riches of the Caspian, but also to the capital of Moscow. Some 200 songs with reference to Razin have been preserved, depicting him not only as a robber on the river, but also as a Robin Hood figure protecting the poor. The fact that these songs could even be collected in the nineteenth and twentieth centuries is testimony to their long-lasting influence. By the 1820s, such songs had even gained popularity among educated circles in Moscow. Rocks, hills, ravines and burial grounds along the river were named after Razin. A hill south of Saratov bears Razin's name and reveals the secret of 'class war' to anyone who climbs it at night.[44]

It was not only Russians or Cossacks who remembered Razin, or who thought there would be a second coming. A popular Chuvash song entitled 'Suddenly Stenka Razin' refers to Razin's corpse being laid in the church, whereupon he suddenly 'Raised his head and sat upright'. And the Chuvash poem 'In the Year of Seventy-Three' commemorates Pugachev:

They seized him, they caught him, they bound him.
Ai they seized Pugach,
Seized him, hung him on the cruel aspen,
On the very top.[45]

Razin and Pugachev were evoked by both radicals and conservatives in the nineteenth century as either an opportunity or a warning that the peasant masses could rise again. When Napoleon entered Moscow in 1812, he asked for documents relating to the Pugachev revolt to be sought out in the archives and public libraries for him to study. He expected peasant delegations to come to him to seek his support for freedom, although this never materialized. In fact, Napoleon feared a peasant *jacquerie*, because it would disrupt supplies and hinder the conclusion of a peace with Alexander I.[46] Fears of another Pugachev-like revolt after the emancipation of the serfs in 1861 seemed to be realized in – where else? – the Volga region. There, the leader of the rebellion, Anton Petrov, an Old Believer, came to be portrayed after his execution as a Christ-like martyr who had sacrificed himself for the poor, like Razin and Pugachev. A fire was said to have broken out upon his tomb and an angel appeared to announce his imminent resurrection.[47]

In the 1870s, radicals, so-called Populists, went into the countryside to try to attract peasants to their revolutionary cause by linking their activities with the anniversaries of the Razin and Pugachev revolts. The anarchist Mikhail Bakunin cited both Razin and Pugachev in the 1880s as an inspiration for social revolution. But Razin and Pugachev were heroes not only of the left: in 1911, the charismatic, fervently anti-Semitic and right-wing monk Iliodor (Sergei Mikhailovich Trufanov) organized a pilgrimage on two steamers up the Volga from Saratov to Kazan, supposedly to return a miracle-working icon to a hermitage in Kazan province. It was a bizarre episode, not least because it transpired that the icon in question had been destroyed; but it is significant that Iliodor, as a native of the Don Cossack region, saw this as a step to a mass movement on the Volga, and directly associated himself with the exploits of Razin: 'The Don is the river of popular anger', he claimed.[48] Nor is it surprising that when cinema came to Russia in the early twentieth century, the film *Stenka Razin* (produced in 1908 by Aleksandr Drankov) should have proved so popular.[49]

Russian Marxists regarded peasant and Cossack revolt as an essentially backward-looking, pointless outburst of violence again feudal oppression. But after the Bolshevik Revolution, and the ideological interpretation that in Russia at least the poorest stratum of the peasantry could be seen as a revolutionary class, peasant revolt came to be portrayed more favourably. Collections of folksongs relating, in particular, to Razin were collected by Soviet scholars in the 1920s. The Historical Museum in Moscow featured both Razin and Pugachev in a room dedicated to social revolt (now, sadly, replaced by other objects). Glorifying the revolt of the 'people' against the state could, however, set a dangerous precedent. In the 1960s, the writer and film-maker Vasilii Shukshin wrote a novel about Razin entitled *I Came to Give you Freedom*; but his efforts to make this into a film failed, because the subject matter was thought by the Soviet authorities to be too provocative. The novel was only published after his death in 1974 – while on a steamboat on the Volga, making another film.

Towns, villages and streets were named after both men in the Soviet period. In the town of Tver, on the upper Volga, there are two embankments: the left bank is named after Afanasii Nikitin, the Tver merchant who travelled down the Volga to Persia and India in the 1460s; the right bank is still named after Razin, who had no connection with the town at all! A number of memorials to Razin were erected, some temporary and some, it has to be said, in rather unlikely places (although, given the mass slaughter that occurred in, say, Astrakhan and Kazan, it would possibly be inappropriate to honour the perpetrators there). On May Day 1919, Vladimir Lenin opened a temporary monument to Razin in Moscow and identified him as 'one of the representatives of the rebellious peasantry'.[50] A memorial to Razin was opened (more appropriately) in Rostov-on-Don in 1972, and shows him on a boat with some Cossack companions and his princess, presumably just before she was cast into the waves of the Volga. There is a statue of him on a plinth in Sredniaia Akhtuba, a small town of some 14,000 people in Volgograd province. A large and very solid statue of Pugachev stands on a plinth in Saransk (a town which opened its gates to him). A wooden house in the town of Oral (formerly Iaitsk, which Pugachev did not take, but which is the main town of the region where his revolt started), now in Kazakhstan, is a museum dedicated to Pugachev;

it contains, among other things, his throne, a bronze bell, a rebel flag, guns and (a replica of) the cage which took him to Moscow. The images of Pugachev (looking, it has to be said, very much like Ermak) appeared on Soviet postage stamps in 1973 to celebrate 200 years since the outbreak of his revolt.

Razin, as a river pirate who targeted the rich, lives on in the detective novel *Pelagia and the Red Rooster*, written by Boris Akunin in 2003 and set on the Volga at the turn of the twentieth century. The story starts on a Volga steamer, where a petty thief is described as a 'razin'. The author explains:

What is known about 'razins'? A small group of river folk, inconspicuous, but without it the River would not be the River, like a swamp without mosquitoes. There are experts at cleaning out other people's pockets on shore as well . . . they aren't paid much respect, but the razins are, because they've been around since time out of mind. As for the question of where the name came from, some claim that it may have come from the word 'razor', since the razins are so very sharp; but the razins themselves claim it comes from Ataman Stenka Razin, the river bandit, who also plucked fat 'geese' on the great River-Mother. The philistines, of course, claim that this is mere wishful thinking.[51]

The memory and commemoration of Razin and Pugachev cannot disguise the fact that the revolts were complete failures. Razin's forces were able to achieve some striking military successes, but they could not be sustained. Pugachev suffered a series of shattering defeats when faced with regular troops. Both men relied on a mixture of popular support and fear for immediate success: garrisons which surrendered without a fight; townsmen who opened the gates for fear of reprisals; gullible peasants and non-Russians who convinced themselves that the revolts were just. Although thousands of men were prepared to fight, and die, for Razin and Pugachev, only Cossacks (and in Pugachev's case, Bashkir tribesmen) were skilled and well disciplined. And when the Cossacks closest to Razin and Pugachev realized that they were facing defeat, they attempted to save their own skins by betraying their leaders. Furthermore, the rebels were fighting for a lost cause. Russia was a dynamic, expansionist, modernizing state, and a return to sometimes real and sometimes mythical old 'freedoms' for Cossacks, nomadic peoples and peasants was never going to happen. The revolts, however, exposed the weaknesses of the Russian Empire in the borderlands in general, and on the Volga in particular. In the second half of the eighteenth century and in the early nineteenth, the Russian Empire pursued a series of policies to extend control over these regions, and it is to this activity that we now turn.

5
TAMING THE VOLGA
The Russian Empire on the borderland

As was fitting, Volga, you used to flow fiercely,

As when the Terrible tsar travelled along you for his victories over the Tatars

You terrified the south with the noise of your waters

Like the thunder Peter brought through you on the proud Persians.

But it is fitting for you to flow in silence

Catherine has brought with her peace to all.

Derzhavin, 'On the passage of the Empress to Kazan' (1767)[1]

Gavriil Derzhavin's ode contrasted the 'behaviour' of the river Volga during the events in the reigns of Ivan IV, Peter I and Catherine II, showing that by the time of her journey at the beginning of her reign, the waters of the river had been quietened, 'tamed' in effect, by the Russian state. The Russian Empire, like all empires, had to establish control over its new territories and its new borders, but the economic and strategic significance of the river Volga made it all the more important to establish control over this region.

We have already seen that the Russian government moved swiftly after the conquests of Kazan and Astrakhan to attempt to control and settle the new frontier in the south. The revolts and disturbances of the late seventeenth and the eighteenth centuries nevertheless exposed the lawlessness of the middle and lower reaches of the river, and showed how difficult it was to suppress major revolt. It was only in the second half of the eighteenth century that the government was able to establish a firmer imperial, Russian control over the whole region – and in particular, over the stretch of river from Kazan to Astrakhan. The policies of the government demonstrated both the determination and the flexibility of the Russian Empire. On the one hand, the government physically increased its *Russian* presence in the region through greater numbers of armed forces, garrison troops and Russian officials, and asserted its authority by determining the movement and settlement of peoples in the region. On the other hand, it permitted a degree of institutional autonomy to non-Russian peoples, while also pursuing cultural policies that asserted Russian dominance, but that were supposed to benefit non-Russians as well. The tolerance by the Russian Empire of cultural distinctiveness that did not threaten central power has led it to be termed the 'Empire of difference'; but it was also a practical way of managing the periphery with a small bureaucracy and limited recourse to military force. The result was that by the early nineteenth century, the Volga lands from the north to the south of the river were fully incorporated into the empire. That is not to say that disturbances did not still break out in towns and the countryside (even in the twentieth century) or that non-Russian nomadic peoples were fully absorbed into settled society; but the era of great Cossack and peasant revolts on the Volga was over. The new unsettled southern border of the Russian Empire was no longer the Volga, but the north Caucasus.

❖

The immediate response on the ground to the defeat of the Razin and Pugachev revolts was to carry out savage reprisals on the Volga, as we saw in the previous chapter. In fact, Catherine II, conscious that the brutal suppression of previous revolts had stoked resentment in precisely the areas that were most receptive to Pugachev, asked local authorities to restrict the number of executions; but her plea went unheeded. In other ways, however, Catherine was ruthless in wiping from the map all reference to the instigators and the main areas and settlements that had supported the Pugachev revolt. The name 'Iaik' was removed from the river, the town and the Cossack host and replaced with 'Ural' (the river Ural retains that name). The Cossack settlement that had been the birthplace of both Razin and Pugachev was moved to the other side of the Don and renamed Potemkinskaia, after Catherine's favourite, Grigorii Potemkin. The Volga Cossack host, mainly located in the Cossack settlement of Dubovka, on the right bank of the Volga, north of Tsaritsyn – whose members had not instigated the revolt, but many of whom had joined Pugachev's forces when they were on the Volga – was abolished and the remaining Cossacks (540 of them) transferred to the north Caucasus, where they became part of the line of defence on the river Terek.[2]

Repression of this nature could not, however, provide a permanent solution to disturbances on the Volga. Razin, in particular, had exposed the weakness of the defences of the towns on the banks of the Volga. Pugachev had been defeated each time he faced regular troops, but it proved impossible to crush his revolt swiftly or to prevent him from causing the destruction of most of the town of Kazan. In Astrakhan and Kazan, an orgy of bloody attacks not only on army officers but also on senior officials demonstrated the vulnerability of the agents of the imperial state. Military control of the region had to be strengthened. This was achieved through a combination of curbing the independence of Cossacks and other irregular forces, increasing the presence of regular forces (mainly the army, but also naval forces on the river) and strengthening garrisons.

Razin and Pugachev were Don Cossacks. By the time of the Pugachev revolt, however, the Don Cossack region had already become well integrated into the Russian Empire, which was why Pugachev appealed to Iaik Cossacks further east, who were resisting the extension of state control into their region as well. After the Pugachev revolt, any lingering sense that Cossacks were genuinely 'free' was swiftly extinguished. The most obvious manifestation of this was the way Cossacks could be disbanded, created or moved to another part of the empire. The abolition of the Volga host was not the only casualty of what was regarded as Cossack disobedience. In 1775, the much larger Zaporozhian host, situated on the river Dnieper in present-day Ukraine, was disbanded and its Cossacks relocated to the north Caucasus. Cossacks were moved from the Don, as well as from the Volga, to the river Terek in the north Caucasus, and it became the new river frontier in the south. Cossacks were also forcibly resettled away from the Volga and the Don to Orenburg province and Siberia. New Cossacks and new Cossack hosts could be formed by the state, as well as abolished, which served to diminish their independence and their sense of separate identity. In 1788, one thousand coachmen – that is, state peasants with a particular responsibility to maintain horses on the major post roads in Russia – were simply made into Cossacks;[3] and in 1824, over 8,000 sons of regular soldiers were transferred to the Orenburg Cossack host and became Cossacks.[4]

At the same time, any semblance of democratic decision making among the Cossacks was eliminated and they were put firmly under the authority of local Russian administrators in whatever province they resided. Cossacks had always been a valuable addition to the regular Russian forces as light cavalry and scouts. By the Napoleonic Wars in the early nineteenth century, they had been fully integrated into the imperial army and had become an important element in almost all the major battles; they played a crucial role as cavalry at the battle of Borodino and in harassing Napoleonic troops during the retreat from Russia in 1812–13. By this stage, the Cossack General Matvei Platov was able to call on Cossacks to 'sacrifice everything for the defence of the . . . Throne'.[5] Cossacks had, within a generation, moved from being potential rebels on the Volga to become the greatest servants of the tsar and bastions of the tsarist regime.

At the time of the Razin revolt in the late seventeenth century, towns on the Volga and elsewhere were defended by members of a separate military caste, normally termed 'musketeers', who had been created by Ivan IV and had fought for him in the conquest of Kazan in 1552. By the seventeenth century, the musketeers were already becoming militarily obsolete. Many of those posted in remote towns on the Volga were poorly armed and had become ill-disciplined, and were occupied in local trades as much as in military defence. Some of the musketeers who were supposed to defend Astrakhan against Razin and his band in 1670 went over to the rebels. Both Tsaritsyn and Astrakhan were given larger complements of troops in the 1680s, after the Razin revolt – over 10,000 men in Astrakhan and over 2,000 in Tsaritsyn.[6] An early travel guide noted that Astrakhan in the 1690s was defended by 80 guns and 'Fortified with Wooden Ramparts'.[7] The strengthening of forces did not, however, prevent a further revolt by the musketeers in Astrakhan in 1705 (described in the previous chapter).

In the 1680s and 1690s, musketeers in Moscow were involved in a series of disturbances and revolts. After these uprisings, most of the remaining musketeers were abolished as a separate military caste and became part of Peter I's new regular army. Peter modernized the army by introducing new, modern military tactics and new training schools for officers, and by developing industries which made the Russian army self-sufficient in arms. At the same time, the non-privileged members of society – peasants and ordinary townspeople – were liable for conscription and were selected from their communities by ballot (a process described more fully in Chapter 8). During the eighteenth century, the Russian army became the largest in Europe. According to the main Soviet scholar on the Russian army, it numbered 104,654 men at the beginning of Catherine's reign in 1762 and reached 180,879 in 1774 and 279,575 by 1791.[8] These numbers are the formal 'establishment' of the army rather than the reality: many recruits, in fact, deserted or died on their way to join its ranks; but it was nevertheless a large force which was always able to defeat Pugachev's men when they fought a pitched battle. The inability of the army to crush the Pugachev revolt after relieving Orenburg and Kazan was, in part, due to other factors: the deployment of most of the Russian army in the Balkans in the Russo-Turkish War (which ended only in July 1774); some complacency on the part of the army and the government after their victory at Orenburg; sheer luck, as Pugachev escaped across the steppe and down the Volga; and simply the extent and remoteness of the Volga countryside, which allowed Pugachev to regroup. By

1801, the Russian army numbered over 440,000 men; by the end of the Napoleonic Wars this figure had risen to over 700,000.[9] Furthermore, this was a standing army. The unfortunate peasant and urban recruits were conscripted for life (25 years after 1793, but in effect that was life for most conscripts). This meant that by the second half of the eighteenth century, permanent and trained troops were available to put down disturbances and were less likely to go over to the rebels. Troops were used regularly to suppress disturbances on serf estates, on the Volga and elsewhere, and in the factories in the Urals.

The army was primarily an ethnic *Russian* and *Christian* army. Most of the officers by the second half of the eighteenth century were ethnic Russian, although there was always a considerable number of Baltic Germans (mostly Lutheran) in the officer corps, as well as foreign mercenaries. The army that beat the Bashkir rebels at the siege of Orenburg and subdued the population after the revolt was primarily a Russian army, serving the needs of the imperial state, but non-Russians on the Volga were obliged to serve as conscripts or as forced labour in this army. In 1718, Mordvin and Chuvash peasants were forced to cut down trees by the banks of the Volga to construct the Russian fleet.[10] In 1722, conscription to the regular army was extended to Tatars, Mordvins and Mari (and young Tatars aged between 10 and 12 could be recruited as sailors) on the same basis as Russian peasants.[11] Nomadic tribesmen in the Volga region, including Bashkirs and Kalmyks, were used as irregular forces in the course of the eighteenth and early nineteenth centuries.[12] In the Napoleonic Wars, Bashkirs formed 28 regiments, each of 500 men, and fought at the battles of Borodino, Dresden and Leipzig.[13]

Being a Muslim soldier in the Russian imperial army was not considered a problem in this period, in part because service for (in effect) life meant that loyalty was to the army, or the regiment, rather than to the community from which the soldiers came. In fact, Muslims from the Volga region enrolled in the Cossack hosts of the Terek in the north Caucasus. In 1826, the Chernigov infantry regiment, which served in the Caucasus, included a number of non-Russians and Tatars from Kazan and Simbirsk among its ranks. They included: 28-year-old Platon Egorov, a Mordvin from Simbirsk province; 21-year-old Iakub Timoraleev, a Tatar from Kazan province; 28-year-old Kozma Mikhailov, a converted Mari from Kazan province; 22-year-old Semen Simonov, a converted Chuvash from Simbirsk province; and 21-year-old Mustai Baiazitov, a Tatar from Kazan province.[14]

The problem for Russia by the early nineteenth century was not a shortage of troops to subdue revolt and control the population, but the difficulty of deploying them within the empire in peacetime, and then of moving them across vast distances to points of disturbance. The poor quality of roads remained a problem; troops could be moved from the Moscow region on a major road to Nizhnii Novgorod and Kazan, but the regions of the middle and lower Volga could only be reached by river or by minor tracks. The majority of regular troops were deployed in the central provinces and western regions, in order to be ready for campaigns in central Europe, and not in the Volga region.[15] There was an attempt to improve major roads in the eighteenth century – not only to move troops and for ease of trade, but also so that officials from Moscow could reach the provinces and check on their performance; but communications remained poor south of Nizhnii Novgorod.

The government, as we have seen, also established additional 'lines' of garrisons to defend and control border areas. These garrison troops were in addition to regular forces and numbered some 65,000 men in 1764 in the whole empire, reaching almost 90,000 by 1774, but falling back to about 77,500 by 1800.[16] As the Russian Empire expanded further south and east, so the 'lines' were extended or new lines built to control the newly acquired lands. In 1731, the so-called 'Tsaritsyn line' was constructed, linking the rivers Don and Volga, and 500 Cossacks were moved from the Don to man the new line. After Bashkir revolts in the 1730s, another line was constructed from Samara, on the left bank of the middle Volga as it turns east, to the river Iaik. Two hundred Cossacks were settled there, and by 1840 nearly 16,000 people (men and women) were living on the line. These fortifications were set up to defend against the 'other' in the Volga region: Tatars in the middle Volga region; Kalmyks in the south; Nogais and Kirgiz in the east.

In 1737, the Astrakhan Cossack host was founded to defend the city and its environs as far down as the fort of Chernyi Iar, south of Tsaritsyn on the Volga. At first, the host was small, but more men were settled from the 1760s; and by 1770, six regiments of Cossacks and Tatars were stationed in Astrakhan. There was still a fear of raids, and Kalmyks were not deployed in the town, because, according to one British traveller, 'the Kalmyks are not, in Astrakhan especially, much to be trusted'.[17] The garrison was sufficiently strong that it was avoided by Pugachev, who sensibly targeted less well-defended forts. By the 1830s, the Cossack host in Astrakhan numbered over 18,000 men.[18] The garrison of Saratov was also strengthened after the Razin revolt, but this did not prevent the town from falling to Pugachev without a fight. It is no coincidence that Pugachev bypassed major, well-defended forts like Iaitsk, but was able to overrun smaller and more remote forts. In Pushkin's *The Captain's Daughter*, a fictional account of the Pugachev revolt written in 1836, the hero arrives in a remote garrison and is shocked by what he finds:

I looked around me in every direction, expecting to see menacing bastions, towers and a rampart; but all I could see was a little village surrounded by a thick wooden fence. On one side of it stood three or four haystacks, half-concealed beneath the snow; on the other a dilapidated windmill with idly-hanging bark sails. 'But where is the fortress?' I asked in surprise. 'There it is,' replied the driver, indicating the little village.[19]

The increase in garrison troops enhanced the capacity of the Russian Empire to control the Volga borderland. It also meant that troops could better defend towns against rebels and could be mobilized more easily to crush disturbances – urban, rural or industrial – in the Volga region. It did not mean, of course, that the borderlands were now completely secure and tamed. The distances covered by the 'lines' were simply too great to be manned adequately at all points. Even in larger forts the garrison troops did not impress. The traveller George Forster described the garrison of Astrakhan at the beginning of the nineteenth century as comprising 'about fifteen hundred men, who have more the appearance of militia than regular troops, and are conspicuously deficient in military order'.[20] The reality was that garrison troops often comprised older and less physically able troops. The garrison became a form of 'outdoor relief' for old soldiers who had nowhere else to go. Garrison records which I have examined included a considerable number of soldiers who were over the age of the 60, and some who were in their 70s! Other elderly soldiers found some contentment in their later years by marrying local women (often non-Russians) and having, almost certainly, a new family.[21] Garrisons may well have provided more security than the alternatives in a country that had no retirement homes for old soldiers – veterans could be found begging in Moscow in the early eighteenth century. The garrison forces were more reliable and numerous than during the Cossack revolts, but could not in themselves ensure the stability of the middle and lower Volga region.

Nor could regular troops and garrisons be expected to curb the piracy on the river Volga, described in the previous chapter. Astrakhan itself was protected by the Caspian flotilla, which had been established by Peter I in 1722 primarily for his campaign against Persia in 1722–23. By 1750, this small flotilla comprised three ships manned by sailors.[22] The Pugachev revolt, however, exposed the weakness of state control over the river, and the few ships that did exist to protect the Volga towns were taken over by Pugachev's men. A naval force to protect the river was established by Tsar Paul I (reigned 1796–1801) only in 1797. He ordered that nine armed oared ships should be constructed to guard the river – three positioned between Astrakhan and Tsaritsyn, three between Tsaritsyn and Kazan and three for the river upstream from Kazan. In 1800, the number of these protective boats was increased to 12 – still a small force to control the middle and lower Volga.[23] In the early nineteenth century, some of the responsibility for constructing defensive fleets passed to the Volga towns themselves.

Robber gangs on the Volga, however, persisted. In the 1820s, travellers on the river to Astrakhan were warned to take pistols with them because of the danger of thieves and bandits.[24] In 1821, the government received a report on brigandage 'on the lower stretches of the Volga, from Kostroma to Astrakhan, especially in the provinces of Kazan and Nizhnii Novgorod'.[25] In 1829, Nicholas I tackled the issue more seriously and issued special regulations for the protection of the major rivers, including the Volga (and several of its tributaries, including the Kama and Oka). In Kazan, for example, he instructed that 18 ships staffed by 90 soldiers and 9 officers were to be stationed in the town to defend against river pirates. The officers had considerable legal powers to try captured thieves and to punish them by flogging or 'running the gauntlet'. The result was that by the 1850s, the dangers on the river Volga had been curbed. By this stage, robbers had become far less of a threat to trade, allowing tourism on the river to develop.[26]

❖

The state presence on the Volga in the late eighteenth century was increased not only through more soldiers, but also by appointing more government officials. The Pugachev revolt had exposed the fact that Russia was both badly and under governed. Pugachev had attracted support by his condemnation of corrupt officials, including judges who took bribes. The provincial governors in Volga towns were unable to defend towns or countryside from Pugachev's men or from his propaganda. In Catherine II's view, the initial success of the revolt was due to the 'weakness, laziness, negligence, idleness, disputes, disagreements, corruption and injustice' of local officials.[27] It was not as if the Russian government had not been warned before the revolt about the inadequacies of provincial administration. One of Catherine II's first acts on coming to the throne was to call a large Legislative Commission of deputies from all over Russia (142 deputies from nobles, some 200 from state peasants and other social groups, including 44 deputies from the Cossacks and 54 deputies from non-Russian tribes). The elections were conducted somewhat haphazardly; Astrakhan and Saratov sent five deputies, but Kazan only two (one Russian and one Tatar).[28] The assembly met in 1767 and was immediately presented with a long

Instruction from Catherine – much of it copied from Montesquieu's *The Spirit of the Laws* and other recent works by European writers – which was supposed to guide the debate.

Before the assembly met, groups of electors were invited to submit 'instructions' of their own to their elected deputies. These bottom-up 'instructions' were very different in tone from Catherine's set of principles in her *Instruction*. Instead of looking forward to possible new forms of government, judicial practice and social order, they looked backwards and demanded, above all, that their traditional privileges should be protected. This was as true of Cossacks and non-Russians, who considered that they were losing their 'freedoms', as it was of Russian nobles, who insisted on their sole ownership of serfs. In this respect, the views of local deputies were not so very different from those of supporters of Razin or Pugachev. The local 'instructions' were also, however, critical of corrupt judges and officials, the lack of doctors in the provinces and the general inadequacy of administration in the countryside. The 'instructions' from the Volga region were no different in this respect from those from other parts of the empire, but they did reflect local, non-Russian concerns. There were complaints, for example, from Tatars in Simbirsk province that their exemptions from poll tax had been overridden; from the Volga and Astrakhan Cossack hosts concerning the need to protect their fishing rights on the Volga; from newly converted Chuvash in Simbirsk province on the failure of the courts to protect them from serfs on neighbouring noble estates, on the problem of raiding Kalmyks and on the slowness of judicial processes; and from Tatars in Kazan province about disorders in conscription and the 'malice' of the local courts.[29]

Catherine was therefore fully aware of the limitations of local administration, but from 1768 to 1774 she was too bound up with foreign affairs to address these issues. It was only after peace was concluded with the Ottoman Empire in 1774 that she was able to turn her attention to local administration – a matter that had become more urgent after the experience of the Pugachev revolt. In 1775, a statute established a new structure of provincial administration. The empire was divided into provinces (*gubernii*) and then subdivided into districts (*uezdy*). New and much more elaborate structures were put in place for courts, financial institutions and other forms of administration. New posts were created in the provinces, some for full-time officials and some for elected, paid representatives of different social groups (the lower courts were based on social class, in that there were courts for nobles, townspeople and state peasants). A statute setting up police forces in provincial towns was issued in 1782. In 1785, Catherine issued a Charter to the Towns, which established a structure of dumas (or councils) within towns, comprising representatives from the urban population, who were now categorized into six groups, to handle administrative and financial matters.

It may seem odd that the complaints about the poor quality of officials were addressed by creating yet more officials, most of whom had no legal or administrative training, but were simply elected to positions of authority by fellow members of the same social group. Russia was severely under-governed in the eighteenth century, with far fewer officials (relatively speaking) than other European countries: in 1763, there were some 16,500 officials in central and local administration, while Prussia, with less than 1 per cent of Russia's land area, employed some 14,000 officials in a far more structured civil service.[30] Some of the resentment which Pugachev (and, even more so, Razin) played on was the image, often all too accurate, of powerful governors (called *voevodas*) who acted corruptly and/or oppressed poor urban and rural inhabitants with impunity. One governor, for example, in the region inhabited mainly by Mari peasants, was said to accept bribes in the period 1762–63 of rye, oats, honey and other goods.[31] The number of local officials rose from an estimated 12,712 in 1774 to 27,000 by 1796 throughout the whole empire. By significantly increasing the number and range of local government officials, the state hoped to curb the power of any one individual. In practice, this could not prevent corruption occurring among senior officials in the provinces, but at least blatant corruption was now more likely to be reported. A governor in Kazan was investigated in the early nineteenth century, after reports of the use of torture in courts. A new governor, Ivan Borisovich Pestel (father of Pavel Pestel, leader of a major revolt in 1825), was investigated in 1803, after reports of disorders and bribes on a massive scale.[32]

There were further advantages in simply increasing staffing in the provinces. Taxation could be collected more easily in towns and the countryside, and by more regular tax collectors. Fugitive peasants, who were particularly likely to be lured into gangs of robbers on the Volga, could be dealt with more effectively. By the end of the eighteenth century, elaborate procedures had been put in place to track down fugitive peasants, hold them and arrange for them to be either returned to their owners, or sold (at a lower price) to local landowners.[33] Urban administration became more efficient and gave salaries and status to at least the wealthier merchants. More courts, and more court officials, meant some inroads into the massive backlog of unresolved cases, which had been a source of complaints in the 'instructions'. Catherine II's legislation also led to the beginnings of a police force in Russia. The 1775 statute established land captains (*ispravniki*), elected from the local nobles, in each district; and a town provost in all towns, with broad responsibilities for maintaining law and order. In 1782, police were established in towns with a whole range of duties, from the arrest and punishment of petty thieves and preserving order, to upholding morals by ensuring the separation of the sexes in bath-houses!

No one claimed at the time that local administration suddenly became effective and incorruptible. The town of Iaroslavl, on the northern Volga, filled all its posts at all levels with able people, according to an official review conducted in 1778;[34] but this was not true of posts in more remote towns or on the middle or lower Volga, which had been subjected to attacks by Cossack rebels or tribesmen. In Astrakhan province, for example, at the other end of the Volga, it was impossible to find enough nobles to fill the local government posts that had been opened up for them, simply because there were fewer Russian noble landowners in this province.[35] Legal training in Russia and the growth of what we would call a legal consciousness emerged slowly. The first ethnic Russian professor of jurisprudence at Moscow University was appointed in 1773; he had been trained at Glasgow University. 'As for justice 'tis a joke. No such thing exists & its shadow covers vice enough to make me shudder', commented an Anglo-Irish lady traveller at the turn of the nineteenth century.[36] Bribery was pervasive, not only in the courts but also among tax collectors and police officials.[37] Russia only began to develop a proper civil service in the nineteenth century. Most of the judges and land captains appointed from the nobility in the provinces after 1775 were retired army officers with no legal training or administrative experience.

Gradual progress was being made, but it all took time. Peasants in Saratov composed a lament in the late 1780s that could have been written for them by Razin or Pugachev:

Now the whole world knows
How the land captain and secretaries make life impossible for us.
By the command of their chief, thieving hundredmen
Constantly make requisitions.
They treat us inhumanely,
Something which has never been heeded.
Does this anger you, heavenly Tsar.
That You have
An incorrigible fool,
A thief as an assessor,
A rogue as a secretary,

Who have ruined us completely?

They did not stop at a chicken or a sheep.

In the past tyrannies detested the Christian faith,

The present one tortures us if we do not give money or measure of oats.

And all our profits and incomes

Have been used up on the expenses of the [lower] land court.[38]

The new officials in the provinces were predominantly *Russian* officials (or Russified Baltic Germans), asserting the authority of the Russian government over non-Russian peoples, as well as Russians. The main elective posts in the new local government institutions, particularly the courts, were filled by Russian landowners, which is why some posts went unfilled in provinces where there were few noble landowners (non-Russian peasants in the Volga region were predominantly state peasants, and not serfs on noble land). The Russian government in this period was content, however, to allow non-Russians to have their own low-level institutions, including courts, and, to an extent, to use their own legal systems. This was not so much an indication of tolerance by the state, but simply a practical and cheap way of governing subjects of different ethnic and confessional groups. We will see, for example, that Armenians in Astrakhan had their own urban institutions and that Tatars in Kazan had their own courts (Chapter 9). These courts also used their own languages, although Russian translations were often required. It was only in the late nineteenth century that the state thought it necessary to impose its control through the standardization of institutions and law, and insisted that Russian should be the language of government.

❖

The Russian Empire (and, for that matter, the Soviet Union) also established control by moving potentially troublesome subjects at will. Restriction of move-ment was the basis of serfdom (see Chapter 3), institutionalized in Russia from 1649. But inhabitants of towns were also tied to their place of residence and needed formal permission to relocate. This was true throughout European Russia (there were very few serfs in Siberia, but other forms of restrictions on movement still applied), but the nature of the Volga region made it more difficult to control the settlement and movement of people: there were nomadic herdsmen on both sides of the river; the multi-ethnic, multi-confessional peasant and urban population had traditionally enjoyed more autonomy than Russian serfs or townspeople; the river and region were havens for fugitive peasants, who could merge into the river workforce with few questions asked. In the eighteenth and nineteenth centuries, the state asserted far greater control over the location of the population of the Volga, and in the process changed the social and ethnic composition of the region – and, to an extent, ensured a greater loyalty of the settled population.

Cossacks, as we have seen, could be moved at will by the Russian government, irrespective of the ties they considered they had to particular regions (and rivers). In the seventeenth century, other forms of military servitors were also moved by the government to the borderlands to assert Russian control of these lands. Musketeers, as we have seen, were stationed in Astrakhan. So-called single-homesteaders (*odnodvortsy*) were a separate category of military service men, who had been given small plots of land on the southern borderlands in return for military duties, which primarily meant defending the Russian borders from raiders. Their service, like that of the musketeers, became obsolete in the eighteenth century; but they still resided in the countryside and the towns on the Volga, and were involved in agriculture and petty trades. In 1780, 447 single-homesteader males were registered in Kazan province alone.[39]

The government also attempted to move new people, who it assumed would be loyal to the regime, into and around the Volga regions. In the 1730s and 1740s, retired (mostly Russian peasant) soldiers were encouraged to settle in 'empty' lands on the Volga, with a number of inducements to do so – a piece of land and 5–10 roubles per family.[40] In 1764, the same inducements were offered to retired soldiers if they were prepared to settle in Astrakhan province.[41] Few, however, took up the offer. By 1763, only 3,480 retired soldiers and their families had resettled in Kazan province.[42] In 1739, the town of Stavropol (the name means the 'city of the cross') was founded on the river Volga west of Samara as a military settlement for Kalmyks who had converted from Buddhism to Christianity. Two thousand Kalmyks were forcibly settled there from the Saratov region. The town, however, grew slowly, and some of the Kalmyks joined the Pugachev revolt. After the revolt, the settlement had to be re-formed because it had been left almost depopulated.

Catherine II took the matter of settlement in the Volga into her own hands when she directly set up colonies of German immigrants on land on both sides of the river near Saratov. Catherine was influenced by current thinking that an increase in population would lead to an increase in the wealth of the country, and by the belief that Russia's backwardness stemmed in part from its low population density. In 1762, she issued a manifesto encouraging settlement in Russia's 'many empty places' by any foreigners (except Jews) who could benefit the crafts and agriculture in Russia. A manifesto the following year specifically encouraged settlement in Saratov province and offered a number of inducements, including land, support to establish settlements, the right to practise religion freely, exemption from taxes for a number of years and freedom from conscription. The manifestos did not specifically target immigrants from German lands, but in practice those proved the best recruiting grounds for new settlers (especially Hesse). Catherine (a German, of course) was probably not too unhappy about this: she and her advisers assumed that the hard-working German peasants would benefit the country economically, and perhaps also provide an example to work-shy Russian peasants. By 1764, lands in Saratov province had been allocated for settlement. Some 30,000 colonists arrived in 1766, often finding themselves in empty fields, with no provision for housing, so that they had to dig holes in the ground to survive the first winter. Between 1764 and 1768, 104 colonies had been founded on both sides of the Volga.[43] Numbers dropped to 23,000 in the wake of disease, drought and lack of seed, but the colonies gradually developed; by 1798, the number of colonists had grown to 39,000, and by 1811 to 55,000.[44]

The settlement of German colonists on the Volga was, however, based on several assumptions by the Russian government. The first was that the settlements would improve agriculture and prosperity on the Volga. In practice, as we shall see in Chapter 8, the German colonists were restricted in what they could do and where they could go, which militated against agricultural or commercial developments. The second assumption was that the lands they were granted by the Russian state were 'empty', whereas in fact they were used by the nomadic tribespeople for grazing. The colonies had a significant impact on the nomadic peoples, whose way of life was already under threat by the encroachment on their lands of noble landowners and peasants and the circumscription of their traditional privileges. German colonies on the eastern bank of the Volga disrupted the movement of the nomadic Nogai and Kirgiz tribesmen, which was why they were so hated and were subjected to devastating raids.

Kalmyks who had settled on the western bank of the river and between the rivers Don and Volga felt hemmed in by the Tsaritsyn 'line' of garrisons and the new settlements of German colonists. As one of them eloquently expressed matters:

> Look how your rights are being limited in all respects. Russian officials mistreat
> you and the government wants to make peasants out of you. The banks of the
> Iaik and the Volga are now covered with Cossack settlements, and the northern
> borders of your steppes are inhabited by Germans. In a little while, the Don,

Terek, and Kuma will also be colonized and you will be pushed to the waterless steppes and the only source of your existence, your herds, will perish.[45]

The choice offered was stark: 'either to carry the burden of slavery, or to leave Russia and thus end all your misfortunes'. This is precisely what happened. In 1771, some 150,000 Kalmyks from the lands between the Volga and the Don simply packed up their tents and moved with their families and herds back to their ancestral homelands on the border with China. Catherine II was furious: her settlement plans for the Volga did not envisage 'losing' subjects. She was helpless, however, to prevent their long march east. It was a tragedy for the Kalmyks: attacked by hostile tribes *en route* and unable to withstand the harsh winter, up to 100,000 of them perished. Those who survived were put into the service of the Chinese emperor, and thus swapped one imperial oppression for another.

Other potentially disruptive social groups were also carefully monitored. Indeed, monitoring of the population in response to imperial decrees and reporting on their location, legal status and occupation seemed to occupy an inordinate amount of the time of provincial officials in the empire, possibly because it required little in the way of initiative on the part of often second-rate officials. In 1843–44, local authorities listed the number of Roma in their districts, in response to a directive from the Senate. Roma had been categorized as state peasants by Catherine II, but were still regarded as a separate social category, treated with suspicion and suspected of thieving and illicit horse trading. It has been estimated that in the mid-nineteenth century there were some 47,000 Roma in the whole empire. Their movements were controlled and monitored. In response to a Senate directive, it was noted in the 1840s that in Kazan province there were 25 Roma families in Spassk district and 7 families in the district of Tsarevokokshaisk (now called Ioshkar-Ola, and the capital of the Mari El Republic). In the town of Kazan itself, Roma numbered 44 males and 39 females; the town asked if they could be resettled from the town to villages.[46]

There are very few descriptions of Roma on the Volga. One exception is the account of a voyage down the Volga in the middle of the nineteenth century by Charles Scott. He came across an impressive singing ensemble in Nizhnii Novgorod 'descended from a race of gypsies':

They were in number about twenty of both sexes, the women predominating. It would have been difficult, by costume or manner, to have traced any connection between these people and those vagabonds who are to be found as well in

Russia as other parts of Europe. The men were dressed in evening coats and shiny boots, and evidently prided themselves upon the tie of their white cravats; and the fine figures of the women were decked in silk and satin dresses, the cut of which would not have injured the reputation of 'Palmire' [St Petersburg]; but the dark complexion, regular features and the black flashing eyes at once proclaimed their origin. In their songs there was a wild and thrilling harmony; and they acted, in face and gesture, with admirable truth the passions intended to be displayed in the words.[47]

In 1854, the Samara civil governor received a request for details of the location of all Kirgiz who lived in the province without written permission, including the names of all the villages and the number of horses that they possessed.[48] In 1861, Samara province listed all the Kalmyks living in all the towns and villages of the province.[49] In the later nineteenth century, the numbers of Jews and Baptists were also recorded by district. The situation of converted pagans and Muslims is discussed in the next chapter. The multi-ethnic and multi-religious composition of the Volga region continued to be regarded as potentially disruptive and was closely monitored.

There were other, rather more subtle ways in which a Russian imperial identity could be stamped on a largely non-Russian region. One way was through architecture. Kremlins were an early manifestation of Russian domination, and were built in all major towns that were seized from the Tatars or that were newly constructed at key points on the river. The kremlins were built on an elevated position, not only for defence, but also to demonstrate visibly the authority of the Russian state over the urban population. Within their walls were the key buildings symbolizing imperial power – the Orthodox cathedral and the governor's house and other major administrative buildings. When Pugachev sacked Kazan, he occupied, and largely destroyed, the lower part of the town, including the Tatar quarter, but was unable to take the kremlin.

In the second half of the eighteenth century, Catherine II encouraged major towns to replace their wooden buildings with new stone buildings in the classical style. But this was not a free-for-all: she specified that towns had to submit proper plans, which she personally approved. Streets were to be straight and broad, and constructed on a grid or a radial plan; government buildings were to be prominent, usually positioned around a central square; provision had to be made for schools and other welfare institutions.[50] In other words, the towns had to be 'European', not 'Asiatic' – something that asserted Russian cultural dominance. In truth, this was as relevant to Moscow and other towns in central Russia (but not St Petersburg, of course, which was a planned city), and to towns on the Volga founded by the Russians (such as Saratov, Samara, Tsaritsyn), as it was to traditionally non-Russian towns such as Kazan and Astrakhan.

The first town that was centrally planned in this way was Tver, on the upper Volga, where the centre had been almost destroyed by fire. Catherine also made 200,000 roubles available to rebuild Kazan after the destruction wreaked by Pugachev's men.[51] The new towns were not simply intended to be more sanitary and less susceptible to fire: their uniformity demonstrated that they were all part of the same empire and served the same administrative purpose as representatives of the power of the state. Other monuments could assert Russian dominance. After the defeat of Napoleon, a memorial was erected on a hill a little distance from the kremlin in memory of those who had died in the subjugation of Kazan in 1552 on a hill a little distance from the kremlin.[52]

Another way in which the state imposed a uniform, and largely Russian, European culture was through education and the arts. The syllabuses of national schools, to which non-Russians were admitted as well as Russians, were determined by the state. Catherine II's schools required the study, among other topics, of a specially written textbook, *On the Duties of Man and Citizen*, which encouraged pupils to perform civic duties to the state. The text was withdrawn by Alexander I in 1819, when it was deemed to contain revolutionary ideas, and the syllabus then became more conservative, with greater attention devoted to the study of Russian Orthodox religion. The Russian government, however, was content that schools should be set up for non-Russians – Armenians in Astrakhan and Tatars in Kazan – and indeed encouraged the establishment of such schools. The government only asserted control over Muslim schools on the Volga attached to mosques in the late nineteenth century. It was only at this time that the use of the Russian language in schools became an important tool of Russification of the non-Russian peoples of the Volga.

Catherine II was acutely conscious of her image and the need to assert herself as the 'Russian Empress' to all her subjects – not least because she had come to the throne by usurping it from her husband, Peter III, the legitimate tsar; she was, of course, German by birth and had also been brought up as a Lutheran. Her sensitivity about this made her understandably furious that Pugachev should have portrayed himself as the 'true tsar'. One way to assert her legitimacy was to visit her varied lands, with all the elaborate ceremonial this entailed, and show herself to her 'people' – or at least to an appropriate selection of them. There was a long western European tradition of kings and queens undertaking 'royal progresses' to force local barons to swear fealty and, in effect, to assert royal control over their lands. Tsars had traditionally undertaken pilgrimages to holy sites, thus forging their association with Orthodoxy, but they did not undertake secular journeys of this nature. Peter I, however, visited forts and towns in Russia that were of significance for his military campaigns. In June 1722, when he was in Astrakhan in preparation for a campaign against Persia, he visited the kremlin and examined the fleet.[53] Catherine's journeys were, however, deliberately chosen to assert her authority over her new subjects and lands. It is no coincidence that the first of these journeys was to the river Volga in 1767, and culminated in a visit to multi-ethnic Kazan.

Catherine set off in May 1767 with a flotilla of four galley ships specially built for the journey (significantly named the *Tver*, *Volga*, *Iaroslavl* and *Kazan*), plus another 10 to 20 ships to transport her entourage of some 2,000 people.[54] The journey of just over six weeks took her to villages, monasteries, historical sites (including the ruins of the ancient town of Bolgar on the river Volga, where she expressed concern that the monuments had been vandalized by the local clergy), and she visited all the main towns on the upper and middle Volga: Tver, Uglich, Rybnaia Sloboda (renamed Rybinsk in her reign), Iaroslavl, Kostroma, Nizhnii Novgorod, Cheboksary, Kazan and Simbirsk. Everywhere she went she was met with elaborate displays of loyalty from local nobles, merchants and clergy. She hosted events for elite members of the towns, visited churches and monasteries and witnessed popular celebrations for the ordinary people.

At one level, the journey was intended to enable Catherine to find out more about her empire on the eve of the Legislative Commission. She noted, for example, that the merchants of Nizhnii Novgorod were hampered by their lack of capital. She saw towns on the upper Volga, such as Tver and Iaroslavl, which were vital for the supply of grain and other goods to St Petersburg and Moscow. She was also able to appreciate the beauty and the potential of her realm, and to express these to a perhaps sceptical audience outside and inside Russia. She wrote to Voltaire on 5 May that 'hour by hour the banks of the Volga become better'. She also lectured her chief Russian advisers in St Petersburg on the same subject. To Nikita Panin she wrote: 'here the people on all the Volga are wealthy and comfortable', and to Mikhail Vorontsov, 'the river Volga is better by far than the Neva [in St Petersburg]'.[55] She was able to present herself as a devout member of the Orthodox Church by attending religious ceremonies, including taking part in a procession of the cross in Tver. At the same time, by meeting Old Believers at the small, prosperous, ship-building town of Gorodets, on the Volga north of Nizhnii Novgorod, she could present herself as an enlightened, tolerant ruler in her attitude towards non-Orthodox Christians. Ordinary people eagerly expressed their gratitude to the empress, not least because her visits were accompanied by public holidays and fireworks.

The visit to Kazan, however, was of the greatest significance (her voyage was reported in the *St Petersburg Gazette* as 'The cruise of Her Imperial Majesty to Kazan', although in fact the journey ended at Simbirsk). This was the city that had been taken in conquest by Ivan IV from the Muslim khans and was a mixed Russian and Tatar population (although predominantly Russian at this time). Catherine met Muslim leaders in Kazan, as well as Orthodox clergy, as she had done in Gorodets. Vladimir Grigorevich Orlov, the younger brother of Catherine's favourite, attended a service at a mosque and was the guest of Tatar merchants. He was at the time director of the Academy of Sciences and, as a typically enlightened scholar, wanted to categorize and identify different peoples during the journey (he later sponsored the ethnographic travel accounts of Russia, including the Volga, by the German naturalist Peter Simon Pallas, described in Chapter 7).

To Catherine, this was the border of her empire between Europe and Asia. 'I am in Asia', she wrote to Voltaire. She further enthused that 'there are in this town twenty diverse sets of peoples who in no way resemble each other'.[56] She made a point of visiting the Tatar quarter in the town, where a public holiday was held in her honour. In what seems to the modern reader a rather grotesque event, the representatives of different ethnic groups – Tatars, Chuvash, Mordvins, Udmurts, Mari – were presented to her and her entourage at the governor's house in national costume and danced for her.[57]

The progress of Catherine down the river Volga was commemorated in a number of odes. These were written in the style of the time, but are significant for the way in which they depict the Volga. By the mid-eighteenth century, rivers and seas had come to feature as popular motifs in poetry, but the Volga was still regarded as simply one of the major rivers in the empire, alongside the Don and the Dnieper. In the wake of Catherine's voyage, however, the river became associated with key historical events that created the Russian Empire.

The poet Vasilii Maikov wrote two odes on Catherine and the Volga. In his first ode, in 1763, when Catherine had already visited Iaroslavl but had not yet embarked on her Volga voyage, he described the river as flowing reluctantly south to the Caspian Sea, when it would rather be at Catherine's side. The river asked:

TAMING THE VOLGA

'Why am I [the Volga] not allowed by fate
To check my flow and be with you
O Sovereign, light of Russia?'

In his second ode, written in 1767, the Volga had in effect become the obedient subject of the empress, and was so in awe of her that it had tamed its waters to ease her passage:

I see before me beautiful meadows,
I see cities, raised up like cedars
Burdening the steep banks
In which the Volga flows; it [the river] marvelled in rapture
Seeing Catherine's progress along it
And restrained its flow and stopped
And thus multiplied its swiftly flowing waters
With which, covering all the rocks and shoals
Made for its sovereign lady a free passage.[58]

Derzhavin's ode, quoted at the beginning of the chapter, commemorated her visit, but also asserted the control now exercised over the river by the Russian state. Of course, poets were writing to honour and please the empress and to assert the power of the Russian Empire. In the process, however, the Volga begins to appear for the first time as a *Russian* river, as an obedient subject of the Russian empress.

The taming of the river Volga was easier to achieve in poetry than in reality. Nevertheless, by the early nineteenth century, the region had become more integrated into the Russian Empire, and the traditional freedoms of Cossacks and nomadic tribesmen had been curbed. The military and bureaucratic presence of the state had been strengthened, as the autonomy of Cossacks and tribesmen weakened. The borders of empire which had to be guarded had moved to the south (the north Caucasus) and the east (Orenburg and Siberia). This did not mean that the Volga had become homogenized. The 'twenty peoples' who had entertained Catherine and Orlov in 1767 represented the multi-ethnic and multi-confessional composition of the middle and lower Volga. Imperial control also had to be established over these non-Christians – pagans, Muslims, Buddhists. The process was complex and sometimes involved contradictory policies, but was an essential part of maintaining a well-functioning and ideologically coherent empire. It is to this that we now turn.

6
ORTHODOXY AND ISLAM ON THE VOLGA

The core identity of the Russian Empire was Orthodox, Christian and Russian. The empire was, however, multi-ethnic and multi-confessional, and the middle and lower Volga regions were the first major non-Orthodox territories conquered by the Russian state. The number of non-Christians was significant, in both the towns and the countryside. Although the majority of non-Christians were Muslim (and they are the main subject of this chapter), the region was also home to Finno-Ugric people with shamanistic beliefs and Buddhists. The Russian Orthodox Church was also challenged from within by Old Believers and by members of dissident religious sects who found refuge in the remoter parts of the Volga region. By the late eighteenth century, the Volga was no longer the border of the Russian Empire, but the issues faced, and the ways in which they were tackled, are fundamental to our understanding of the way the imperial Russian government attempted to convert, manage and control its non-Orthodox subjects. The river Volga divided west from east: it formed, as we have seen in the way Catherine II expressed it to Voltaire, the border between the Christian 'West' and non-Christian 'Asia'.

Government policy towards non-Christians fluctuated over the eighteenth and nineteenth centuries, determined partly by the personal attitudes of individual tsars, partly by the extent of the activity of Orthodox clergy on the ground and partly by the degree of success of conversion policies. The experience of the middle and lower Volga regions showed that the Russian government was consistent in its determination to monitor and control the movement of non-Christians in the empire, but that at the same time its policies on conversion could be flexible and pragmatic. Above all, the government had to accept the existence of a large number of non-Christian, and non-Russian, subjects within the empire and ensure their loyalty.

It is not possible to give precise figures for the ethnic and religious breakdown of the population of the Volga in the eighteenth century (a regular census – of male

106

subjects – started only at the end of the reign of Peter I, and was introduced so that taxation could be allocated per head, the poll tax). In the empire as a whole, ethnic Russians dropped below half the population by the end of the century,[1] although Slavs – Russians, Ukrainians and Belarusians – always constituted a significant majority. The ethnic and religious composition of the Volga region was distinctive. One set of figures states that by around 1860 there were 418,504 Tatars in the province of Kazan, 152,908 in Samara, 85,412 in Simbirsk, 59,897 in Saratov and 31,950 in Nizhnii Novgorod (at the same time, Tatars in Orenburg province numbered 858,695 and in Perm 71,965).[2] Other figures for Kazan province at around the same time put the number of Chuvash people at 338,440, and 89,728 Mari.[3]

Ethnicity did not always, of course, correspond to religion. Not all Tatars in the Volga region were Muslim: those who had converted to Christianity were divided between so-called 'old baptized' (*starokreshchennye*), 'newly baptized' (*novokresh-chennye*) and 'apostates' and are discussed further below. The majority of Chuvash and Mari peasants had converted, at least nominally, to Orthodox Christianity by the end of the eighteenth century. As a result, Christians consistently outnumbered Muslims. It is estimated that in Kazan province in 1861 some 74 per cent of the population were Christian and 24 per cent Muslim.[4] It was the interest, if not obsession, of the local and central authorities to know precisely how many adherents of which faiths there were in each region. In 1854, a survey was undertaken to record the number of Muslims, Jews and 'idolaters' (pagans) in Kazan province by district. It found that there were 380,601 Muslims and 8,246 pagans in the province, and that there were 794 mosques.[5]

❖

Russian government policy towards the conversion of non-Christians was a mixture of persuasion and force. In effect, those at the top and the bottom were the easiest to convert.[6] Elite members of society – landowners – could be assimilated by being offered the same titles and rights as Russian nobles, including the right to own serfs, if they converted. It was always easier to convert pagans – that is, those with animistic or shamanistic beliefs – than Muslims. The pragmatism, and even tolerance, of rulers like Peter I and Catherine II ran counter to the conviction that pagans were 'uncivilized' and 'wild savages' who could only benefit from Russian, European civilization, manifested through conversion to Orthodox Christianity (something that was not unique to Russian thinking, of course, being also true of the multi-ethnic empires of, for example, France and Britain). The pagans were mostly Chuvash, Mari and Mordvin peasants on the middle and lower Volga who lacked the support of a religious leader or the social organization to resist proselytizing priests. It was also easier to convert peasants in small settlements – both serfs and state peasants – than urban communities, which were more complex and more ethnically mixed. In fact, trading communities of non-Christians in the Volga towns were left alone by the tsars – an obvious example of this is the town of Astrakhan, which had probably the most diverse ethnic composition of any town in the Russian Empire.

Incentives were offered to shamanistic peasants, but even more so to their noble landowners, to convert to Christianity. Peasants were threatened in the late sixteenth century if they did not convert, but were also given cash and clothing if they did.[7] In the major Law Code of 1649, which institutionalized serfdom, it was decreed that converted peasants could not be given away as slaves, that is, as people with no rights at all.[8] In 1681, landowners were given financial rewards for Tatars who had converted to Christianity – five roubles for a converted male and two roubles for a female.[9] There was, however, no systematic conversion policy by the Russian government in this period. Much depended on the activity of some of the large monasteries that were established on the Volga. The state also decreed in 1685 that 'force' should not be used in conversions,[10] but the fact that this stipulation was repeated several times in the eighteenth century strongly hints that in practice force was used, particularly towards Chuvash and Mari peasants. In the late seventeenth century, Chuvash peasants in Simbirsk province complained that they were being forced to convert and were 'punished without mercy' if they refused.[11]

Peter I (reigned 1682–1725) took a more active role in the policy of conversion. His policy was due not so much (if at all) to personal faith, as to the fact that he saw Muslims as a potential 'fifth column' within the Russian Empire during his campaigns against the Ottoman Empire and Persia. Peter feared that there would be converts to, in his words, the 'disgusting faith of Muhammad' not only among pagans, but also among Orthodox Christians.[12] Peter was particularly concerned about the loyalty of the Muslim landowning elite, and threatened to confiscate lands if the Muslim owners did not convert to Christianity within six months. He was, however, primarily pragmatic and was prepared to tolerate different beliefs, provided his subjects paid for those beliefs. Old Believers, for example (discussed more below), had double taxation imposed upon them in 1716. Incentives for conversion were also offered, in particular by exemptions from conscription and from taxation for a number of years.

Peter's adviser Ivan Pososhkov nevertheless compared Russia's policies unfavourably with those of Catholic empires, and made his views on the 'childlike' character of pagans clear:

> And these peoples have been the subjects of the Russian Empire for two hundred years, but they did not become Christians and their souls perish because of our negligence. The Catholics are sending their missionaries to China, India, and America. [Despite] the fact that our faith is the right one – and what could be easier than converting the Mordva, the Maris, and the Chuvash – yet we cannot do this. And our non-Christians are like children, without a written language, without a law, and they do not live far away, but within the Russian Empire, along the Volga and the Kama rivers; they are not sovereign, but the subjects of Russia.[13]

A sharp difference in tone and policy was introduced in the reign of Elizabeth (reigned 1741–61), the daughter of Peter I. In part, this reflected her own dislike of all non-Christians, and in part was a consequence of her desire to curb Islam after Muslim Bashkirs revolted in the Orenburg region in the 1730s. Missionary activity was much more active during her reign, and was promoted personally by the empress, who founded a special Office of Converts to take this process forward. Kazan province was the focus of this activity, in part because the first head of the Office of Converts, Father Dmitrii Sechenov, characterized the province as a centre of 'all the non-Christians residing in south-eastern Russia'.[14] It was claimed that there were over 100,000 converts in the year 1747 alone in Kazan province, the majority among Chuvash, Mordvin and Mari pagan peasants. Incentives were still in place: in 1747, Chuvash, Tatars, Mari and Mordvins were promised exemption from conscription and from the poll tax for three years if they converted.[15] Criminals could also be excused punishment if they converted – an attractive incentive, given the savagery of punishment by the knout in eighteenth-century Russia. In 1743, for example, three Chuvash peasants accused of theft took a practical decision to convert rather than undergo savage punishment: they 'did not admit to stealing the sheep, but accepted holy baptism' – and some cash!'[16]

Incentives, however, were mixed with threats and violence during Elizabeth's reign. Pagan peasants were threatened with resettlement if they did not convert, and pagan sacred sites and cemeteries were destroyed. Mosques were demolished if they were located near Orthodox churches or settlements of Christians. In the period 1740–62, some 418 of the 536 mosques in Kazan province were pulled down. The remaining 118 mosques were only spared because they had been built prior to the conquest of Kazan in 1552. In Astrakhan province, 29 of the 40 mosques were destroyed. Tatars were only allowed to retain or build mosques in the provinces of Kazan, Voronezh, Nizhnii Novgorod, Astrakhan and Siberia in villages that were populated exclusively by Muslims or where there were at least 200 male Muslims.[17] Towards the end of Elizabeth's reign, the government's policy softened a little, possibly influenced by concern that resentment at their treatment had contributed to a further Bashkir revolt. In 1756, it was decreed that new mosques could be built, but only with the explicit permission of the tsar; however, this resulted in only two new mosques opening, both in the town of Kazan. It proved harder to convert Muslims on the upper Volga: it is estimated that by 1749, only 226 Muslims had been converted in Nizhnii Novgorod province.[18]

Tatars who became Christian in this new wave of conversions in the eighteenth century were termed 'newly baptized'. They were identified separately from Tatars who had been converted earlier, in the sixteenth and seventeenth centuries, and who were termed 'old baptized'. That distinction became important in the late nineteenth century, because it was the 'newly baptized' who were more likely to reconvert to Islam (see Chapter 12). In general, the 'newly baptized' were often from poorer and more vulnerable backgrounds – such as barge haulers on the Volga, labourers in factories and those who wanted to avoid conscription. Whole Tatar villages were also targeted for conversion in this period.[19]

There was a clear change of policy towards Muslims in the reign of Catherine II (reigned 1762–96). Tolerance was an essential part of the Enlightenment, which Catherine espoused at least in principle (and often in practice) at home and projected to the rest of Europe; but her policy was also based on pragmatism. The forced, and often violent, conversions by Elizabeth were deeply resented, and this was expressed in the 'instructions' to deputies for Catherine's Legislative Commission in 1766 from Mari and Chuvash peasants on the Volga.[20] Catherine's own personal faith, it has to be said, probably rested lightly on her: she had seemingly experienced no problems in converting from Lutheranism to Orthodoxy as a condition for becoming betrothed to her husband (the future Peter III, whom she replaced in a coup in 1762). As a result, she was able to take a pragmatic approach to non-Christians, including Muslims and Jews. One of her first acts was to abolish Elizabeth's Office of Converts. We have seen that she met Muslim clergy during her visit to Kazan in 1767 and visited the Muslim quarter of the town. In 1773, Catherine allowed mosques to be built again, although formal permission was still required. She was as keen as Peter to assimilate the Muslim elite. In 1788–89, she established a central administration for Muslims – the Muslim Spiritual Assembly – in the town of Orenburg, where there were many Muslim Bashkirs (some of whom had partici- pated in the Pugachev revolt a few years earlier). The assembly was made responsible for all Muslims in the empire, as a parallel institution to the Holy Synod for Russian Orthodox Christians. It was given the authority to determine matters of dogma, marriage and divorce, appoint mullahs and supervise Muslim schools. The head of the assembly and other senior Muslim officials were given noble status and assimi- lated into the imperial elite.

Catherine's policies were pragmatic and more tolerant than those of Elizabeth, but that did not mean that non-Christians were treated equally or that they could not be penalized in particular circumstances. Catherine made no attempt to change those regulations that subordinated non-Christians (and, for that matter, non- Orthodox Christians) to Orthodox Russians. Conversions were, of course, only allowed one way. Muslims were not allowed to proselytize among Christians, and to do so had been specified as a criminal offence in the Law Code of 1649. Marriage in the Orthodox Church was a sacrament and Muslims or pagans (and indeed Protestants and Catholics) had to convert and be baptized in order to marry Orthodox Russians.

Moreover, Catherine could be vengeful against Muslims, or at least sanction as much brutality as had Elizabeth. She fought two wars against the Ottoman Empire in her reign and projected these victories, in a none too subtle way, as the triumph of Christianity over Islam. A cameo of the Russian victory at the battle of Chesme in 1770 showed a kneeling Turk laying down his arms before a bust of Catherine II, who was crowned with a laurel wreath with an eagle flying overhead, a laurel leaf in its beak as a sign of peace . . .[21] During her reign, the Russian Empire acquired more Tatars when she annexed the Crimea in 1783. Many Tatars left the Crimean peninsula for the Ottoman Empire (as Kalmyks had left in 1771, as we saw in the previous chapter) – at least 35,000 and possibly as many as 100,000. This was in part a result of the deliberate demolition of mosques by Russian forces (although some of that destruction may have been in revenge *after* their departure). The result was a devastating and deliberate despoliation of sacred sites. As a British traveller commented in 1800, the Russians:

> . . . have laid waste to the country; cut down the trees; pulled down the houses; overthrown the sacred edifices of the natives, with all their public buildings; destroyed the public aqueducts; robbed the inhabitants; insulted the Tartars in their acts of public worship; torn up from the tombs the bodies of their ancestors,

casting their reliques upon dunghills, and feeding swine out of their coffins; anni-hilated all the monuments of antiquity; breaking up alike the sepulchres of Saints and Pagans, and scattering their ashes in the air.[22]

Tatars in Kazan must have been aware of the fate of less cooperative Muslim subjects.

Alexander I (reigned 1801–25) embraced Catherine's policy of tolerance. Missionary work among Tatars in the Volga region, however, increased in the 1820s, in part as a response to Alexander's own spiritual development. The tsar had a spiritual 'experience' during the dark days of the Napoleonic invasion of Russia in 1812, which was strengthened in the period after 1815. In 1817, he established the Ministry of Religious Affairs and Public Instruction, which was responsible not only for Christians, but also for non-Christians. The tsar also encouraged the work of the British and Foreign Bible Society in Russia, whose missionary activity not only involved dissemination of bibles, but also led in some cases to the first written form of native languages. By the end of 1818, the society had published 371,000 copies of the bible in Russia, in 79 editions and in 25 languages and dialects.

Russian missionary work continued under Nicholas I (reigned 1825–55), and was particularly active in the 1830s. The incentives for conversion increased with inflation: by 1849, converts were exempted from the poll tax for life and from other taxes for six years, and individuals were rewarded with sums of between 15 and 30 silver roubles per person.[23] By the end of the nineteenth century, it was estimated that almost 500,000 subjects had been converted to Christianity, including 336,911 Chuvash, 88,272 Mari, 18,670 Mordvins and 7,751 Udmurts.[24] Catherine II's policy of allowing mosques to be built was continued, and by 1858 there were 430 mosques in Kazan province alone.[25]

❖

The genuineness of many of the conversions was questionable. This was particularly true of the shamanistic Chuvash and Mari peasants on the Volga, many of whom had converted under duress or for petty incentives, and often had little or no under-standing of, or sympathy for, the beliefs they had apparently adopted. This could lead to conflicts between converted pagans and priests. It was more common, however, for pagans simply to continue in their old beliefs and only to adopt the outward forms of Orthodoxy. In the 1770s, the German botanist Johann Gottlieb

Georgi travelled through Russia and noted that the Chuvash only 'outwardly make profession of Christianity' and were more pagan than Christian, while the Mordvins had been converted but were 'still attached to their ancient idolatry'.[26] Prince Mikhail Shcherbatov, a councillor in Catherine's reign, lambasted the Orthodox Church for its crude approach to the conversion of pagans, which:

> . . . brought them to baptism in the same way they would have been brought to a bath; gave them a cross, which in their ignorance, they considered some kind of talisman, and an image of Christ, which they regarded as an idol; and forbade them from eating meat on fast days, a prohibition which they did not follow, while priests took bribes from them for overlooking this.[27]

The situation did not improve in the nineteenth century, despite the long period in which pagans had been Christian. In 1827, the land captain reported from Tsarevokokshaisk district that the Chuvash and Udmurt peasants were disrespectful to the priests and were hiding 'pagan habits'.[28] In 1828, Archbishop Filaret considered that the Chuvash would remain pagan and ignorant of Christian beliefs as long as they lived separately from Orthodox Russians, and that the Udmurts had continued with pagan beliefs, including ancestor worship.[29] These fears were not unjustified. For example, in 1837, a converted Chuvash called Nataliia Petrova (that is, a Russified name) in the village of Khochasheva in Iadrin district (south-west of Cheboksary) was accused of enticing several converted Chuvash to 'idolatry' through her 'sorcery'.[30]

The 'newly baptized' Tatars were acknowledged to be particularly susceptible to returning to Islam. The Kazan Clerical Consistory reckoned in 1829 that of the estimated 12,129 newly baptized Tatars in Kazan province who had converted in the eighteenth century, 10,526 had now reverted to Islam.[31] Concerns about the activities of reconverted Tatars were not groundless. For example, in 1866, the land captain in Chistopol district (south-east of Kazan, bordering the rivers Volga and Kama) reported that a Tatar in the village of Verkhnii Nikitkin had not only reconverted to Islam, but had built a prayer house with 'the appearance of a mosque without a minaret'; he recommended that it should be sealed up, so that other peasants did not attend prayers there.[32]

Government policy towards apostate Tatars was to resettle them either individually or as whole villages. This was an extension of the traditional government treatment of groups of people who were troublesome: for example, individual and whole hosts of Cossacks could be moved in the seventeenth and eighteenth centuries, as discussed in the previous chapter. In the process, local authorities employed an extraordinary amount of effort to monitor the conversion and reconversion of individuals, families and whole villages. Such cases generated a vast amount of paperwork for local police and officials in local courts and higher institutions – not untypical, it has to be said, of the way local administration operated in the empire, when a decree from St Petersburg could lead to a vast paper trail but little change in policy. Cases, sometimes involving only one or two individuals, could not always be settled at a lower level, and would reach the Governor's Office and the Kazan Clerical Consistory. For the apostates themselves, of course, resettlement was almost certainly a traumatic experience. Individuals were uprooted and expelled to other villages, which were often deliberately located at some distance and not necessarily even in the same province. Parents were separated from children; husbands from wives. Resettlement could easily be to an underdeveloped region where conditions for agriculture were harsh. It was a cruel fate, which hardly accords with a stated enlightened policy of 'tolerance'.

Cases from Kazan and Samara provinces illustrate the human cost of the government's policy on apostasy, but also the complexity of unravelling the situation in villages where Muslims, converted Tatars and Christian Russians lived together. In 1854, an incident was investigated which dated back to 1835 and involved numerous reports from witnesses and local officials. The case started with the allegation that an elderly 'newly baptized' Tatar peasant, with the perhaps unfortunate name of Aleksei Romanov, had reconverted; but it soon broadened to investigate the alleged activities of 132 converted Tatars in villages in Orenburg province who lived among Christian Russians. After investigation, it transpired that 22 families of 'newly baptized' Tatars had reconverted to Islam, and the decision was taken to resettle these families at least 100 *versty* (just over 100km) away in Kazan or Orenburg province. Children were resettled with their parents if they were under 15 years of age, and older children were allowed to stay behind only if they could show that they were practising Christians.[33] The case dragged on into the 1850s as the peasants resisted resettlement; at least one Tatar objected on the grounds that he had never converted to Christianity in the first place. Such policies continued. Some 1,200 converted Tatars who had reverted to Islam were resettled from Chistopol district in 1843.[34] In 1860, Muslim Tatars from the same district were resettled because of their allegedly 'fanatical beliefs'.[35]

The more sensitive Russian missionaries were aware that the heavy-handed policies of the Russian government only served to alienate converted Tatars. In 1865, one missionary summed up the failure of government policies as follows:

From 1827 to 1864, parish priests and consistories, for all their admonitions, have done nothing to make baptized Tatar apostates listen to reason, because they acted without love and compassion for the apostates, without prior preparation . . . Police intrusion only annoyed baptized Tatars, turning them away from Orthodoxy; measures such as resettlement brought Tatars to ruin and demoralized them rather than fortifying them in the Christian faith and piety. It is not surprising that conversions of Tatars to Christianity were scarce and apostasies numerous.[36]

Cases of resettlement did not always simply involve apostate Tatars. Indeed, Tatars who had converted to Christianity could suffer the same fate. In the 1840s, the 'newly baptized' Tatar Dmitrii Garilov and his family were resettled from a Muslim village to a Christian one,[37] presumably so that they would be in less danger of reconverting. There was no evidence that Garilov or any fellow Muslim villagers had acted inappropriately in any way.

The 'old baptized' Tatars were generally assumed to be more stable in their Christian beliefs than 'newly baptized' Tatars, but this belief was shaken in 1866, when there was a serious outbreak of apostasy on the Volga, which drew in 'old-baptized' Tatars. Rumours started in Chistopol district that the tsar had passed a law allowing Tatars who had been baptized now to practise Islam freely. This was a time of uncertainty in the countryside, and there had been violent reactions to the edict emancipating the serfs a few years earlier. The movement spread rapidly from village to village on the Volga and led to a spate of petitions requesting reversion to Islam (accompanied symbolically by the shaving off of beards for men and the adoption of Islamic dress for women). The initial reaction of the government was to reject the petitions and to arrest nine Tatars who were thought to be the ringleaders. When this did not stop the revolt, the government became very concerned and arrested and investigated a large number of converted Tatars. In the end, 21 of the instigators were sentenced to hard labour, 20 sent to prison and 260 subjected to intensive religious instruction, along with their families.[38] Most of the other accused Tatars were released from prison, which only served to encourage further outbreaks of apostasy over the next few years.

Fundamental for the assimilation of converted Tatars were questions of language and education. One criticism of the policy of the government and the Church was their failure to produce materials in native languages. The first translation of the catechism into Tatar only appeared in 1803. Schools for converted Muslims had been set up in the eighteenth century, but these schools were not seen as a way of inculcating Russian 'values'. Indeed, one of the main aims of the schools for converted Tatars was to provide translators for the Russian army and state – that is, entirely practical and not ideological.[39] It was only in the late nineteenth century, under the leadership of the academic and missionary Nikolai Ilminsky, that the emphasis changed to attempt to promote the values not only of Orthodoxy, but also of 'Russianness' through the education of Tatar youth (discussed further in Chapter 12).

❖

Foreign travellers depicted Russia as a deeply religious Christian country. Indeed, the extent of Christian piety struck some as excessive and bordering on the uncivilized. In the late seventeenth century, Samuel Collins, an English physician to the tsar, commented, 'upon some great Vigils they stay all night in their Churches, at certain times prostrating and crossing themselves, and knocking their heads against the Ground'.[40] Some 130 years later, Catherine Wilmot, a member of the Anglo-Irish minor gentry and a Protestant, remarked:

> At the rising & setting of the Sun & on other occasions they begin to cross themselves, but so *obstreperously* that the operation does not finish under qrtr of an hour. They bow their heads down almost to the ground, & then not only *recover their ballance* but throw themselves proportionably back again, crossing themselves at arms length.[41]

The Orthodox Church, however, faced challenges not only from outside – from pagans and Muslims – but also from within. These problems occurred throughout the Russian Empire, but were more acute on the middle and lower Volga, simply because the inaccessibility of much of the region meant that abuses went unchecked and dissident sects could flourish.

The ordinary parish clergy had a poor reputation in Russia. The clergy was, in effect, a closed social estate, in that parish priests had to marry and their sons were expected to become clergy in due course, if there were posts for them. Pay for the parish priest was low, and this – combined with the fact that the families of clergy were often very large – meant that parish clergy often lived no better than the peasant communities they served. Seminaries existed, but the level of education provided in them was often very low, and the schools in any event had a reputation for brutality rather than for learning. The result was that the parish clergy were often poorly educated, impoverished and sometimes lacking a genuine vocation for their work, with the result that inappropriate behaviour and drunkenness were all too common. Petr Levshin, the bishop of Tver, had parish priests defrocked in the 1770s for drunkenness and debauchery and for 'behaviour unbecoming their station in life'.[42] Bishops were drawn from the monastic orders and not from the parish clergy, and were normally of a higher intellectual and moral calibre, though there were exceptions. John Parkinson, who travelled through Russia at the end of the eighteenth century, remarked caustically from Astrakhan that:

> The Archbishop's disorder was indolence and Drunkenness. When in the country he devotes himself to those vices. He [is] obliged in the Town to be more on the reserve and therefore prefers the country.[43]

The poor calibre of many of the parish clergy often made them ill-equipped to win the respect of converted pagans or Muslims on the Volga. In the 1880s, there was a report concerning the alleged 'fermented minds' of baptized Chuvash in Cheboksary district (Simbirsk province). They had disrupted a religious holiday and refused to go to church, although in this case their protest might have been not so much a rejection of Christian beliefs as an expression of dissatisfaction with the local clergy, and in particular with the way they conducted marriages.[44] Parish clergy often depended for their survival on fees for services such as baptism, marriage and funerals, and that often caused resentment, particularly in poor villages.

The Orthodox Church, however, faced a more fundamental internal threat than the calibre of its parish clergy. In the middle of the seventeenth century, the Orthodox Church underwent a schism that has never healed. Patriarch Nikon attempted to introduce a series of reforms to church texts and liturgical practice. Those who did not accept the reforms became known as Old Believers, led at the time by Archpriest Avvakum. The changes seem on the surface to be minor matters of ritual and outward forms (unlike the split within the western Church in the Reformation): for example, the simplification of certain rites, the number of fingers used when making the sign of the cross, the number of alleluias and the spelling of certain words in holy texts. The change to the form of crossing oneself became a matter of passionate opposition among many believers across social classes because it was the most common ritual in an Orthodox service. To many believers, including members of the church hierarchy, this was, however, more than an attack on much-loved rituals, and was perceived as a fundamental and heretical attack on the unique purity of the Orthodox Church. There was a deeply held belief that Russian Orthodoxy was the only pure form of Christianity, with Moscow as the 'Third Rome', after the fall of Constantinople in 1453. Nikon's reforms were seen by many believers as an attempt to impose Greek forms on the pure Russian Church, to the extent that they were seen as the sign of the devil (and it was not lost on Old Believers that the reforms commenced in 1666, a date including the three sixes, which the Book of Revelation had identified as the sign of the 'great beast' – that is, the Antichrist).

In the late seventeenth century, many Old Believers fled to the more remote parts of the Russian Empire. Some staged dramatic resistance to regular troops; this almost always ended in a tragic mass self-immolation, in which thousands of Old Believers perished. Self-immolation continued into the eighteenth century and was especially prevalent in the reign of Elizabeth, when the persecution of Old Believers increased. Old Believers also fled to the southern borderlands. Some became Cossacks and others simply formed small, hidden communities in the less accessible territory, including the upper, middle and lower Volga. Archpriest Avvakum, the leader of the resistance, claimed that there were 'tens of thousands' of adherents in Volga villages and towns who would 'take up the sword against the Antichrist'.[45] The number of Old Believers across the country is impossible to gauge accurately, because many would not admit to it in the census. It has been estimated that in the mid-seventeenth century their numbers could have been anything from 800,000 to over a million; by 1911, it has been suggested, there were over 2 million Old Believers across the empire.[46]

Old Belief posed a genuine problem for the Church and the state, because it meant that a significant number of the Orthodox population did not accept the authority of the official Church. By the late eighteenth century, the Old Believers had divided into two types of communities: 'priestist' and 'priestless'. The former accepted priests who had been ordained in the Orthodox Church and the latter did not; but both communities undermined the authority of the Orthodox parish clergy, who could be completely excluded or only accepted on sufferance. Parish priests lived among peasants almost as peasants themselves, and depended on parishioners for fees for services. They could be reluctant to report on Old Believers, since they depended on them for their livelihood. Furthermore, some parish priests probably sympathized with their beliefs. For the state, Old Belief could be a rallying cry for all those who resented change. We have seen that the Razin and Pugachev revolts appealed to Old Belief in their manifestos; this was to garner support not only from Cossacks, but also from local peasantry on the Volga. The situation therefore was that at the same time as missionaries were proselytizing, and the local authorities were checking on all the activities of Tatar apostates, the Orthodox Church was not able to control its own clergy, and the clergy had limited authority over parishioners.

Local authorities recorded the existence and location of Old Believers in the same way as they recorded non-Christians, converted Tatars and apostates. In particular, they recorded the number and gender of Old Believers, and their age, in order to assess the taxes to be collected from each household. These records demonstrate that Old Believers could be found all the way up the Volga, and not simply in the regions of the middle and lower Volga where there were non-Christians. In 1766, for example, 53 Old Believers were recorded in Cheboksary district (Simbirsk province), the majority of whom (29) were unskilled town dwellers, 7 were guild members (that is, more skilled) and 3 were serfs.[47] In 1778, Old Believers were listed on lands owned by nobles in Iaroslavl province, on the upper Volga. For example, on the Sheremetev estate there were 213 male and 321 female Old Believers, and on the estate of a nobleman called Ivan Lobanov there were 164 male and 280 female Old Believers. In total, 1,912 male and 3,571 female Old Believers were recorded on six noble estates in Iaroslavl province.[48] In 1856, the traveller William Spottiswoode noted Old Believer merchants and traders in Saratov province.[49] Kostroma province had been home to Old Believer settlements from the seventeenth century, and by the beginning of the twentieth century they numbered some 31,000 of both sexes and included wealthy merchants.[50]

Catherine II was more tolerant of Old Believers than her predecessors. She made a point of visiting the town of Gorodets, on the Volga north of Nizhnii Novgorod, and its Old Believer community during her voyage down the Volga in 1767. Gorodets had become almost entirely an Old Believer town and played an important role in ship-building and the production of wooden handicrafts in the eighteenth century. Alexander I continued this policy and met Vasilii Obraztsov, a prominent Old Believer and the town head of Rzhev, during a visit in 1824.[51] In 1830, there were over 2,000 Old Believers (some 40 per cent of all the Old Believers in Tver province) in the town of Rzhev, a number that had risen to over 6,000 by the mid-century.[52] Policy towards Old Believers changed significantly in the reign of Nicholas I. The tsar regarded Old Believers as subversive – as did his minister of interior affairs, Dmitrii Bibikov – and there was a deliberate attempt to eradicate them by force and to undermine their successful economic activity in Moscow and on the upper Volga.[53]

The recording of the existence of Old Believers shows that the authorities regarded them as a potential threat to law and order. Their greatest concern was that more Orthodox peasants would become Old Believers, often with the encouragement of existing Old Believers. In 1724, a priest reported from Nizhnii Novgorod province that a certain Ivan Ievlev used 'improper words' on Sunday, had refused to let children be christened in the sun or weddings be conducted, and had even threatened to kill the priest.[54] The suspicion was that this was an attempt to lure peasants into accepting the Old Belief. In Nicholas I's reign, Old Believers came to be regarded as criminals. The tsar and Bibikov wanted to eradicate Old Belief as part of a broader determination to eliminate any dissent and challenge to the regime; but their conviction that Old Believers were subversive was strengthened by the activities of a criminal gang on the Volga that had been hidden by peasants in the Old Believer village of Sopelki, in Iaroslavl province, and also by the exposure of a particular Old Believer sect on the Volga, called the *beguny* (runners), which denounced the tsar as the Antichrist.[55] This, in turn, led to an even more detailed investigation of the location of Old Believers on the Volga and elsewhere.

The death of Nicholas I in 1855 reduced the immediate threat to Old Believers (in particular, his prohibition in 1853 on Old Believers registering with Russian trade guilds was not implemented by the new tsar), but suspicion of Old Believers continued in the second half of the nineteenth century.

In 1856, a secret report was sent to the Kazan Governor's Office concerning a state peasant, Zinovii Rizhev, who had left Old Belief in 1849 for the Orthodox Church but had allowed his daughter to marry an Old Believer and was suspected of reconverting, with his whole family, back to Old Belief.[56] Three years later, the government investigated an Old Believer sect that had formed in the village of Korobova in Kostroma province, on the upper Volga; it had attracted deserters and vagrants, and was allegedly causing disorders due to 'religious fanaticism'.[57] In 1862, the Samara town police and provincial office received reports concerning two peasants – Petr Kovolev and Aleksei Avunichev – who had allegedly become Old Believers.[58] The following year, the Ministry of the Interior asked for details of the location of Old Believers in Saratov province, noting that the local police had the right to disperse them to other locations 50 *versty* away (just over 50km).[59] In 1868, the priest reported from the village of Murasa in Spassk district (Kazan province) that some 200 peasants had rejected Orthodox rituals at the Easter service, and had stolen the icon, because they had become Old Believers. The instigators were said to be two retired soldiers who had returned to the village after military service.[60]

The main concern of the government was that Orthodox Christians should not become Old Believers. There were also, however, cases of Old Believers who reverted to, or simply became, Orthodox. In 1852, the Iaroslavl Governor-General's Chancellery was given the names of 215 Old Believers from some 30 villages who had, in the words of the report, 'returned' to Orthodoxy. It included whole families, including children aged 11 and 12, and men and women of all ages – from their twenties to their sixties. No reason was given.[61]

There were other dissident religious sects in Russia that were also attracted to more remote parts of the empire. Local authorities also took an interest in the members of such groups and their location on the Volga. One such sect was known as the 'milk-drinkers' (*Molokany*) – so called because they drank milk during Lent, i.e. they rejected the fast, which was itself a very important part of Orthodox popular practice. The sect originated in Tambov province in central European Russia, but 'milk-drinkers' could also be found on the Volga. In 1856, the Samara Provincial Chancellery received a secret report about 'milk-drinkers' appearing in the village of Ivanovka. Upon investigation, it transpired that eight families were involved, although it is not clear what action, if any, was taken against them.[62] By the late nineteenth century, another challenge to the Orthodox Church came from Baptists. The authorities were concerned about the rise of this sect, and carefully monitored its members. The police were required to report on the presence of Baptists in Kazan province in 1881.[63] In 1893, it was reported to the Kazan Clerical Consistory that 13 Baptists were active in one village in Tetiushii district.[64]

For all the concern and careful monitoring by the local authorities, sometimes cases of conversion could occur which defied all logic and simply showed that personal relations could lead to strange outcomes. One bizarre case occurred in Kazan in 1867. Two men petitioned the Kazan Clerical Consistory for permission to convert from Orthodoxy to Judaism – Stepan Matveevsky, a townsman, and a junior officer, Dmitrii Iusupov, who wanted to convert with his wife and his three children. It transpired that Matveevsky lived in the same building as a Jew and had been influenced by him. The two men were investigated by the Kazan police, who determined that Iusupov had not in fact wanted to convert, but that Matveevsky did. The upshot was that Matveevsky's petition was refused and he was forcibly resettled. The Jewish neighbour was also expelled from Kazan.[65] Such was the fate of those who broke all the rules!

7
THE VOLGA
Disease and science

It seems fully natural, that every year in July the information comes some-where from the Volga that cholera cases were reported, then it is noted, that in other towns so and so many people got caught by the disease, and so and so many died. Every day the news is spread that cholera cases were reported in the south, and in the central provinces and at the periphery.[1]

The river Volga was a carrier of disease – mainly plague and cholera – into Russia, primarily from the south, through Astrakhan. It then spread north, following the river. Disease was carried on the boats on the Volga – by rats and fleas in the cargoes or in the clothing and on the bodies of ship workers. Volga boatmen and other ship workers and traders played a key part in spreading disease.[2] Serious epidemics on the Volga posed a significant challenge both for local authorities in Volga towns and for the Russian government – a challenge that was met only slowly. It was recognized that disease spread on the river, and at the time of major epidemics shipping was banned on the river and cordons set up to restrict the movement of people from infected Volga towns. In time, the experience of epidemics also led to a better medical understanding of the causes of infectious diseases and to improvements in the way they were treated and prevented. The Volga was also a region of scientific interest in its own right. The eigh-teenth century in particular was a period in which peoples, flora, fauna and soil were described and classified as a scientific study in the Russian Empire, as elsewhere in Europe. Expeditions to the Volga were sponsored by the tsars or by leading state academic institutions. Knowledge is power, and the process of acquiring knowledge of new lands, and of non-Russian peoples, was a means of asserting control by the tsars and the Russian government over this multi-ethnic empire.

❖

Plague is a bacterial infection, caused in humans by the bite of an infected flea. The fleas are the main transmitters of the disease, which they pick up from rats infected with the bacteria. Once the infected rats die, the fleas look for a new host, which may be other rats or humans (particularly if the rats have died), thus spreading the infection. Bubonic plague is spread by fleas from infected rat to rat or from infected rat to humans, but it is uncommon for the disease to spread between people via fleas, because humans rarely produce enough of the harmful bacteria to infect fleas in turn. The bacteria attack the lymph nodes, particularly in the groin (the term 'bubonic' is derived from the Greek term for 'groin' and the term 'buboes' is used to refer to the swollen lymph nodes). The symptoms include severe headaches, fever, vomiting, muscle cramps, rapid pulse, bloodshot eyes and delirium. Dr Gustav Orraeus, a military doctor from Russian Finland, described the appalling impact of the disease on Russian troops in the northern Balkans in 1770, as follows:

> The disease always begins with a great pain in the head. At the same time almost everybody infected by it feels nauseated, yet with few does vomiting happen by itself. Soon the sick fall into great despondency, feel anxious, and have a very great heat, to which delirium is joined with the majority. Buboes and carbuncles sometimes come out from the very beginning of the disease, and in this case the disease does not continue long; yet sometimes they show themselves within 12 to 24 hours, counting from the onset of the disease. On the third, fourth or at most the fifth day the buboes suppurate or dissolve, while the carbuncles begin to separate themselves from the healthy part, in which cases there is the best hope for recovery or, on the contrary, the sick die. A large part of the sick completely come to their senses several hours before the end, say they feel better, even go so far as to ask to eat. At death, or soon afterwards the places around and on the very buboes and carbuncles become bluish and darken, while under the skin dark spots quickly spread in great profusion.[3]

The mortality rates for bubonic plague vary, but can reach 60–90 per cent of those afflicted during epidemics. If untreated, those affected can die within five to seven days of the onset of the disease. In the fourteenth century, it is estimated, some 50 million people across Europe were killed by the plague, known as the Black Death. The only treatment at the time – not only in Russia but in Europe generally – was the isolation of infected victims and the speedy burial of the dead, although bleeding was also used. Plague bacteria are very sensitive to temperature and flourish in warm, moist conditions; and so in reality the onset of cold weather was the main way in which the disease was checked, without human intervention.

The Black Death, which devastated the Golden Horde territories, Kiev and the Rus principalities, as well as central and western Europe in the mid-fourteenth century, probably spread from Persia. Popular opinion sometimes blamed the Tatars, however, because the disease was seen to have come from the east. A second wave of plague hit the Golden Horde territories in 1365–66 and probably came up the river Volga from the south. Further outbreaks occurred in the late fourteenth century; the severity of an outbreak in 1396 was said to have prevented a Tatar raid on Moscow. Outbreaks of plague continued throughout the fifteenth and sixteenth centuries. In 1654, plague spread up the Volga from Astrakhan, reached Kazan in July 1654 (claiming some 48,000 lives) and then moved on to Nizhnii Novgorod, and from there to Moscow and central Russia and to Vologda in the north. Plague died out in central Russia in early 1655, but reappeared in the southern Volga in 1656–57. In 1692–93, plague reoccurred in Astrakhan, this time claiming up to 10,000 lives.

The following decade was almost free of the disease, but plague broke out again in 1727–28, and again originated in Astrakhan. It had been a particularly warm and damp winter, but this was also the time when Terek Cossack troops were marching back to Russia from the Persian campaign, and the soldiers could easily have carried the infected fleas. A total of 411 persons died from 'buboes, malignant fever, and petechiae' in the town in April 1728 alone, a month that was described as moist and misty; the situation was exacerbated by a hot June, during which a further 1,300 died. The order was made by the local governor to evacuate Astrakhan, and at the same time all shipping and other transport was banned for six weeks – a policy that apparently caused a food crisis in the town. Pickets were set up in and around the town of Tsaritsyn, on the lower Volga, to protect inhabitants north of Astrakhan. In all, it is estimated that some 3,000 people died in the town during this epidemic.[4] The Russian government recognized the danger of the river: it banned shipping movements on the waterway and set up cordons and quarantine areas in Volga towns and villages in and on the borders with neighbouring provinces.

Although neither contemporaries nor the Russian government realized that plague was carried by fleas infected from rats, they were all too well aware that plague often originated in the south and spread through Russia up the waterways, and up the river Volga in particular. Astrakhan was often the place where plague first manifested itself and from where it spread. The town, as we will see in Chapter 9, was a meeting place for traders, ships and ship workers from the south and the east. Crowded conditions, poor sanitation, stagnant water in the town, a canal system that ran throughout the town and into which raw sewage was dumped, a warm and often moist climate, described as 'sickly' by a British visitor in the middle of the eighteenth century[5] – all this made Astrakhan a perfect breeding ground for plague. In the winter, the raw sewage lay on the ice and then melted into the canals and wharfs in the spring.

Further outbreaks of plague occurred in the eighteenth century, but were now caused mostly by the movement of troops. The major outbreak in the late eighteenth century was in Moscow, in 1770–71; it culminated in a horrifying riot in September 1771, when Archbishop Amvrosii was savagely beaten to death by an enraged crowd, who believed he was about to remove a supposedly miracle-working icon; no one knows how many casualties there were in the riot, but it could have been up to 1,000. The Moscow plague was caused by the movement of troops, and their infected baggage, from the northern Balkans, where the disease had broken out in 1770. Up to 70,000 people could have died in Moscow as the epidemic raged. There were particular conditions which made Moscow so vulnerable to the epidemic, including filthy waterways which ran through the centre of the city, the overcrowding, and the numbers of temporary residents and soldiers.

The epidemic spread eastwards to the towns and villages of the northern Volga, helped by people fleeing the city without any effective government control and carrying the infected fleas. It was reported from Moscow in August that:

Many inhabitants living here are departing, and sick persons are leaving Moscow . . . even at night many ride and walk away, wherefore one can see from reports this disease has also appeared in several villages and hamlets.[6]

A Moscow merchant, Ivan Tolchenov, recorded in his memoirs that his grandfather had died in Tver province during the height of the Moscow plague. He recalled that 'people began to scatter to nearby towns and villages and some of them were already infected, while others carried off clothing and other things impregnated with the deadly poison, and so the infection multiplied across many towns and villages'.[7] It was estimated that some 100 deaths occurred in the town of Iaroslavl, but that almost 2,000 also died in the villages in the province, with deaths also occurring in the Volga towns of Kostroma and Uglich. It is also likely that ship workers from Nizhnii Novgorod helped spread the disease from Moscow into central Russia. The plague reached Nizhnii Novgorod in September 1771, when a coachman died of the disease. His body was buried by three wretched female convicts, who were then quarantined, and the disease did not spread further. In the end, cold weather, rather than effective cordons and quarantine or any treatment, almost certainly led to the subsidence of the disease.[8]

Outbreaks of plague continued to occur in the nineteenth century. In 1808, the government established a cordon around Saratov and Tambov provinces to protect Moscow from a further outbreak.[9] The next major epidemic, however, was of cholera, which devastated the population of the Volga and spread to central Russia. Cholera is a bacterial infection of the small intestine, spread mainly by unsafe water and food containing the bacteria. Symptoms start with dizziness and anxiety, followed by violent diarrhoea and vomiting and muscle seizure, followed by lethargy, low blood pressure and poor pulse. The disease is sometimes known as the 'blue death', because dehydration causes the skin to turn bluish, especially around the lips and finger nails. At this point, the affected person is almost certain to die. It is estimated that without treatment, cholera kills about half of those infected.

Cholera, probably originating from India and the Middle East, broke out in Astrakhan in 1823, and by 15 September some 29 people had died. The local authorities set up cordons around Saratov and Kazan provinces, and at first the disease seemed to have been isolated.[10] The following year, however, the epidemic started up again in the town of Orenburg, on the Ural river, and this time barriers were less effective at stopping trade east to west – in part because local authorities were reluctant to damage the lucrative market at Makarevo, near Nizhnii Novgorod (see Chapter 10). By the following year, cholera was rife again in Astrakhan – by May 1830, there were 3,633 cases in the town and 2,935 in the countryside (at the time, the town's population was around 37,000 and the population of the province was almost 330,000). Panic and violence broke out in the town, as people tried to flee and victims were forced into the quarantine hospital, from which few emerged alive.

Cholera then spread up the river Volga, affecting the towns of Saratov, Kazan and Nizhnii Novgorod, from where it spread west to Moscow and Novgorod and then north to Tver. Kazan was described by the rector of the university as a 'besieged town' as the disease took hold.[11] Cordons were set up between Tver and Rybinsk in 1830, and this probably helped to contain the disease; but in 1831, it returned to Tver as it spread from Iaroslavl further up the Volga. Official records state that a thousand people died in Tver province in 1831, although that was probably an underestimate. There were riots in Tver in June 1831, after the death of some prominent figures, including Amvrosii, the bishop of Tver.[12]

The cholera devastated German colonies in Saratov province,[13] and recruit levies were temporarily suspended in Astrakhan, Saratov and Orenburg provinces.[14] The local authorities set up a special orphanage in Tsaritsyn to cope with the unfortunate infants and children whose parents had died.[15] Astrakhan was particularly badly affected; it was described a few years later by a British traveller as still 'melancholy' as a result of the cholera.[16] The disease nevertheless weakened as it followed the river north. It killed some 90 per cent of affected persons in Astrakhan (perhaps as many as 10,000 people died in total in the town), but the figure declined to 70 per cent in the town of Saratov, 60 per cent in Kazan and 50 per cent in Moscow.[17]

Government policy was to try to contain cholera by quarantining infected people and goods. Individuals crossing quarantine cordons were isolated for 14 days. The process was, however, always incomplete. Cordons were mainly set up on major post roads, but people and goods could still move on minor roads. Nor was shipping on the river Volga restricted, despite the fact that the boatmen at Nizhnii Novgorod had contracted the disease. Within towns, individual houses or whole quarters of towns were quarantined. The government instructed that the bodies of those who died should be buried deeper than normal, and be placed in special parts of cemeteries or in special cemeteries. There were concerns, however, about disrupting the important Volga trade routes from the south and from Central Asia, which weakened potential quarantine measures.

Epidemics – mainly of cholera – continued to ravage Volga towns during the middle of the nineteenth century. Kazan suffered epidemics in 1847–48, 1853–54, 1857, 1859 and 1860.[18] Several thousand people died of cholera in Saratov in 1848. Dr Nikolai Tolmachev recorded his travels through Kazan province in 1848, and noted the cholera deaths in the villages he passed through. He observed, for example, in April 1848 that in the village of Kirmeni (an Orthodox village of some 400 peasants) 7 persons had died from cholera and 2 had recovered; in the nearby village of Zmeevo, 13 peasants had died.[19]

Nearly 10,000 people died in Tver province in the cholera outbreak of 1848, 922 of them in the town itself, after the disease came up the Volga. There was little improvement in the treatment of the sick in the mid-nineteenth century, despite the establishment of more hospitals. The records of the town hospital in Tver at the time show that many died within hours, if not minutes, of arriving at the hospital. Father Timofei Gorodetsky recounted that when ministering to those dying of cholera in a Tver hospital, he 'often had to walk nearly to his ankles in the excrement of those suffering from cholera'. Ivan Lazhechnikov wrote to the poet Fedor

Glinka, warning him to avoid the town because 'cholera is raging around us here in Tver'. Two local priests had died only a few hours after dining with him, and 'the plaintive tolling of the bells for the dead priests woke you up at night in the midst of already uneasy sleep and did not even let your heart rest during the day'. Religious processions in the town drew vast crowds, and in themselves increased the risk of the disease spreading.[20]

The 1870s saw a number of serious outbreaks of cholera on the lower and middle Volga. Between June and August 1872, there were 4,211 cases of cholera in Saratov and over 2,000 deaths.[21] A plague epidemic broke out in the village of Vetlanka, some 200km from Astrakhan, in 1879.[22] According to a report presented to the Privy Council in London, the village was quite prosperous, but had some of the conditions which helped the spread of plague: pigs revelled 'in the refuse of the village' and were found in herds 'in the swamps in the bed of the Volga'; troops had passed through the region from campaigns against the Ottoman Empire, and this could have spread the disease. Most of those affected died within two to three days of contracting the disease, including the first three doctors who were sent out to the village.[23]

In 1891–92, a further cholera epidemic ravaged the Volga towns and spread from there into central Russia and Ukraine. The impact was so great that it contributed to a severe famine in the region. Yet another outbreak occurred in June 1891 in Astrakhan, despite improvements to the water supply system that had been made in the 1880s. Some 200 people died each day in Astrakhan, and in all 3,500 perished.[24] Panic gripped the town and a mob attacked the doctors who were attempting to isolate patients. The disease then spread very rapidly up the Volga, through Tsaritsyn, Saratov, Kazan and Nizhnii Novgorod, to reach Tver by August. The river was still the crucial route for the spread of cholera, and the development of steamships could account for the very rapid progress of the disease during this epidemic (although by this date, disease could also spread across the railway and road networks).

An excellent study of the town of Saratov in this period allows us to see how and why cholera took hold there and demonstrates that the local authorities were unable to deal with the epidemic. The town grew very quickly in the second half of the nineteenth century – the population doubled between 1850 and 1900. This rapid expansion put pressure on housing and facilities and also attracted a large number of itinerant and casual workers, including boat workers on the Volga. The 'normal' death rate per thousand   that is, in times when there was no major epidemic – was worse in Saratov than in many other towns of the Russian Empire (higher, for example, than Tver on the upper Volga).[25] As the port flourished, more boats arrived and brought infection to the town, particularly in the high season of June and July. Saratov was constructed around a number of narrow ravines, through which untreated sewage reached the river Volga. Some 4,000 people lived in crowded shanty towns along the ravines, where sanitation was very poor and clean water almost completely inaccessible, and from which steep, muddy and foul streets led to the waterfront. These were the areas where the highest death rates occurred during the cholera epidemic of 1892.

The epidemic started in Saratov in June 1892. The disease probably originated with a barge worker on the Volga, although cholera could have reached the town earlier and not been recorded. Panic broke out all along the river, and refugees tried to escape from the affected towns, so spreading cholera further up the river. Refugees arrived in Saratov from Astrakhan, making it even more difficult for the people of Saratov to escape by steamer. The *Saratov Journal* noted on 23 June 1892 that:

Lately, we are told, huge numbers of passengers are noticeable on the ships coming from Tsaritsyn. Rumour has it that these are *Astrakhantsy* moving up-river and to other provinces. All classes on the ships are overcrowded by them, so that in Saratov almost no places are left for those who want to take a place at the local docks.[26]

The town authorities closed the small shops on the river bank the following day, in an attempt to stop the sale of contaminated food. But this not only affected the livelihood of shopkeepers, but also made the conditions on the ships even worse, because as a result the unfortunate passengers could not feed themselves. The administration of Saratov proved unable to deal with the disease. The town police tried to separate the sick from the healthy in the town, but the forcible removal of victims to overcrowded hospitals, combined with rumours of further removals, led to disturbances. Some 2,000 people rioted, and three people were killed – not quite the scale or savagery of the Moscow riot of 1771, but the events exposed the inadequacy of the way in which the local administration handled epidemics, as well as the medical failures. Conditions in Saratov hospitals were appalling, as one worker recorded:

The hospital was overcrowded, in fact too overcrowded, with too few nurses, nobody could look after the sick. I cannot remember this dark barracks, where masses of sick were piled up, without horror. The place was terribly crowded.

Here a patient vomits, there another has diarrhoea, all this falls directly on the floor, the same floor, on which their nearby neighbours roll in agonizing cramps. And the groaning, why all the groaning? What terrible agony must the sick, from whose mouths these terrible groans are heard, experience![27]

In all, it is estimated that 10,980 people died in Astrakhan; 21,091 in Saratov; 18,115 in Samara; and 3,703 in Kazan.[28] As before, the number of deaths decreased with the distance from the source, Astrakhan, although that was of little comfort to those who died from this painful and distressing disease or to their families.

Epidemics continued to plague Volga towns in the late nineteenth and early twentieth centuries. Outbreaks of cholera occurred in Astrakhan in 1902, in Samara in 1904, and in Saratov in 1907. In Tsaritsyn, there were cholera epidemics in 1879, 1892, 1904, 1907 and 1908.[29] By this time, the authorities were at least conscious that the railway network could spread disease, and disinfected stations and carriages whenever cholera struck. In 1908, physicians noted a particularly dangerous point where the railway line crossed the Volga, in Volsk district (Saratov province), where 'cholera, each time it is present, weaves a nest for itself within the outer boundaries of the government [province]'.[30]

Compared with the German states, France or Britain, Russia was slow to develop a medical profession. Training of a medical corps only really started in the reign of Peter I. His new armed forces needed doctors, most of whom were German in origin (apart from anything else, doctors needed to know Latin, which was rarely taught in Russia). By 1803, there were still only 2,000 doctors in the whole empire[31] – some progress at least (there had been fewer than 200 in 1700), but well below the figures for central and western Europe. Progress continued to be slow in the nineteenth century. It is estimated that there were only 37 doctors in the town of Kazan in 1861,[32] and Samara had only two hospitals in the mid-nineteenth century.

The medical profession only slowly came to understand the causes of cholera, but this was the case world-wide and not simply in Russia. By the late eighteenth century, there was a realization that dirty living conditions and poor diet were contributory factors, but the normal treatments – bleeding, bathing, opium – did little to tackle the causes of the diseases. The main treatment, in Russia and in other countries, was to try to isolate the victims and wait until the disease played itself out, through lack of contact with the victims or with the onset of colder weather which killed the bacteria. By the late nineteenth century, however, there was a growing awareness among Russian and other doctors that better sanitation, clean water and more medical provision all helped to prevent cholera and other diseases occurring and spreading. It had become clear that the poorest parts of town were more susceptible to epidemics,[33] and that this was linked to insanitary conditions and overcrowding. It was also recognized that it was particularly dangerous when large groups of people came together, such as at the fair at Nizhnii Novgorod, where the 'throng' led to a 'prevalence of epidemics', as one traveller commented.[34] Medical provision improved slowly in Volga towns; by 1875, for example, Samara had a new town hospital that could house over 2,000 patients.[35] Improvements to water supply and sewerage were made in many Volga towns in the late nineteenth century. At the same time, Russian physicians conducted serious analysis into the cause and spread of cholera,[36] something which contributed to a better understanding of the disease and to later Russian and Soviet attempts to control it.

The river Volga and shipping were identified as a major cause of the spread of disease. By the early twentieth century, more regulations were in place to control any outbreak. Ships had to be disinfected, sick passengers had to be isolated and ships had to have a doctor on board. Special cholera isolation barracks were built on the shores of the Volga, and were provided with proper equipment and staff. This was, however, also a time of further population pressure, as the towns grew and social conditions in many cases worsened.

❖

Astrakhan may bear the unwelcome label as the source of many of the epidemics which ravaged the Volga and central Russia from the seventeenth to the twentieth centuries. It was also, however, a region that fascinated the Russian government and botanists, because of its unique fauna and flora. Some of this fascination was pure scientific enquiry; but the government was also interested in the potential economic and medicinal value of the products of the new lands that had been acquired by Ivan IV. Such interests were not unique to Russia, of course, but the government had a particular need to understand and exploit its recently acquired lands. In the late seventeenth century, it sent an expedition to Astrakhan to examine mulberry trees, with a view to taking them back to Moscow, in order to develop a silk industry. The members of the expedition were instructed to

 . . . in the gardens of the residents of Astrakhan and of every government worker, count every mulberry tree, both young and old, and write down whether they have fruit, and from these fruits collect seeds and send them to Simbirsk.[37]

The expedition also went to Tsaritsyn and Simbirsk and examined vineyards and the cultivation of watermelons, which were grown in the region and were said to be especially sweet and delicious.

In the reign of Peter I, the Englishman John Perry was sent to Astrakhan to explore the possibility of canal construction between the Volga and the Don (the waterway was only finally opened in 1952). Perry wrote an account of Astrakhan, in which he noted the tsar's interest in developing a trade in wine from the Caspian after he had experienced grapes from Astrakhan:

> The *Czar* has Thoughts of planting Vineyards, and improving the making of Wine on this Side of the *Caspian* Sea, in *Terki* and *Astracan*, where the Grapes, both red and white, are very large and good, and are brought from thence every Year to *Mosco*, with great Quantities of that delicious Fruit a Water Melon.[38]

The enterprise, however, was halted by a revolt in Astrakhan in 1705.

Peter I visited Astrakhan in 1722, but even before his visit he decreed, in 1720, that an apothecary garden should be established there, so that medicinal plants could be imported into the Russian Empire via the town from Persia, and instructed that a doctor should be sent there with medical supplies.[39] The Medicinal Chancellery duly established a small pharmacy and a garden 'to cultivate medical herbs'. The pharmacy and garden were intended to be of military benefit to the state, because they would act as a medical outpost that would supply Russian troops in the Astrakhan garrison and more generally along the empire's new southern border in the north Caucasus. In 1731, these gardens were visited and described by Johann Jacob Lerche, a German field doctor attached to an Astrakhan regiment. He found: 'there are so many trees and other strange plants, that a lover of botany would take the greatest pleasure to see it'. He also noted that the inhabitants of the town kept their own gardens, with a system of water channels to irrigate them.[40] In 1740, another pharmacy was set up in Astrakhan.[41] This was an age when medicinal gardens were becoming popular throughout Europe, and the Volga delta was exotic, both to Russians from St Petersburg and Moscow and to foreign travellers.

In the eighteenth century, several expeditions were dispatched by the Academy of Sciences in St Petersburg, with government backing and approval, to Siberia and the Crimea in particular, to describe and catalogue the new lands.[42] The middle and lower Volga were areas of scientific interest, not only because the non-Russian peoples could be regarded as rather exotic (it was seen in Chapter 5 that Catherine II was entertained at Kazan with dances and music performed by native peoples), but also because the region and its inhabitants needed to be better understood, given their potential disloyalty to the regime, as seen in popular revolts in the seventeenth and eighteenth centuries. In the early eighteenth century, John Perry made a point of describing the appearance and habits of Kalmyks and their relationship towards Russians, and in his words, the '*Hordes of Tartars*' in the vicinity of Astrakhan, noting that:

> Some of these *Tartars*, particularly the *Cullmick* [Kalmyks], own Protection from the *Czar*, and others live in good Amity with the *Russes*, and come every Year on the East side of the *Wolga*, and trade with the *Czar's* People.[43]

In the reign of Catherine II, Peter Simon Pallas, German botanist and scientist, and professor at the St Petersburg Academy of Sciences, was commissioned by Catherine II to undertake expeditions to the furthest parts of the empire to record peoples, flora, fauna, climate, minerals, weather and fishing. He was instructed to note the state of agriculture, in particular, with a view to ascertaining where developments and improvements could take place – within that region and elsewhere in the empire – in order to improve the economy of the empire as a whole. His first expedition to the Volga took place in 1769–70. In 1793 and 1794, Pallas travelled from St Petersburg down to the Crimea by way of Tsaritsyn and Astrakhan (travelling on the Volga in winter on the ice). He recorded the features of the countryside for the Academy of Sciences, including the quality of the soil, the nature of the rocks, the fauna, the flora and birds. Plants on the Volga were drawn, catalogued and sent as samples to St Petersburg.

Pallas noted that the river constituted a climatic as well as an agricultural divide, and that the 'desert' between the Volga and the Iaik rivers (later renamed the river Ural) would be better if it 'was not [made] so very droughty by the Hills on the Westside of the Wolga attracting every cloud, that hardly one year in ten the Barley comes up'.[44] He noted the harmful effects of the hot, dusty winds from the east in the summer, which could devastate crops; but he also commented on the rich quality of the black soil in the middle Volga region. In Astrakhan, he made a special point of recording the different types of fish in the Volga delta, the mulberry trees (for silk) and the nature and quality of the grapes, noting the attempts of the tsars to develop a wine trade.[45] In this way, the Russian government had a record of the nature of its territory and its people in the Volga region, and an assessment of its potential for economic development. The work of Pallas and the leaders of the other expeditions also made a significant contribution to scientific knowledge in the Academy of Sciences.

The Pallas expedition was sponsored from the top, by the tsar. Educated Russians were not unaware that the country had resources that could be exploited. In 1765, the Free Economic Society had formed in Russia to promote economic and agricultural developments in the country. The Russian Imperial Geographical Society was founded in 1845, and is one of the oldest geographical societies in Europe (the French Geographical Society was the first such society, founded in 1821; the British Royal Geographical Society was founded in 1830), and rapidly became a very important and distinguished institution. By this time, Russian academic institutions were no longer dominated by Germans, and Russians led further expeditions to the Volga and elsewhere. In 1876, the Free Economic Society commissioned Vasilii Dokuchaev to conduct research into the fertile black-earth region. Dokuchaev and his team covered much of the steppe area, including Saratov province, in 1877, 1878 and 1881, making meticulous notes and calculations; these led to the publication of *The Russian Black Earth*, a founding text of genetic soil science.[46]

The Volga, along with other parts of the empire, had by now become an area of interest for ethnographers, who recorded, among other things, the customs and costumes of the region. By this stage, local historical journals and publications of the local bishoprics recorded daily life and published some serious studies of the Volga and its people, including non-Russians. The Ethnographic Museum in St Petersburg has a good collection of peasant dress from the Volga region, including the costumes of the Udmurts, Mari, Chuvash and Tatars.[47]

By the mid-nineteenth century, the Volga region itself had become a centre for scientific knowledge and dissemination. The University of Kazan, established in 1804, was the centre of the educational district that incorporated the towns of the middle and lower Volga, the province of Orenburg and Siberia. In addition to the university, there was a grammar school, and by the late nineteenth century there were teacher training colleges for converted Tatars. Boys who wanted to study at the university came from all those regions, and teachers who had been trained in Kazan were employed in local schools. By the late nineteenth century, towns on the Volga had established their own literary and geographical societies. By the dawn of the twentieth century, advances had been made in education and science which should have benefited both the Volga region and the empire as whole. Despite these advances, the Volga region remained an area that suffered terribly from disease and famine in the 1920s and 1930s. These tragedies were, however, due largely to actions by the Soviet government, rather than to natural events, and resulted in an enormous cost in human life, as we shall see later.

Part 3

The Volga in the Russian Empire

Life and identity on the river

8
THE VOLGA VILLAGE

The opening words of the novel *My Children* by the contemporary Tatar writer Guzel Yakhina contrast the left and the right banks of the river Volga through the eyes of German colonists on the left (eastern) bank, and present the river as a divide:

> The Volga divided the world in two. The left bank was low and yellow, and flat stretching out into the steppe, from beyond which the sun rose every morning. The earth here was bitter to taste and dug up by gophers, the grass was dense and high but the trees were stunted and sparse. The fields disappeared beyond the horizon, multi-coloured melons like a Bashkir's blanket. The villages clung along the water's edge. The wind blew hot and spicy from the steppe – from the Turkmen desert and the salty Caspian. What it was like on the other bank no-one knew . . . A cold wind always blew from the right bank, from over the mountains from the distant North Sea . . .[1]

The sheer breadth of the Volga meant that peasants rarely crossed it even when, as above, the other side was home to fellow ethnic and religious settlers. The world of the peasants centred on their village – on whichever bank of the river – but it was not a completely closed world. The Volga countryside was home not only to Russian peasants, but also to Tatar, Chuvash, Mari and Udmurt – and, from the second half of the eighteenth century, German colonists. The Russian Orthodox community included not only ethnic Russians, but also converted Tatars and pagans, some of whom retained their original beliefs. Peasants of different ethnicity, language, culture and religion for the most part lived separately, but they also had to exist together in the same region.

The river Volga flowed through a number of distinct agricultural zones: mixed forest zone on the upper Volga; the steppe south of Samara; rich black-earth soil in parts of Saratov province; and then semi-arid steppe east of Tsaritsyn, extending down to Astrakhan in the south.

The river Volga also divided terrain in its middle and lower reaches: the western bank, known as the 'mountain' side was hillier, but had better arable land; the eastern bank had poorer soil and less vegetation.

Nevertheless, in many ways life was very similar for all peasants on the Volga: they grew roughly the same crops in the same seasons and used similar implements; they ate similar products from the earth and from the river Volga and its tributaries; they experienced much the same climatic conditions; they were subjected to very similar fiscal and other obligations; they built their houses and made their clothing and footwear from much the same materials. Communities also, however, had their own customs and habits, depending on their ethnicity, faith and origins. Some of these were unique to specific communities, but others spread and influenced peasants throughout the region. The multi-ethnic and multi-confessional composition of the rural population made the Volga countryside distinctive, and the complexities of these relationships will be explored in this chapter.

❖

Russia was a peasant country – over 90 per cent of the population were peasants in 1719, and the proportion remained between 80 and 90 per cent up to the First World War. The Volga countryside was primarily inhabited by peasants and it was predominantly a peasant economy, although, as we have seen, there were nomadic peoples in the south and east. Peasants, not just on the Volga but everywhere in the Russian Empire and in Europe, shared many of the same customs and ways of life; but there were legal distinctions between peasants on the Volga, which in part related to ethnicity, and which in turn affected their obligations. Serfdom was institutionalized in Russia under the Law Code of 1649, and the legal status of serfs has been discussed in Chapter 3; but not all peasants were serfs. The other main category of the peasantry consisted of 'state' peasants, who lived on state land and were administered and taxed by the state. There was also a category of 'church' peasants, who lived on church or monastic lands, but these were assimilated into the category of state peasants when church lands were secularized in 1764. In the early 1760s, over half of the ethnically Russian population in the empire was classified as serfs, but by 1795 that had fallen to under 50 per cent. The proportion of state peasants in the whole empire rose from 19 per cent in 1724 to 40 per cent in 1781–82, before falling back to 35 per cent in 1816 (the fluctuation largely depending on who was classified formally as a state peasant).[2]

The proportion of serfs within the peasantry was highest in the central provinces of the Russian Empire; it dropped the further away one went from Moscow, and was almost non-existent in Siberia. This was due in the main to the quality of land, but as a result peasant status was linked with ethnicity. The vast majority of the *serfs* who lived along the Volga were ethnically Russian, while the Tatars, Chuvash and Mari were more likely to be *state peasants*. In Kazan province, for example, on the middle Volga and on the eastern side of the river, state peasants comprised 75 per cent of the rural population from the early eighteenth to the early nineteenth century.[3] It has been estimated that 86 per cent of the Chuvash rural inhabitants, mainly located on the western side of the river Volga, were state peasants in 1763.[4]

Land was plentiful in Russia, but labour to work the land was not. The wealth and status of the Russian nobility were therefore determined by the number of serfs they owned. As one English visitor noted, 'the Wealth of a great Man in Russia, is not computed by the extent of the Land he possesses, or by the Quantity of Grain he can bring to Market, but by the number of his Slaves [serfs]'.[5] Only a few nobles in the empire, however, were fabulously wealthy, with vast lands, tens of thousands of serfs in many estates all over the empire, and great houses in the country and in St Petersburg; some 60 nobles owned between 5,000 and 30,000 serfs.[6]

There were some great noble estates on the upper Volga, which had been historically ethnic Russian territory and had been incorporated into the Moscow principality by the late fifteenth century, and on the western bank of the middle Volga in Saratov province, where the land was richest. The wealthiest family in the Russian Empire was that of the Sheremetevs. Count N.P. Sheremetev owned 185,610 serfs and over 10,500 square kilometres of land in the whole empire in the late eighteenth century, and had an annual income of over 600,000 roubles. One of the Sheremetev estates was Molodoi Tud, located in Tver province on the upper Volga, and comprised over 800 square kilometres inhabited by some 15,000 serfs.[7] In Iaroslavl province, property was owned by the great families of the Sheremetevs and the Golitsyns, as well as by Prince Bariatinsky, and the Lobanov family.[8] The Orlovs, another prominent family (two Orlov brothers had helped Catherine II to seize the throne in 1762), owned an estate called Usolsk in Samara province on the middle Volga, comprising 36 villages and some 26–27,000 peasants.[9]

The size of the noble estates, however, varied considerably. In Simbirsk province, where the land was not as rich as in Saratov province, one of the wealthiest noblemen was A.N. Zubov, who owned 193 peasant households (869 peasants); but many holdings on the middle and lower Volga were of between 100 and 200 peasants (or fewer), and there were very few noble estates at all in the delta region.[10] When serfs joined the major Cossack revolts on the Volga in the late seventeenth and eighteenth centuries, it was often to take revenge on particularly cruel landowners or their agents. Much of the violence of the Pugachev revolt on the Volga was said to have been directed at small noble estates, where peasants came into more day-to-day contact with their owners and their families; elsewhere in the empire, managers ran the estates of the wealthiest nobles, who thus had less personal contact with their serfs.

The state peasants – Russian and non-Russian – were considered by the serfs to be in a more advantageous position, because they were not so subject to oppression (or at least interference in their lives) from noble landowners or their representatives. In the case of serfs, by contrast, the landowners or their agents determined how many days they should work on noble land; they could also arbitrarily move peasants to other estates or force them to become house serfs and remove them from the village. When serfs claimed to be seeking 'freedom', this often meant the freedom enjoyed by the state peasants to run their lives without interference from a noble landowner or his representatives. Siberia was perceived as 'free' because it had almost no serfs; before 1649, serfs from the Volga region fled across the Urals to become state peasants in Siberia and to seek a better and 'freer' life. This did not mean that the lives of the state peasants were necessarily materially better – not least because the noble estates normally had the best-quality land, and could provide some protection in the event of crop failures. As an Anglo-Irish visitor noted, it was in 'the Master's interest to treat them [his serfs] kindly. His population constitutes his riches, & he who neglects or oppresses his subjects becomes their victim & sinks himself.'[11] State peasants were also tied to their village through the collective obligations it owed to the state, in the same way as serfs (see below).

After the conquest of the khanates of Kazan and Astrakhan in the mid-sixteenth century, pagan and some Muslim peasants in the Volga region were subjected to a tax called *iasak*, or tribute, which could be in kind (furs or grain) or cash. The payment of tribute (which was often extracted with violence by soldiers or Cossacks) was regarded as subjugation, and so-called 'tribute-paying' people, who included tribespeople in Siberia, as well as non-Russians in the Volga region, were considered to be inferior to the ordinary, Russian, peasants. Peter I equalized taxation when he introduced a 'poll tax' in 1718 (calculated by Peter initially at a level that would cover the costs of his armed forces), which had to be paid by every male non-privileged subject of the empire (peasants and ordinary townspeople). This had the effect of equalizing not only the obligations, but also the legal status of non-Christian peasants on the Volga and in other borderlands. The former tribute-paying peasants simply became state peasants, and non-Russians on the Volga were now taxed in the same way as Russians. In the eighteenth century in the Volga region, one of the incentives to convert to Christianity was a promise of exemption from the poll tax for a number of years. And in the 1760s, German colonists were recruited with the promise that they would be exempted from the poll tax for 30 years.

Slavery was officially abolished in Russia when the poll tax was introduced. Slaves had included military captives, who could be Tatars, although the majority of slaves in Russia were not prisoners of war, but rather ethnic Russians who had fallen into slavery through debt (and in effect became permanently indentured labour). It was estimated that 10 per cent of the population of Russia in the mid-seventeenth century were slaves. In the course of the eighteenth century, slavery disappeared in the Russian Empire, although in practice debt servitude continued, and some Chuvash, Mari and other non-Russian peoples effectively remained enslaved, by being made 'bondsmen' tied to the houses of rich noblemen.

German colonists in the Volga region were technically 'free' people. In practice, however, they had very little more freedom of movement or autonomy than serfs or state peasants. Colonists had no right to move to other colonies or into the towns. This meant that they remained where they had originally been settled on either bank of the river Volga, mainly in Saratov province. They had been recruited as 'farmers', and that is what they remained – at least until the very end of the nineteenth century, when wealthy Germans set up grain businesses in Saratov and agricultural reforms allowed greater movement. They were not allowed to practise trades or professions, but had to farm, despite the fact that over half of them orig- inally came from a non-farming background. A special office was set up in Saratov to deal with the affairs of the German colonists.

The organization and obligations of the peasant communities in the Volga region, and elsewhere in the empire, were similar, irrespective of legal status, ethnicity or religion. The peasant commune (called the *mir* in Russian) was the standard form of self-government not only for serfs but also for state peasants. It was the commune that divided, and re-divided, lands within the village, so that a rough equality was established between the available labour in the household and the land worked (normally according to the number of able-bodied males); deter- mined the use of communal woods and ponds; allocated the poll tax and other dues among peasant households; determined which boys would be dispatched as military recruits; and acted as a court for minor offences. The commune met in a general village assembly, but the lead was almost always taken by the older and wealthier male peasants, who were almost always the heads of large households. Large noble estates established a chain of responsibility from the landowner through his stewards to the assemblies; but in practice, a great deal of autonomy was exercised by the senior peasants in both serf and state peasant villages. The German Volga colonies had their own form of commune, called the *Dusch*, which also distributed land and taxes.

In addition to the poll tax, which was a state tax, all state peasants – Russian and non-Russian – paid a quitrent (called *obrok* in Russian). The level of the quitrent rose more rapidly than the poll tax in the eighteenth century although it was the latter that was often the cause of more resentment during peasant revolts, possibly because it was seen as a 'new', additional tax. Serfs either paid a quitrent direct to their owners, or they provided labour service (called *barshchina* in Russian), which meant usually three days a week working on the nobleman's land – or, very commonly, a combination of the two obligations. By the late eighteenth century, there was a move by noble landowners away from quitrent to labour service, because the price of grain increased, which made peasant labour more valuable. This was particularly true on the fertile black-earth lands, including land in Saratov province, where yields were highest. On the Borisogleb estate of the wealthy Kurakin family (in Saratov province), for example, serfs were obliged to provide labour duties three days a week in the 1780s. When an attempt was made to increase this to four days a week, a number of peasants fled from the estate.[12] Serfs resented this labour, and it was a major reason why they envied state peasants, who were seen to be in charge of their own land, even if that land was of poor quality.

By the eighteenth century, all peasants were subject to conscription into the Russian army, irrespective of ethnicity and religion. The only exemptions were for German colonists (supposedly in perpetuity, but that was overturned at the end of the nineteenth century) and for converts to Christianity (for a number of years). The exemptions provided a significant incentive, because in the eighteenth century – and right up until the late nineteenth century – conscription was for life (25 years after 1793 – but in effect that was life). The conscripted peasant was effectively 'dead', and his departure was accompanied by funeral laments. If he was married, his wife was often treated as a widow and, in practice, frequently allowed to marry again. The 'take' of peasants varied according to the pressures of warfare: over the whole empire, some 4 million men were recruited in the period 1705 to 1825.

Peasant communes – serf and state peasant – selected the recruits. It was usually done by lot, but on some serf estates provision was made to protect the families of married men and small households from the levy. Some noble owners tried to protect families – sometimes for genuinely humanitarian reasons, but primarily because their wealth depended on the number of serfs, and so it made no sense to lose married men who could have (or already had) families. The nobleman V.I. Suvorov, for example, rebuked an official on his estate in Kostroma province, on the upper Volga, who had 'maliciously' conscripted a married man without the permission of the peasant commune.[13] Wealthy peasants were able to 'buy' substitutes from their poorer neighbours – a process that was sometimes supported by noble landowners, who obviously wanted to retain all their able peasants. On the Manilov estate in Tver province, owned by the Gagarin family, wealthy peasants spent 34,000 roubles in the period 1812–57 on buying substitutes.[14] This was an expensive time to buy substitutes, as it coincided with the most demanding years of the Napoleonic Wars and the Crimean War.

The practice on noble estates described above concerned serfs in the Volga region who were overwhelmingly ethnically Russian. All peasants, however – serfs and state peasants, Russian and non-Russian – were subject to conscription, and in all cases the peasant commune played the most significant part in selecting conscripts. It was in the interests of all peasants that the least productive members of the community should be conscripted. This could mean the drunk, the idle and the dissolute; but it could equally mean peasant boys from poor households that could not contribute to the collective tax burden of the village. In reality, peasant communes – state and serf – could be much more brutal than noble landowners, because their economic survival depended on the ability of the village to meet their collective financial and labour obligations. In 1788, for example, the serf commune in the village of Molodoi Tud, in Tver province, resolved to 'dispatch as recruits 71 men for negligence in ploughing, for not paying dues, suspicious [characters] and landless peasants'.[15] In the village of Baki, in Kostroma province, three peasants were conscripted in 1819 on the grounds of 'dissolute and drunken behaviour and failure to pay taxes'.[16] In fact, serf communes sometimes came into conflict with their noble landowners, if the latter tried to preserve family units that were too small to contribute much to the collective dues.

Other peasant obligations included carting and labour. In the reign of Peter I, both Chuvash and Tatars were forcibly conscripted to work as labourers in the Admiralty and had to provide carts to transport salt from Lake Elton. This was at a time when Peter was trying to establish a fleet in the south of Russia. Mari peasants, along with others, were forcibly sent to build St Petersburg and the port of Azov in the early eighteenth century.

Taxes and other obligations were based on the size of the peasant household, which often included the extended family, including brothers and their families. Household sizes were largest in the Russian Empire in Siberia and in the rich, black-earth territory (where households of 7–9 people were the norm, but could be as high as 20–30 people). On the middle Volga, household size was smaller – averaging almost seven members in Kazan province.[17] But there were exceptions. The families of Orthodox priests were often very large (Orthodox parish clergy had to marry, and clerical posts were reserved for their sons), as were the families of German colonists, among whom it was normal to have 10 or 12 children. It was in the interests of noble landowners that their serfs should marry early and have large families, since that increased their wealth. Serfs on the Sheremetev estate in Tver province had to get formal permission from their owner to marry outside the village – and then had to pay a fine for the privilege of as much as 200 roubles in 1803.[18] On the Sheremetev estate of Voshchazhnikovo (in Iaroslavl province) in the first half of the nineteenth century, serfs not only had to pay a fee if they married outside the estate, but peasants (including widows and widowers) were also fined if they remained unmarried – an additional two to six roubles a year for every year over the age of 20 that they remained unmarried![19]

❉

Russians and non-Russians on the Volga shared many of the same obligations, but were their everyday lives very different? And how much contact did these different groups have with each other? For the most part, peasants of different ethnicity and religion lived in separate villages. Inter-marriage between different religious groups was rare, and was not encouraged either by the government or by the local communities (Muslims had to convert to marry an Orthodox subject, and Muslims were forbidden to proselytize). As one Russian peasant remarked about Muslims in 1864: 'We are not close with them; and we are not in conflict; there's nothing to argue about. We have our own environs, our own estate, and our own land. The Tatars, too, have their own possessions.'[20]

In practice, Tatars and Russians lived in the same villages on the Volga, but usually only when the Tatars had converted to Christianity. The village of Alkeevo in Chistopol district (Kazan province) was home to 151 Tatars and 60 Russians in the eighteenth century, and the nearby village of Mamykovo was home to 182 Russians and 42 Tatars,[21] though these were almost certainly converted Tatars. In the 1830s, the Samara Chancellery received a report from Orenburg province on the number of villages where converted Tatars and Russians lived together. The main concern of this report was the number of Tatars who had reverted to Islam; but in the process, it revealed that Russians and converted Tatars and Chuvash lived together in the same villages.[22] Dr Tomachev travelled through Kazan province in the middle of the nineteenth century. He reported whether the village was populated by Russians, Tatars or converted Tatars, but noted villages where Muslims and converted Tatars lived together. In the Tatar village of Berdebiakovy Chelny he recorded 7 Muslim families and 12 'Tatar-Christians'.[23]

Peasants of different ethnicities and religions came into contact with each other at markets and fairs, although these were usually close to the village and rarely involved crossing the Volga. Some peasants knew some words of each other's language. In the mid-nineteenth century, it was reported that on the middle Volga 'Russian peasants spoke Chuvash very well, and Mordvin as if it was their native tongue',[24] but this was almost certainly exceptional or an exaggeration. By the 1860s, it was said, economic integration had taken place in some villages in Nizhnii Novgorod province, with the result that Russian peasants spoke Tatar – and Tatars Russian.[25] But this was very rare, and may have been because there was only a small number of Tatars in Nizhnii Novgorod province. Most ethnic groups lived separately and probably only spoke a few basic words of each other's language (in the early years of the Bolshevik state there was criticism that government instructions had not been translated into local languages on the Volga, so that peasants did not understand policy). Words were borrowed from other languages, of course, particularly for foodstuffs; and place names and geographical features often remained in the original Tatar or Chuvash language. Some of the vocabulary of Chuvash and Mari is identical, despite the former being a Turkic language and the latter Finno-Ugric.

Russian government policy was to keep different faiths apart. We saw in the previous chapter that converted Tatars who had reconverted from Christianity to Islam were moved away from their villages, and converted Tatars were moved to Christian villages. According to the legislation of 1756, Tatars were only allowed to retain or build mosques in villages that were exclusively populated by Muslim Tatars, and where there were at least 200 male Muslim Tatars.[26]

Government policy could, in part at least, have been influenced by the clashes which took place between Christian and Muslim Tatars. In 1770, because of 'quarrels and fights', the Senate approved the resettlement of converted Tatars from Muslim Tatar villages in Kazan province, noting that in 56 villages there were only

601 converted Tatars among the almost 5,000 inhabitants.[27] The problem became more acute towards the end of the nineteenth century, when Tatars were becoming more conscious of their separate, Muslim identity. In Kazan province in 1873, the local teacher of a school for converted Tatars reported on the pressure ('tyranny' as one of the victims called it) put on his pupils by Muslim Tatars in the same or neighbouring villages, including by parents and close relatives. He presented a number of personal testimonies, including that of a 14-year-old boy who had been sworn at and beaten by his parents and other relatives in an attempt to persuade him to reconvert to Islam. Crowds of Muslim Tatars had put pressure on Christian Tatars attending church, telling them that the law now allowed them to reconvert to Islam, blocking their path to church and being abusive and attacking them. Christian Tatars had water and stones thrown at them at the market. The result was that many of the teacher's pupils had indeed reconverted to Islam.[28] Another report in the same year told the same story of worshippers being prevented from getting to church, and reported the case of a man who had beaten his wife when she refused to convert to Islam.[29] In 1897, a religious procession of Orthodox Chuvash peasants in Simbirsk province went through a Muslim Tatar village, whereupon the Tatars blocked the road, threw stones and disrupted the procession. The police had to intervene and accompany the procession, although it was reported that the peasants displayed some sensitivity, refraining from singing as they passed through the village and covering the cross.[30]

German colonists were both Protestant and Catholic: initially, 71 Protestant and 31 Catholic settlements were created (plus 2 mixed settlements). They had little cultural contact with either Russian or non-Russian villages – or even with colonists of the other faith. German colonies were divided not only by religion, but also by the river. There were settlements on both sides of the Volga, and communication between them was almost impossible: 45 colonies were set up on the western side and 59 on the eastern. Many colonies became the size of small towns, with some 10,000 to 15,000 people, and so there was little incentive to move outside them. The mathematician and physicist William Spottiswoode commented in 1856, when he travelled through Russia, that it was:

. . . remarkable to see how little these colonists amalgamate with the Russians. They retain their own language, customs, and habits; and it must be added, that the Russians also retain theirs. Neither the neater habits of the former, nor their ingenuity, nor their thriftiness, seem to have had the slightest effect on the people among whom they dwell.

'They live as strangers among a strange people', he concluded.[31] German colonists were slow to learn Russian, and really only did so when conscription was introduced in the 1870s. The colonists in any event retained their local dialects, and had to be taught High German in addition to Russian.

Foreign visitors to German colonies often praised the industriousness of the colonists, and often contrasted it with that of Russian villages and peasants. The colonists were, however, restricted in their occupations, made very few innovations in agriculture, and were slow to adopt any mechanized equipment. Indeed, as time went on, and the number of colonists grew, they became more impoverished, as the land they had been allocated had to be subdivided between them. In addition, a large colony of Moravian Brethren from Saxony was established in Sarepta, just south of Tsaritsyn on the lower Volga, quite separate from the other German colonists. This colony was plundered by Pugachev's forces (although the colonists escaped), but it recovered quickly. Peter Simon Pallas, himself a German, found the colony in the 1790s to be 'beautiful, and in a state of increasing prosperity'.[32] The population of the Sarepta colony numbered between 400 and 500 until the end of the nineteenth century, when it grew substantially.

Peasant villages in the Volga region were distinguished most by their place of worship: Orthodox, Protestant, Catholic churches or mosques. Cemeteries, which were normally located at some distance from the village, were also distinctive according to faith and pagan symbols; carvings and offerings appeared in some nominally Christian cemeteries of the Chuvash and Mari, alongside crosses and other forms of Christian symbolism. Most villages were laid out in a linear fashion, with the place of worship often some distance from the centre. In the Tatar village described above, where there was conflict between Christian and Muslim Tatars, the Muslims were able to block worshippers by occupying the bridge that led from the village to the church.

There were fishing villages on the banks of the Volga, but most settlements were located some distance from the banks of the river. The western bank was steep and could be cliffy, particularly on the middle Volga, and so villages were constructed on the hills, with access to the river by steep paths or ravines. On the eastern side, the land was low-lying, but the risk of flooding meant that construction had to take place some way away from the water's edge. Villages needed a water supply, of course, but it was easier to make use of the small tributaries of the Volga, or lakes or pools, rather than the river Volga itself, which was often difficult to access and navigate.

Peasant huts were built of wood, but timber was less readily available on the eastern side of the Volga, so that huts there tended to be smaller. It was normal for the peasant household to have a smaller log cabin in the yard, often functioning as a separate bedroom for married children or as a store room, and a number of barns. The yard could be enclosed or open, but Tatar villages tended to have their outbuildings separate from the main house, and some Russian villages copied this style. Tatars also often had large, decorated gates at the front, leading into the yard, and sometimes decorated fences linking the various wooden buildings; and this style was also copied by some Russian villages. Russians and Tatars used different carvings around the windows and on the eaves of houses – Tatars used stars, for example, and they also preferred brighter colours on their gates and houses than did Russians. Indeed, Tatar villages in Kazan province and elsewhere are still distinctive today, not only for the mosques, but also for their brightly coloured houses and large gates, as opposed to the Russian style of low fences. By the nineteenth century, wealthy Tatar peasants were building two-storey houses with elaborate carvings. Some of these styles of carving merged on the Volga, although each province, and each village within a province, could develop its own style, irrespective of the ethnicity of the inhabitants.

Within the main hut were wooden benches, tables and sleeping shelves; but by the second half of the nineteenth century, these basic items were being supplemented by new styles of furniture, beds and clocks. The main feature of all peasant huts was the stove, which could occupy up to a quarter of the room and was used for cooking and warmth. People would also sleep around and on top of the stove during the winter. In the 1720s, Friedrich Christian Weber reported from Tver:

> They have extraordinary large Stoves, which take up one fourth Part of the Room . . . At *Tweer* [Tver] I met a whole Family of twenty Persons, Master and Mistress, Children married and unmarried, with the Servants, lying thus together on Heaps on the Stove, and the Shelves above; and upon my asking them, whether they lay easy, and had Room enough to sleep, they answered perfectly well in such a warm Place, and wanted no Beds.[33]

Russian stoves were traditionally fixed in the corner of the hut (usually diagonally opposite the icon corner); in Tatar huts, moveable stoves were more often found in the centre, away from the walls of the hut. The style of stove was known locally as either a 'Russian stove' or a 'Tatar stove' and was an indication of the ethnicity of the inhabitants of the hut, although some Russians copied the Tatar style in the nineteenth century.[34] Stoves provided the only heat, and the poorest peasants only had a hole in the roof for the smoke to escape, although more prosperous peasants used brick flues.

Foreign travellers to Russia often commented on the poverty and dirt of Russian huts, although it has to be said that many travellers were predisposed to find the lot of serfs miserable, because they assumed that their lack of legal rights inevitably led to poverty and despair. Despite this caveat, the general view seems to have been that Tatar homes were cleaner than Russian ones. John Ledyard, who travelled across Russia in the late eighteenth century, found the 'Tartars universally neater than the Russians, particularly in their houses'.[35] Baron Haxthausen, a German agriculturalist and scientist, commented on Tatar homes that 'Everything in the house is very clean.'[36] Tatar villages were also said to have more open, green spaces than Russian villages, which were often very stark.

The German colonies were distinctive in appearance. William Spottiswoode commented in 1856 that the Moravian colony of Sarepta was:

> . . . thoroughly German; it has not a vestige of the Russian form or fashion about it. The houses are neat, and conveniently, but not rigidly, arranged. There are trees, walks, and gardens, such as we had not seen for many a hundred miles.[37]

Baron Haxthausen, writing at much the same time, found the colonies well kept and planned, but noted that the interiors 'presented a mixture of German and Russian', at least in terms of the material goods on show. There was a samovar and, oddly, a Russian saint in the icon corner, but all the furniture and utensils were German.[38]

Russian agriculture was, for the most part, peasant agriculture. Few noble estates experimented with agricultural practices, and were instead content to let the peasants cultivate their fields in their own way, although there were a few exceptions on the Volga and elsewhere in the empire. Russian peasants who settled on the middle Volga brought agricultural practices and implements from central Russia. The major crop in Russia was rye, which was harvested between July and August, with oats and barley as a spring crop; millet, flax and wheat were also grown. Land was divided into strips, and all peasants used the three-field system, whereby one field remained fallow in rotation. Peasants on the Volga and in central Russia used wooden ploughs: the *sokha* was a light wooden plough with metal shares which needed only a single horse, but produced only a shallow furrow; however, in parts of the Volga region the traditional Tatar plough, the *saban*, was used by Tatars and by at least some Russian peasants. This was a heavier plough, and more effective, but needed four horses or up to eight oxen to pull it. The main hand tools were the scythe for mowing hay and the sickle for reaping.

Yields were low by the standards of central and western Europe. This was true of land worked by both serfs and state peasants, by Russians and non-Russians. A lot, of course, depended on climatic conditions. The upper Volga experienced climate typical of north European Russia. The middle and lower Volga region typically experienced hot, dry summers, wet autumns and cold winters. When Peter Pallas travelled down the Volga in 1769–79, the temperature fell to as low as minus 27 degrees centigrade in Simbirsk. In 1803, it was said, crows froze to death not only on the steppe, but also in the towns.[39] In contrast, the summers on the middle and lower Volga could be hot and dry, and winds from the east could parch the fields, leading to crop failure and famine. John Perry commented that 'in the Height of Summer, from the latter End of *June* to the Middle of *August*, when the Heat has taken its full Power on the Continent, then an Easterly Wind brings more sultry Hot.'[40] The drought of 1833 was particularly severe, and affected the Ukraine, north Caucasus and parts of Saratov province. A German colonist wrote: 'In general 1833 was very unfavourable. The very unusual drought lasted from spring to autumn: the failure of grain and grass harvests was complete.'[41] Further droughts occurred in every decade for the rest of the nineteenth century (1873 was especially bad in Samara province). At the same time, the region was prone to brief but violent thunderstorms and downpours in the summer, which could also destroy crops.

As a generalization, the quality of land was better on the western bank of the Volga than on its eastern bank, which was why German colonists felt aggrieved if they were settled on the eastern shore. Conditions could, however, differ enormously within a province, or even within a district. Kazan province, for example, had poor-quality soil in the north; was heavily forested in the north-west and swampy in the north-east; but in the south, the soils were far better, and the south-eastern corner had very rich, black earth.[42] Saratov province included rich, black-earth land in the west, but also contained poorer-quality land in the east.

All peasants kept cattle, sheep, chicken and geese. Russians also kept pigs, while Muslim Tatars and Chuvash were more likely to keep goats (many Tatars believed goat's milk was better for infants than cow's milk). Tatars had a reputation for bee-keeping and for producing excellent honey. In the mid-nineteenth century, Baron Haxthausen found the Tatar peasants to be 'very industrious, and excellent bee-keepers'.[43] Some tobacco was grown in the south in the nineteenth century, in particular by German colonists (including in the Sarepta colony). Peasants grew a variety of fruits, and the Astrakhan region had an excellent reputation for watermelons. Potatoes were introduced to the Russian Empire by Peter I in the early eighteenth century. They were called the 'apple of Eve' or the 'Devil's apple' by Old

Believers, who refused to grow them; but they became a staple for Russian peasants on the Volga and elsewhere.

Peasant diet on the Volga, as in other parts of European Russia, reflected the products grown locally and was primarily cereal-based – rye bread and groats. Russian peasants, in particular, ate such a variety of vegetables – onions, beets, garlic, cabbages – that foreign travellers thought the diet of Russian peasants better than that of their peasant equivalents in western Europe. An Anglo-Irish visitor to Russia, Martha Wilmot, was impressed by the dress and diet of the peasants she saw in western Russia at the beginning of the nineteenth century, and was one of the few visitors to recognize that the material well-being of Russian peasants could compare favourably with that of their counterparts at home:

> . . . the peasantry of this country really and truly enjoy not only the necessarys but the comforts of life to an astonishing degree . . . those who imagine the Russ peasantry sunk in sloth & misery imagine a strange falsehood. Wou'd to God our Paddys . . . were half as well clothed or fed the year round as are the Russians.[44]

Russian Orthodox peasants observed fasts and refrained from eating meat, not only during Lent but for up to 30 weeks of the year. The German diplomat Friedrich Christian Weber, who visited Russia in the early eighteenth century, noted 'with Astonishment how zealous they are in observing their Fasts'.[45] He went on to say that you could tell if a Russian peasant was nearby, because of the smell from the amount of onions and leeks they consumed!

Russians were said to eat more vegetables and fruit than Tatars, Chuvash and Mari, although that does not mean that the latter ate no vegetables. Mushrooms were an essential part of the Russian peasant diet, but were less popular with Tatars. Russians ate little cheese, apart from a curd cheese (*tvorog*), and little fresh milk, although sour cream was used in special dishes; Tatars ate more milk dishes and used butter in their cooking. A special clotted cream dish called *kaimak* was eaten by Russian peasants in the southern Volga and was said to have been influenced by Tatar and Bashkir cooking (variants can be found today in Central Asia, Iran, Iraq and the Balkans). All peasants ate a rich variety of fish from the river Volga and its tributaries. In the 1830s, Robert Bremner claimed that half the fish of the whole empire were caught in the Astrakhan region, including sturgeon, sterlet, carp, salmon and pike: 'without this river the Russians could not live', he concluded.[46]

The basic drink of Russian peasants was *kvass*, a lightly fermented beer, which could be brewed from malted rye, barley, wheat, pastry or bread, and flavoured with fruits and berries. Tatars and Russians also made honey-based drinks. Tea was probably introduced to Russia in the seventeenth century, but only became popular in the late eighteenth century. Both Russians and Tatars drank tea, and by the nineteenth century all peasant huts would contain a samovar.

It is hard to establish whether differences in diet led to health differences between ethnic groups, or between the categories of peasant – serfs and state peasants – on the Volga. A study has been made of the health of 20,000 recruits in Saratov province in the middle Volga region from the mid-eighteenth to the mid-nineteenth centuries. It found that state peasants, who would include most of the Tatars, Chuvash and Mari, were on average 2cm taller than serfs, who were predominantly ethnically Russian; urban dwellers were also taller than all rural inhabitants. The study also found that poor harvests in the years 1859–62 and 1875 had a detrimental effect on the height of recruits across all ethnic groups.[47] It was said that Tatar babies had a better chance of surviving the first few years of life because they were suckled for two or three years, whereas Russian babies were given a pacifier – a cloth filled with partially chewed food to suck. Furthermore, death rates among Russian infants were highest in the summer months, when women were working in the fields to bring in the harvest. Chuvash women also laboured in the fields, but Tatar women (and Old Believers) on the whole performed less arduous physical labour.

Peasant dress was made of local and home-spun material and was designed to suit the climatic conditions and occupations. All peasants wore a long shirt, but Tatar shirts – for men and women – were traditionally longer and unbelted (hence a popular proverb: 'One without a cross and a girdle looks like a Tatar'). But there is some evidence that Tatar long tunics may have influenced the dress of Russians (and Mari and Chuvash), because Russian men on the Volga often wore longer, and sometimes unbelted, shirts, in contrast to their countrymen in central Russia. The Marquis de Custine remarked in Nizhnii Novgorod province in the 1840s that 'Some of the peasants in this part of Russia wear white tunic shirts, ornamented with red borders: the costume is borrowed from the Tartars.'[48] Tatars wore loose trousers, and that style also spread to Russian and other peasants.

The reason for the influence of Tatar dress may have been that Tatar tailors travelled between Russian villages and sold Tatar-style clothes and boots to their inhabitants. Tatar looms were traditionally wider than Russian peasant looms and could produce a broader cloth. The most common peasant footwear was made of bast (usually lime bark), worn with cotton leggings; only wealthy peasants wore leather boots. Tatars had a reputation for producing elaborately decorated leather boots, sometimes using colour, but these were prized possessions. Tatar men wore special square-shaped hats, which were also sold to Russians at fairs. Elaborate headdresses were made for married peasant women, irrespective of ethnicity: Tatar women, for example, wore a small cap called a *kalfak*, with a decorated head band. Converted Tatars developed some of their own costumes, but men and women tended to continue to wear long Tatar tunics, and unmarried girls wore the *kalfak*, with the headband often decorated with coins.

Popular festivities for Russian (and Ukrainian) peasants coincided with major Christian festivals. The festival of *Sviatki* (Christmastide) took place in January, from Orthodox Christmas to Epiphany, and involved special songs, dances and games for young people, who would wear their best clothes and used this occasion to meet each other. These festivals have been recorded for the Volga region in the nineteenth century and were drawn from customs in central Russia, although individual villages could develop their own variations on costumes and games. Lent was the other major Christian festival celebrated by Orthodox Christians throughout the Russian Empire. Celebrations in the middle Volga region, which included sledging and other physical activities, special meals and bonfires, were similar in form to those in the central and northern European Russian provinces. Easter celebrations involved special foodstuffs and the exchange of coloured eggs. The Volga region developed a few of its own rituals at this time; for example, a special ball game was played in Chistopol district, and particular local regional dishes evolved, including for major festivals and for weddings and other important family occasions.

It is hard to know the extent to which these festivals involved non-Russians and, in particular, non-Christians. It is particularly difficult to establish the cultural habits of the converted Tatars. In the late nineteenth century, Russian ethnographers recorded many of the customs of peasants in the Russian Empire, including in the Volga region; but they did not always make the distinction between converted Tatars and other Christians. These Tatars shared Christian festivals, but also had a cultural link to Tatar festivals, and it was said that they participated in the major Tatar festival, the *sabantuy* (see below).[49]

Mari and Chuvash peasants traditionally followed shamanistic beliefs, and, as we saw in the previous chapter, their conversion to Christianity was often only nominal. In some villages, particularly in remote areas, their ancient pagan rituals coexisted with Orthodox festivals. A scholar of Mordvin customs, M.E. Evsevev, recalled that in his childhood the Mordvins called upon their old gods at harvest time.[50] In any event, it is sometimes hard to draw a distinction between pagan and Christian celebrations, as Christian festivals often also coincided with the onset of seasons and the solstices, or with particular events in the agricultural year. Religious customs did – and still do, and not only in Russia – overlap with older, pre-Christian rituals. *Sviatki* was an ancient pagan festival that coincided with the Christmas period, and many of the customs in Orthodox Christian churches at Easter reflect the onset of spring and the birth of new life – just as they do in central and western Europe.

All peasants of whatever faith (and indeed not only peasants) believed in 'evil sprites' and the 'evil eye' and held superstitions and magical beliefs to satisfy the most basic human needs, such as finding a spouse, happiness, a cure for sickness, revenge on rivals, etc. Such beliefs could merge to suit the occasion, as a Chuvash priest recorded in the early twentieth century:

> In my parish there is one woman, a zealous Christian, whose husband has epilepsy. She served [requested] a service of intercession in the church, but when it didn't help, contacted a yumza [shaman]. Then, she took her husband 12 versts away to a Tatar mullah. Finally she again came to me.[51]

There is some evidence that Russians and non-Russians – and possibly Christians and non-Christians – took part in each other's festivals in the Volga region. It is said that Russian peasants gave presents to Tatars, as well as to Christians, on Christian holidays. Horse racing during the Lent festivities could involve Chuvash, Mari and Tatars, some of whom might not have been Christian. Fist-fighting also took place at Lent, on a large area of open ground, and could involve organized fights between several villages and between different ethnic groups. Some festivals were not associated with a particular religion or ethnicity. Festivals to celebrate, for example, the name days of the tsars were not associated with any religious group. Chess was played by all members of male society and different ethnic groups, including Tatars and Kalmyks.

Russians could also participate in major Tatar social festivals, such as *sabantuy* and *dzhien*. *Sabantuy* celebrated the beginning of the most important stage of agricultural work, before the sowing of spring crops, and could therefore overlap with Russian Easter festivals. Dyed eggs were given as presents, as at Easter, although some Tatar communities specifically chose other days on which to give their presents, so that the event should be distinguished from Easter. *Dzhien* was celebrated between the end of sowing and the start of harvest, during a lull in agricultural work. It also involved horse racing, wrestling matches, songs and music, as well as traditional feasting. Neither festival had a fixed day. That meant that converted

Tatars could celebrate *sabantuy* in their own way: they often held it on Easter day, so that they could assert their Christian and their Tatar identities together. *Dzhien* was often held in villages of converted Tatars on a Christian saint's day.[52]

The participation, if limited, by non-Christians in Tatar festivals did not mean that Muslims did not also have their own festivals. Tatar Muslim holy days – the fasting holiday (*Uraza gaete*) and Sacrifice Feast (*Qorban gaete*) and key dates in the history of Islam and the life of Mohammed – were exclusively for Muslims. Tatars had their own musical instruments, and special music and songs were performed at major family events, including weddings.

The German colonists were Christian, but celebrated their own festivals separately; they had special celebratory food, which often originated from their villages in German lands (and had sometimes been forgotten in their original homeland). Chuvash, Mari and Udmurts also came into contact with Tatars and often borrowed customs from them, especially in areas where Tatars were numerous. A priest in the Tetiushii district of Kazan province commented at the end of the nineteenth century on the influence of Tatars over their Chuvash neighbours. According to him, the Chuvash sang Tatar songs, adopted Tatar dishes and even wore their moustaches in 'the Tatar way'.[53]

Few changes took place in the lives of peasants – serfs and state peasants, Russian and non-Russian – over the course of the eighteenth and the first half of the nineteenth centuries. Significant attempts to change the obligations and way of life of peasants were only made after 1861. That is the subject of Chapter 12.

9
THE VOLGA TOWN

Was there such a thing as a typical river Volga town? On the surface, there was little in common between the historic Russian towns of Tver, Iaroslavl and Kostroma in the heartland of old Russia, the towns that had been founded as forts to control key points on the river after conquest such as Saratov and Samara, and the non-Russian towns that had evolved in the period of the Golden Horde such as Kazan and Astrakhan. All were linked, however, by trade along the river south to north and north to south, which meant the movement not only of goods, but also of people. All towns in the Russian Empire were moulded by the policies of the Russian government. Volga towns were, however, distinct in their ethnic composition. Their inhabitants included non-Russians, non-Christians and non-Orthodox Christians. This had an impact on almost all aspects of urban life, including the physical appearance of towns, the occupations of the town inhabitants, urban administration and urban culture. Astrakhan was unique in its ethnic composition, and is discussed separately at the end of this chapter.

Towns were built along the banks of the Volga, usually at the confluence with another river: the Volga and the Oka (Nizhnii Novgorod), the Volga and the Kazanka (Kazan), the Volga and the Kotorosl (Iaroslavl) and the Volga and the Tvertsa (Tver). Towns could be built on the higher (right) or the lower (left) bank, but only towns on the upper Volga, where the river was narrower, could be built on both sides of the river. On the middle and lower Volga, the breadth of the river meant that towns were both linked and divided by the river: ships moved up and down, but it was often very difficult for the inhabitants of a town to cross the river, and rare for them to want to do so when the other bank was often sparsely inhabited. In any case, the roads ran along whichever bank the towns were located on. The construction of first rail and then road bridges across the Volga in the late nineteenth and twentieth centuries finally linked the banks, but towns still developed primarily on one or the other side of the river. Even now, there are relatively few bridges across the middle and lower Volga, even in major towns.

The small town of Rzhev (population 22,000 at the beginning of the twentieth century), the first town downstream from the source of the Volga in the Valdai hills, was built on both sides of the river, with the main part on the higher bank and originally connected to the other side by a wooden pontoon bridge. Tver was the first major town on the upper Volga; plans from the early nineteenth century, after it was rebuilt following a major fire, show that it was constructed fairly evenly on both banks of the river, although the major administrative buildings are on the right (southern) bank. The two shores were also connected by a pontoon bridge until the 'old bridge' was constructed in 1898. Tver was a key provincial town in the late eighteenth century, under the reforming governor Jacob von Sievers;[1] it was described in the 1790s by Pallas as 'so much improved, that it may with propriety be ranked among the most elegant and regular provincial towns in Europe'.[2] However, the town declined economically in the nineteenth century, and other settlements on the upper Volga overtook it in population and commercial significance. Below Tver, the towns of Rybinsk, Iaroslavl and Kostroma are built on both sides of the river. Iaroslavl was the largest town on the upper Volga throughout the imperial period, and was of 'grand appearance', according to Baron Haxthausen.[3] In the early eighteenth century, 36 different artisan occupations were listed in Iaroslavl, but the town was particularly known for its clothing and leather industries and from an early date for its ship-building. It was a town of wealthy merchants; this was reflected in the number of private chapels they founded, only a few of which survived the Soviet period.

Nizhnii Novgorod is situated at the confluence of the Volga and Oka rivers. The town is located on both sides of the river but is predominantly on the western bank. It was a major trading centre (and as such will be discussed in the next chapter, on trade and the river). In 1797, over 2,000 merchants and over 3,000 artisans were registered there.[4] In the 1840s, it was described by the Marquis de Custine as 'the most beautiful [town] I have beheld in Russia'.[5] By 1897, its permanent population was estimated at 98,503, but this increased substantially during the time of the fair.

Cheboksary, on the western bank, was a small town until the twentieth century, but some significant merchant families lived there, and some of their villas can still be seen in the town today. Kazan, on the eastern shore of the river, was also a key town for east–west trade. Its population of some 30,000 in the early nineteenth century grew throughout the century – 43,900 in 1843, 63,100 in 1863 – and by

1897 had grown to 131,508. The cosmopolitan character of the town is discussed more fully later.

Below Kazan on the river, Simbirsk, on the western bank, developed slowly (it had a population of only 41,702 in 1897). It was described by Laurence Oliphant, an author and traveller in the middle of the nineteenth century, as of 'a mean and insignificant appearance . . . the streets are deserted, the shops poor, and the *tout ensemble* most uninviting' (although he apparently found the melons to be 'magnificent'!).[6] In 1860, a Russian steamship captain was even more disparaging: 'a *gubernskii* [provincial] town without a governor, a noble assembly without an assembly, a theatre without actors, a [Protestant] church without a pastor, and [Catholic] church without a choir'.[7] Simbirsk was bypassed by the main railway lines, which led to its decline in the late nineteenth century. The Volga then turns to the east, and Samara is located on the 'bend' on the eastern bank. The town developed as a centre of the grain trade and flourished by the end of the nineteenth century (it had a population of 91,659 in 1897). Despite this, the town was dismissed by a traveller at the end of the nineteenth century as 'dreadfully dusty' and 'a very uncomfortable place of residence'.[8]

Saratov, on the western bank, became an important commercial centre in the second half of the nineteenth century (its population grew to 133,116 in 1897). Tsaritsyn, also on the western bank, below Saratov, was originally a small town, but became a 'boom town' in the late nineteenth century (55,186 inhabitants in 1897, rising to 131,782 inhabitants in 1913).[9] The main reason for Tsaritsyn's rapid growth was that it became an essential railway hub for the export of grain after the completion in 1871 of the Tsaritsyn–Riga line. As Tsaritsyn expanded, so the town (former Cossack settlement) of Dubovka, north of Tsaritsyn, declined because it was not connected to the railway hub. In the south, Astrakhan was a major urban centre and port (with a population of 113,075 in 1897).

In Russian, there is no distinction between the words for a 'town' and a 'city', unlike Russian villages, which had several different names depending on size and on whether or not they had a church. The two towns that were exceptional in size, in status, and as the real or potential centre of government were St Petersburg and Moscow – they were the two 'capitals', in effect. In Britain, a city becomes a city by being granted city status by the monarch; and in the United States, a city is a legally defined entity, with powers of administration delegated by the state. But in Russia, the distinction between types of towns was purely administrative and was determined by the state. After Catherine II's reforms of 1775, the Russian Empire was divided into provinces of roughly the same population (300,000 to 400,000 inhabitants), and subdivided into districts of between 20,000 and 30,000. The major town in the province was termed the 'provincial town' (the Russian equivalent of the British county town), where the organs of the provincial administration and higher-level educational facilities would be located, and the major town of the district was termed the 'district town'. The legal status of a town could be anomalous and at odds with its economic significance. Samara became a provincial town in 1851, at which time it had only 20,000 inhabitants.[10] Tsaritsyn was designated a district town in Saratov province after Catherine's reforms in the late eighteenth century, which made sense because at the time it was very small (just over 4,000 people in 1825);[11] but it retained that status throughout the imperial period, even though it grew rapidly and became larger than many provincial towns.

❖

Many towns in Russia were ethnically mixed – Odessa and St Petersburg, for example – but the ethnic and confessional composition of Volga towns reflected their origins and the fact that trade on the river had led to the movement of peoples as well as goods. Kazan was a distinctive town, because it comprised large numbers of both Russian and Tatar merchants and artisans. The proportion of Russians and Tatars is not easy to establish with any precision. In Kazan province as a whole, Tatars were greatly outnumbered by Russians (roughly 75 per cent to 22 per cent), but they were more concentrated in the town of Kazan and among the merchant classes. One estimate suggests that there were 61,714 Tatars in the town of Kazan in 1858;[12] but Tatars were not the only non-Russian residents. Baron Haxthausen visited the bazaar in Kazan in the middle of the nineteenth century, where he found 'Russians, Tartars, Tcheremiss [Mari], Votiaks [Udmurts], Mordvins etc.'; he reported that 'there was much noise and running about, and many stout, ugly women, and good-looking men'.[13] Kazan was exceptional, because it had been a Tatar town and the capital of the Kazan khanate, but Tatars also traded in other towns. In 1858, it was estimated that over 10,000 Tatars lived in the small town of Chistopol (granted status as a district town within Kazan province in 1781).[14] Tatars also lived and traded in Nizhnii Novgorod and Simbirsk, and at least one was an industrial entrepreneur: the Tatar Khasan Timerbulatovich Akchurin owned a paper mill in Simbirsk in the late nineteenth century.[15] Baron Haxthausen also noted the presence of Ukrainians, Chuvash, Mordvins, Kalmyks and Bashkirs in Saratov.[16]

A snapshot of the economic contact between Russians and Tatars in Kazan can be found in the records of one court, the Tatar *ratusha* (a Russian word from the

German *Rathaus*) in the late eighteenth century. This was a court of first instance
– that is, one which dealt with minor disputes, although it also handled administrative and judicial matters. Many cases involved disputes between Tatars over such
matters as unpaid bills and contracts, or cases brought by the local authorities
against individual Tatars – for example, against a Tatar who had counterfeited a
passport. The court dealt with cases relating to behaviour within Tatar families,
such as husbands who abused their wives or children who disobeyed their parents.
There were also, however, records of disputes between Russians of various social
backgrounds and Tatars, almost all of which concerned financial contracts. When
Catherine II set up courts in the provinces in 1775 (see Chapter 5) she seemed to
assume that conflicts would be exclusively between the same ethnic and social
groups (there were separate courts for nobles, townspeople and state peasants); but
this bore little relation to the reality of day-to-day interaction between people
within the town. A few examples will demonstrate the variety of issues: a serf who
accused a Tatar of stealing his horse; a dispute over a contract between an official of
noble background, Stepan Popov, and a Tatar, Mukhamet Isaev; a Kazan merchant,
F. Poliarkov, who disputed a contract with a Kazan Tatar who lived in the new Tatar
quarter in the town; a dispute over a debt owed by a Kazan merchant, A. Savin, to
a Tatar who lived in the old Tatar quarter; the case of a Kazan artisan, Kh. Galeev,
who had borrowed money from a Kazan Tatar from the new quarter.[17]

Christians who were not Russian Orthodox also lived in Volga towns. Nizhnii
Novgorod province was known for the presence of many Old Believers, many of
whom were involved in trade. There were also Old Believers in Iaroslavl – we have
seen that 215 of them 'returned' to the Orthodox Church in 1852 in Iaroslavl
province.[18] Baron Haxthausen found Old Believers in the town of Rybinsk during
his travels. This group was also involved in trade in Kazan: for example, I.S.
Kurbatov was a prominent Kazan Old Believer trader.[19]

Volga towns also included a small number of Protestants and Catholics. In Nizhnii
Novgorod town in the 1840s, there were 364 Protestants and 471 Catholics – small
numbers compared with over 39,000 Orthodox Christians, but more than the Old
Believers (estimated at the time to be as few as 260 in the town).[20] Germans –
not only Baltic Germans, who were subjects of the tsar, but Germans from German
lands – worked in Russia as merchants, administrators and educationalists. The town
architect of Tver in the period 1776–84, for example, was Johann Stengel, a German
from Saarbrücken.[21] Polish Catholics were exiled to remote parts of the Russian
Empire after the Polish revolts of 1830 and 1863: some 20,000–50,000 after each
revolt. Most of the Polish 'rebels' were sent to Siberia, but some ended up in Kazan,

in effect in internal exile, and formed a significant community. In the 1850s, Catholics were allowed to construct a church in Kazan, provided that it blended in with the local buildings and was not too overtly Catholic in architecture (it has now been replaced by a new church). The first Catholic church was built in Saratov in 1805 for German Catholics, and this was followed by Catholic and Lutheran cathedrals in the late nineteenth century (the Catholic cathedral was destroyed in the Soviet period, but has been rebuilt and reopened in 2002).

The Russian Empire acquired a substantial Jewish population after the three partitions of Poland-Lithuania in the late eighteenth century. Russian government policy on Jewish settlement was inconsistent. At first, Jews were confined to a 'Pale of Settlement' in the former Poland-Lithuania, in the western part of the empire; but then in the 1820s and 1830s they were allowed to settle elsewhere. Jewish boys also served as 'cantonists' – that is, as soldiers in the Russian imperial army – during the reign of Nicholas I, when Jews were not conscripted alongside Russian peasants and townspeople. After the completion of their service, these Jews were allowed to register in various towns in Russia. Photographs of Jewish cantonists who settled in Kazan are displayed today in the small museum of the Kazan synagogue. Even in the 1860s, however, only certain categories of Jews could settle outside the Pale. In practice, Jews did come to the Volga, but only in small numbers; and they settled primarily in towns. Their presence was constantly monitored by the authorities: the population statistics noted above for Nizhnii Novgorod included 354 Jews,[22] and in 1855 a report was made of Muslims, Jews and 'iconoclasts' (Old Believers) in Kazan province, which found that most districts had no Jews or very few (11 in Cheboksary, 16 in Chistopol, 4 in Iadrin district, 2 in Tetiushii district).[23] Almost all those Jews were male, which suggests they were probably itinerant pedlars. At the same date, 10 male and female Jews resided in Kazan. There were 541 Jews in Nizhnii Novgorod in 1870, a number that had risen to almost 3,000 by 1913. There were over one thousand Jews in Iaroslavl by the end of the nine-teenth century.[24]

Some urban trades became associated with particular ethnic and religious groups. In Nizhnii Novgorod province, Old Believers were said to dominate trade in several craft areas, such as wooden dishes, felt boots, lambswool hats and nails,[25] but they were also involved in major trading and ship-building (especially in the small town of Gorodets, some 50km north-west of Nizhnii Novgorod on the Volga). In the late nineteenth century, a small number of wealthy Germans, many of them colonists, dominated the grain trade in Saratov. Meanwhile Tatars domi-nated the silk trade in Kazan.

The occupations of Jews are easy to establish, because they were constantly being checked by the police to determine if they had the right to live in the Volga towns. The Jewish military 'cantonists' of the mid-nineteenth century had become merchants and petty traders by the end of the century.[26] In 1875, the Jews in Kazan province included tailors, traders, workers in the distillery, teachers of Jewish children and a musician.[27] Seven Jews lived in Rybinsk in 1882: a teacher of Hebrew, a tailor, a watchmaker, traders in petty goods and butchers.[28] A list in Samara in 1913 showed how Jews had progressed in the professions in the twentieth century, while still retaining traditional occupations: they included several pharmacists, doctors and tailors, two watchmakers, two dentists, two metalworkers, two hat makers, one egg trader, a machinist, a cobbler and, perhaps to cater for the Jewish community, a midwife.[29]

❖

The Russian government made several attempts to reorganize urban administration in the eighteenth and nineteenth centuries. This was in part to tackle widespread bribery and corruption, about which townspeople regularly complained. The reforms were also, however, recognition that Russian towns generally had not developed economically in the same way as towns in central and western Europe. The urban population was categorized in the eighteenth century into privileged and less privileged groups. At the top were merchants, in 1785 divided into three guilds according to the amount of capital they declared; their privileged position was recognized by their exemption from the poll tax and from other onerous burdens, such as having troops billeted on them. Below the merchants were artisans, and below them the lowest category of 'general townspeople', who were employed in unskilled jobs, such as labouring. At the same time, the government introduced a series of administrative bodies, with posts filled by elected members of the town community according to status. At the base of these institutions were the 'societies' – that is, the collective bodies of the merchants and artisans, which conducted elections to minor courts and to dumas that handled administrative and fiscal affairs.

The problem with the reforms was two-fold. First, few towns in the Russian Empire outside Moscow and St Petersburg had enough people of sufficient wealth to become members of guilds and to fill the urban positions. The towns on the river Volga illustrated these problems. In Iaroslavl, an important commercial town, in 1800 there were 4,095 merchants, but only 7 of those had enough capital to be members of the first guild, and only 122 had enough to register with the second guild.[30] Samara by 1853 had become an important town for the grain trade, but had no members of the first guild at all, only 8 in the second, and 701 in the third.[31] In the less developed Cheboksary, only two merchants were registered in the first guild in 1775.[32] Second, and more fundamentally, the main problem in Russia was the overall structure of society, which was dominated by two social groups: nobles and peasants. Noblemen owned factories, dominated the grain trade and distilling and, in general, looked down on merchants. In Iaroslavl in 1810, a merchant called Kuznetsov, with the support of other merchants in the town 'society', challenged the rights of nobles to participate in the fishing trade in the Volga, but it is not known how the case ended.[33] Furthermore, peasants traded in artisan goods without any restrictions on their production (unlike trade guilds in central and western Europe, which controlled quality and production), and constantly undercut the products of urban artisans. If these so-called 'trading peasants' were serfs, they were even encouraged to do this by their noble owners so that they could pay more cash dues.

The position of non-Russians, and non-Orthodox Christians, in the Volga towns only served to complicate the bureaucratic structures and hinder the evolution of confident self-governing urban institutions. Russian and Tatar merchants lived apart from one another and had their own separate institutions and separate merchant and artisan 'societies'. As we have seen, Tatars had their own court of first instance, the Tatar *ratusha*, to handle judicial and economic affairs, and more minor courts – oral courts – to handle minor criminal cases. Some Tatars nevertheless registered with the Kazan *Russian* merchant and artisan societies. In the 1790s, the *ratusha* received petitions from Tatars to register in the Kazan merchantry as artisans.[34]

Old Believers also fitted uneasily into the new urban structures. In 1834, Kazan merchants requested that Old Believers should be eligible for election to urban posts in the town – apart from the post of town head, who was the most important elected town administrator. It may be that the Kazan merchants regarded these urban posts as a burden rather than a privilege. Their petition pointed out that Old Believers already took part in merchant and artisan trade, and so this initiative would, in their view, cause no disruption. It was also noted that many Old Believers possessed considerable capital, and could therefore register in the first and second guilds. They even suggested that this move might encourage Old Believers to return to the Orthodox fold![35] The problem continued, however, and in 1856 the request was repeated, this time pointing out that Catholic Poles were not excluded from these posts.[36] The police interest in the status of Jews is another example of restrictions being imposed from above. Many of the Jews cited in this chapter had testimonials from the merchant and artisan 'societies' in Volga towns stating that they were trading conscientiously.[37]

All Volga towns had fortifications and a kremlin, which housed the main administrative buildings and cathedral. Apart from the kremlin buildings, the towns were originally constructed of wood, which made them very susceptible to fires. Samara, for example, had conflagrations in 1765 (when 418 buildings were destroyed), in 1772 and in 1850, when 521 houses, 126 grain stores and 2 churches were destroyed.[38] Iaroslavl lost 210 homes, 15 churches, 2 monasteries, 583 shops, 13 barns and 37 smithies in a fire in 1768.[39] From the late eighteenth century, particularly after fires, town were planned with broad streets which often led to the kremlin or cathedral, and with classical-style administrative buildings, often constructed round a large square. The new towns could look quite grand: Kazan 'looks exceedingly well', commented William Spottiswoode in the 1850s,[40] after a major fire in 1842. Gardens and parks were laid out at the same time. In the late nineteenth century, impressive, solid merchant houses were built in all the Volga towns. Samara, in particular, still has some fine examples of art nouveau-style merchant houses near the town theatre. In the second half of the nineteenth century, the embankments of Volga towns were developed and often reconstructed, with poor houses and dirty lanes replaced by fine walkways and impressive merchant houses.

That does not mean that Russian towns were comfortable places to live in, especially for the poor. Clean water and sewers were rare. The air quality was so poor in Iaroslavl (where there were leather and paper factories) that in the 1760s it was reported that people could not go out into the streets.[41] Kazan was no better: it was reported that in 1872 people could not open their windows in the summer or autumn because of the stinking air.[42] Part of the problem was the construction of factories on the water front. Cheboksary, at the time a small town, was low-lying and marshy, but the water and air quality were rendered worse by the location of seven leather and four soap works on the banks of the Volga.

The other unpleasant feature of Volga towns was their muddy, unpaved streets. A description of Kazan in 1812 stated that it was 'not possible to walk on even the best streets in wet weather because of the mud'.[43] The artist Ivan Shishkin, who was born in Viatka province but was educated at Kazan grammar school, found Kazan 'tedious' because of the 'cold and the dirt which are rife here'.[44] Conditions in Saratov were also very poor, especially in the ravines which split the town and were occupied by the poor, and on the crowded and insanitary waterfront. A satirical poem indicated that in Saratov goats and pigs could be found in the streets, and 'the dirt is such/ That people were stuck in it up to their heads'![45] Smaller towns could be worse than large ones because there were few stone buildings or paved roads until the late nineteenth century. All Russian towns suffered from the same social problems that occurred in other major towns of Europe. Cheap taverns and eating houses led to drunkenness (as well as formal drinking fraternities at Kazan University); and it was estimated that there were 1,361 prostitutes in Kazan in 1852.[46]

Towns had separate quarters for non-Russians, and could look physically very different – although this might not have been obvious to the casual traveller, who only visited the central, Russian, part of the towns. Catherine II was the exception, because she insisted on visiting the Tatar quarter of Kazan when she travelled down the Volga in 1767. Filipp Vigel, a nobleman of Swedish extraction who served in the Foreign Ministry, visited Kazan in 1805 and commented that it was very like Moscow; but he recognized that he had only been in the ethnically Russian areas. A French traveller at the beginning of the nineteenth century remarked of Kazan that 'there is nothing Asiatic here, but only European'.[47] The two Tatar quarters, or suburbs, of Kazan (the old and the new Tatar quarter) were (and still are) very different in appearance from the classical centre of the town. Tatar buildings in their quarters were of wood, but painted in bright colours, with elaborate carvings and large, coloured gates. There were also mosques, of course, in the Tatar quarters, most of which had been built in the second half of the eighteenth century, after Catherine II cancelled the ban on their construction.

Kazan was a very distinctive Volga town, as was Astrakhan (see below); but other Volga towns could also have separate quarters for different ethnic groups. Chistopol, a small town in Kazan province, had a very distinct Tatar quarter, laid out between two major streets.[48] In 1805, following a report from the governor, Saratov established a special section of the town for 455 'homeless peasants' and 17 Roma.[49] Separate parts of Saratov were occupied in the late nineteenth century by German colonists. The Mennonite colony of Sarepta was located about 20km outside Tsaritsyn, but became a separate part of the city in the twentieth century, as the town expanded rapidly, and narrowly, down the banks of the Volga (and the Sarepta buildings which survived the battle of Stalingrad are now part of an open-air museum within the city boundaries).

The very rich lived similar lives, irrespective of ethnicity or religion. By the mid-nineteenth century, wealthy Tatar merchants in Kazan lived as comfortably as their Russian equivalents, and often decorated their homes with 'western' luxury goods, such as clocks and mirrors.[50] In the 1870s, a traveller found that the houses of the richer Tatar merchants 'were provokingly European-looking'.[51] Baron Haxthausen found the furniture in their houses 'European', but he noted the soft feather beds, which he associated with eastern style.[52] Edward Turnerelli found that in wealthy Tatar houses 'the sofas, chairs, and tables are placed around the room according to the European fashion'.[53] Many Tatars in Kazan, however, still favoured their own style of stoves in their houses, received guests in a special room, had eastern-style sofas and usually had fewer paintings on the wall (and those were possibly pictures of Medina and Mecca). Russian merchants were, in any event, slower to take to western dress than were Russian nobles, and were often perceived as 'eastern' by western travellers. In Rybinsk in the mid-nineteenth century, Baron Haxthausen found a merchant (almost certainly a Russian, because he was bearded) dressed in a kaftan, but noted that 'his house however was luxuriously furnished in the European style'.[54] Turnerelli noted that rich Tatar ladies in Kazan lived a life of luxury and idleness, in what he regarded as an oriental manner:

> A rich Tartar woman, hardly has she left her bed, when she begins her daily task of painting her face rouge and white; then she clothes herself in gaudy vest-ments of gold and silver texture, puts on her various ornaments, rings, necklace, bracelets etc; and this done, she throws herself on the soft Turkish sofa, on which she almost lies buried.[55]

❖

Towns began to evolve as cultural centres in the late eighteenth and nineteenth centuries. The first Volga town to adopt this role was Tver, which had a literary culture and salons from the eighteenth century. Alexander I's sister, Catherine, took part personally in these, and it is said that extracts from the first history of Russia, written by Nikolai Karamzin, were read out at such gatherings. In the 1840s, the town had a literary circle centring on the Russian composer Mikhail Glinka.

In the early nineteenth century, however, Kazan became the leading cultural centre of the Volga. There were several schools in Kazan in the eighteenth century: an ecclesiastical academy (although in 1723, of the 52 pupils at the school – it was said – 14 had fled, 9 had been sent home as too poor, 11 were considered too young, 6 had died and 2 had been expelled from the school; only 5 of the original contingent remained a year later!);[56] a school for baptized Tatars, Mari and Udmurts (established in 1735 for 30 pupils aged between 10 and 15 at a cost to the government of 2,000 roubles);[57] and a grammar school (the *gymnasium*), only the third to be established in the empire, after those in Moscow and St Petersburg.

Until the nineteenth century, the only universities in Russia were in Moscow and St Petersburg (St Petersburg University evolved from the Academy of Sciences, founded in 1724); within the empire, there were also universities in Warsaw and Dorpat (present-day Tartu in Estonia)). Alexander I founded two new universities, in Kazan and Kharkov (present-day Kharkiv in Ukraine), in 1804. Alexander did not explain why he chose Kazan to be the location for a new university. It could have been a conscious assertion of Russian cultural dominance over territories that were ethnically mixed, but Kazan was also simply a very important town, a nexus for east–west trade, and the gateway to the Urals and Siberia to the east. By 1839–40, 250 students had enrolled and the university rapidly became a cultural centre for the whole Volga region. The university stimulated the setting up of literary societies and bookshops, and from 1812 it sponsored a series of public lectures on history, science and technology (440 public lectures were given in the period 1858–61).[58] Students were drawn from other regions of the Volga, and from the Urals and Siberia. The university taught Arabic from 1829 and Mongolian from 1833, and published books in Arabic and Tatar: it had become a genuine cultural link between east and west.

Educational and cultural facilities also developed in other Volga towns. In 1786, Catherine II established a framework for national schools in the Russian Empire, including a 'major school' of four classes in each provincial town and 'minor schools' of two classes in all district towns. In theory at least, the Russian national schools were the most modern of their day, having drawn on the best and most up-to-date models in the Austrian Empire. They taught not only religion and the three Rs, but also geography, history, science and foreign languages. By 1792, major schools had been set up in Tver, Iaroslavl, Kostroma, Nizhnii Novgorod, Kazan, Saratov and Simbirsk, and minor schools in most districts. By this time there were over a hundred pupils in the schools in Tver, Iaroslavl, Kazan and Saratov, and 65 pupils in the less developed Simbirsk.[59] Schools developed more slowly in Samara, and in the late eighteenth century it was reported that 'there is no building for a school in Samara, and there are only 14 pupils'.[60] By the mid-century, however, the town had a grammar school with 53 pupils; that had grown to 424 pupils by 1875 (including 166 children of nobles and officers, 154 children of merchants and artisans, 54 from a clergy background and 50 peasants and 'others'). The number of pupils in Saratov elementary schools increased from 634 in 1865 to over 3,000 by 1881 and over 6,000 by 1905.[61] Education in Saratov, as in some Siberian towns, also benefited from exiles. Nikolai Ivanovich Kostomarov, for example, a distinguished historian, was exiled to Saratov in 1859 and taught at the town grammar school.[62] So, too, for that matter, did Nikolai Chernyshevsky, the son of a priest; he was born in Saratov in 1828, educated at the local seminary, and taught literature at the grammar school, before moving to St Petersburg and becoming the inspiration for revolutionaries through the publication of his novel *What Is to Be Done?* (1862).

It is worth noting in this respect that Catherine II and Alexander I were not 'Russifiers' in the modern sense of the word. Catherine, in fact, specified that the languages of the borderlands had to be taught in national schools – these included Tatar, Mongolian and Chinese. Kazan University was open, at least in principle, to all ethnic groups. In its first hundred years, the university educated 92 Tatar students.[63] In the writer Sergei Aksakov's autobiographical account of his education at Kazan grammar school at the beginning of the nineteenth century, he wrote that he was taught Russian literature and mathematics by Nikolai Ibrakhimov, whose name, Aksakov assumed, was either Tatar or Bashkir. It was only later in the nineteenth century that language and ethnicity became a sensitive issue associated with identity and loyalty to the throne.

Newspapers, literary journals and local history journals were published in major towns like Iaroslavl, Nizhnii Novgorod and Kazan in the nineteenth century. By the end of that century, literacy rates had reached over 50 per cent in major towns like Kazan, though they were lower in smaller towns – 34 per cent, for example, in Chistopol.[64] Clubs for nobles and merchants and for particular interests (for example, literary and geographical societies) cut across ethnicity (Saratov, for example, set up a literary society in the late nineteenth century which existed until 1918; the town also founded a museum dedicated to the writer and radical Aleksandr Radishchev in 1885). Both Tatars and Russians formed philanthropic societies in the second half of the nineteenth century, although these were often separate, if complementary, activities. Wealthy merchants made individual philanthropic donations. In Astrakhan, the Tetiushinov family, which had shipyards that constructed warships, participated in a 'list of donations for the poor at Easter'. Grigorii Tetiushinov also donated annual sums to a town library from 1858 to 1862.[65]

Catherine II consciously tried to involve the provincial nobility in the administration of the provinces, and as part of this process established noble 'assemblies' in the major towns. The main role of these was every three years to elect noble representatives to administrative and legal posts in the provinces, and to determine the eligibility of nobles to vote and to be entered in the 'book of nobility'; but this was also an attempt by Catherine to provide an opportunity for nobles to meet and socialize away from their estates, and, in the broader sense, to stimulate the development of a 'provincial culture' for nobles of middling wealth outside the grand noble houses in St Petersburg and Moscow. In the first instance, it proved hard to persuade impoverished nobles to attend assemblies – or for that matter even to find enough candidates to fill the elected posts in many provinces, especially those where there were few noble landowners, such as Astrakhan. During the nineteenth century, the elected officials from the assemblies began to take a greater interest in local affairs.[66] In 1812, noble assemblies throughout Russia collected donations for the war effort. The most important social side of the assemblies proved to be the ball, which took place after the elections. Assemblies could also make charitable donations: in 1805, the Nizhnii Novgorod assembly gave loans of between 400 and 5,000 roubles to impoverished members of the nobility.[67] Assembly attendance could be very low: in 1862 only 15 per cent of eligible voters took part in the elections in Tver province; in 1902, there were 1,275 noble landowners in Saratov province, and 600 of those were eligible to vote but only 225 bothered to do so.[68]

By the second half of the nineteenth century, more opportunities were available for nobles to socialize in major towns. The main assembly hall for the Kazan nobility could accommodate 800 people. The young Lev Tolstoy attended concerts, balls and plays in Kazan while he was a student there in the 1840s, and particularly admired the actor Aleksandr Martynov, who acted in Russian and foreign productions. Above all, the young Tolstoy enjoyed evenings of *tableaux vivants* – that is actors posing in scenes – which were held in the homes of the local nobility.[69]

The first provincial theatres in Russia were serf theatres of the local nobility, but a number of town theatres opened in the second half of the eighteenth century. In 1760, Mikhail Verevkin directed *School for Husbands* in Kazan and declared 'Molière is now known in Tartary'![70] By the mid-nineteenth century, large (stone) theatres had opened in Tver, Iaroslavl, Kostroma, Nizhnii Novgorod, Kazan, Samara and Saratov. The Samara town theatre still exists and is particularly fine. When Baron Haxthausen visited Nizhnii Novgorod in the 1850s, he noted that the performers were serfs,[71] but the theatre flourished in the 1860s and 1870s because it attracted a wide audience from merchant traders, of various nationalities, to its fair (which was situated nearby). Turnerelli noted that the theatre in Kazan attracted large audiences, although he was scathing about the actor taking the part of Hamlet (!), who, in his words, 'would have constituted an admirable ghost, but made a sorry one of the ghost's son'.[72]

There was some scepticism within the smart social sets of Moscow and St Petersburg about whether 'provincials' could understand the subtlety of theatrical performances. Apparently, the primarily merchant audience in Saratov burst into laughter at what were supposed to be tragic moments in performances; but the journalist who reported this suggested that it was as much to do with the exaggerated gestures and the poor acting of the performers as with the unsophistication of the audience.[73] The town duma subsidized the theatre to the tune of 2,500 roubles a year in the 1870s,[74] demonstrating that the urban elite recognized the value of the theatre. Volga towns could attract top performers: Franz Liszt played in concerts in Volga towns, as well as in Moscow and St Petersburg.[75] And the Volga towns produced their own 'stars', such as Liubov Nikulina, a serf actress and singer who rapidly acquired a reputation and performed in Aleksandr Ostrovsky's *The Storm* (see Chapter 11).[76] The theatre director Mikhail Lentovsky was born in Saratov in 1843, and became a well-known actor who performed there and in Kazan before moving to Moscow, where he established the Hermitage garden theatre. He also set up a drama theatre in Nizhnii Novgorod and went on to organize the official celebrations in Moscow for the accession of Alexander III.[77] In addition to formal theatres, itinerant groups of actors travelled down the Volga by steamer and performed at various venues. In the nineteenth century, I.I. Lavrov took a company of actors from Nizhnii Novgorod down to Tsaritsyn.[78] In the early twentieth century, a company under the direction of Sergei Kusevitsky presented classic plays at various Volga ports.[79]

In towns, as we have seen, there were separate quarters for different ethnic groups. Despite this physical segregation, it was more difficult than in the countryside for non-Russians to live completely separate lives. There was an urban culture in Russia by the mid-nineteenth century which could often, if not always, override differences in ethnicity and religion. Public theatres and concert halls were open to the educated elite of all ethnic and religious groups. In popular culture, street theatre and public festivals provided open-air entertainment for all the urban inhabitants, irrespective of ethnicity or social class. It was said of Kazan that urban classes 'promenaded' around the Black Lake: army officers, officials, merchants, students, fashionable ladies.[80] Chess was popular with Russians and Tatars and cut across ethnicity and social class. It was easier for people to mix in the towns than between separate villages. Tatars could join in the open-air entertainments, such as swings and sliding down ice hills, at *Sviatki* (Christmastide) and Lent. In Kazan, Russians took part in the Tatar *sabantuy* celebration on Lake Kaban. The village came to the town, although the organized fights between villages during rural festivals easily descended into street fights in towns between different social and ethnic groups.

❖

So far, the town that has not been discussed is Astrakhan. In many ways, it shared the characteristics of towns on the middle and upper Volga. But unlike other Volga towns, Astrakhan was recognized as a real 'border town' between the Russian Empire and lands to the south and east: Persia, Central Asia and the Indian sub-continent. It was the emporium for goods that came from across the Caspian Sea and was a major nexus in the trade from east to west. That trade is discussed in the next chapter; the focus in this chapter is on the unique and extraordinary ethnic composition of the Astrakhan trading community and its significance for urban life.

Astrakhan was a trading emporium, and its ethnic composition reflected this. As well as Russian merchants, who were in a majority, distinct ethnic communities of Armenians, Indians, Tatars and Persians grew up from the seventeenth century. At the end of that century, it was noted that 'There are divers Nations inhabiting here, drawn by the Conveniency of Trade, and the sweetness of the Air.'[81] There are 'Orientals of many kinds', remarked a British traveller rather more bluntly in the mid-nineteenth century.[82] Another commented that in the town 'there is perhaps a more diversified assemblage of nations, than on any other spot on the globe'.[83] In the 1740s, non-Russians, or 'Asiatics', comprised some 25–30 per cent of the population of the town – there were some 776 male Armenians, 76 Indians and 109 so-called Agryzhans (children of Muslim Indian fathers and Tatar mothers).[84] Different ethnic groups lived separately in the town, although trade could be conducted in many languages. Astrakhan was divided into quarters for different ethnic groups: one quarter for Tatars and Indians, one for Persians and some Tatars, and one for Central Asians. In addition to mosques, there were Armenian churches and a Hindu temple.

After the Russians, the Armenian trading community was the largest, and many Armenians had grown very wealthy.[85] The first Armenian church was opened in Astrakhan in 1669.[86] In 1779, one estimate was that there were nearly 2,000 Armenians in the town (possibly this referred to male Armenians only), many of whom lived in 'rich stone houses'.[87] By 1867, the Armenian community numbered some 5,000.[88] Armenians dominated the silk trade and were important suppliers of credit.[89] They wore distinctive clothing, and were said to dress 'in the German

171

way'. A community of Indians had existed in Astrakhan from the mid-seventeenth century, when they established their own quarter, or 'court'. They came mainly from the Punjab, Sind and Afghanistan, dominated the transit trade of goods up the river Volga and acted as creditors. Indian traders were particularly active in the 1670s and 1680s – to the extent that Russian merchants complained in 1684 that 'Indians [live] . . . in Moscow and Astrakhan for many years without leaving and falsely call themselves Astrakhan inhabitants although they have wives and children in India.'[90] Numbers dropped after this complaint, but Indians remained a significant community.[91] 'Armenians & Indians are the principal Shopkeepers in *Astracan*', commented a traveller in the 1730s.[92] By 1702, some 260 Tatars lived in the town, and their number increased during the eighteenth century.[93] So-called 'Bukharans' were traders from Central Asia who brought goods from the east to Russia through Astrakhan. In the late 1770s, it was estimated that there were also some 400 Persians in Astrakhan.[94]

Other traders were attracted to the town on a temporary basis. In the sixteenth century, there was even a British community of traders: a major British expedition to Persia had sailed down the Volga to Astrakhan in 1579, and the group stayed there for six months, exploring the possibilities for trade, before moving on to Persia with their goods. The British continued to have a presence in Astrakhan until their right to trade with Persia was cancelled in 1584. A French account of Astrakhan, written in 1606, noted that during the lifetime of Ivan IV 'the English traded there, and from there into Persia'.[95] By the nineteenth century, German traders had established shops in the town. Jews settled there in the nineteenth century, but were expelled from the town in 1835.[96]

Non-Russians in Astrakhan were granted special privileges, which created a constant conflict with Russian merchants. The right to trade freely was granted and confirmed at several points in the eighteenth century for Armenians, Indians and Bukharans, and they were also allowed to become subjects of the Russian Empire and to register with the Russian merchantry; in addition, they were freed from some onerous duties, such as various fiscal dues, until 1836.[97] Persians had the same privileges, but were only allowed to become temporary subjects of the tsar.[98] In the eighteenth century, different ethnic groups – including Armenians, Tatars, Bukharans and Agryzhans – were also allowed to have their own courts to deal with minor crimes and conflicts. Armenians had their own loans bank, set up in 1776,[99] and in 1804 an Armenian school was opened at the initiative of a wealthy Armenian merchant: the success of the school fluctuated, but it was still in operation in 1917.[100]

1. The Bolgar archaeological and historical complex became a UNESCO World Heritage Site in 2014. The Bolgar state is the subject of Chapter 1. The site is significant not only for its archaeological remains and excellent museum but also for the evolution of present-day Volga Tatar identity, as discussed in Chapter 16.

2. Tver (Kalinin in the Soviet period) is the main town on the northern stretch of the Volga. The principality of Tver challenged the dominance of Moscow in the fourteenth century, and is discussed in Chapter 2. In the eighteenth century, Tver became a provincial town of moderate economic significance.

3. The Tatars were the largest ethnic group that the Russians encountered after they had established control over the whole length of the Volga through the conquests of Kazan and Astrakhan in 1552 and 1556. Their interactions with ethnic Russians, from the sixteenth century to the present, and the evolution of a distinctive Volga Tatar identity, are the subject of several chapters in this book, but feature particularly in Chapters 3, 6, 12 and 16.

4. The Chuvash are an ethnic Turkic people who settled mainly in the middle Volga, in the region which is now the Chuvash Autonmous Republic within the Russian Federation, with its capital in Cheboksary. They became subjects of the Russian Empire after the conquest of Kazan. The majority of the Chuvash people had converted to Russian Orthodoxy from animist religions by the eighteenth century. The Chuvash are discussed in particular in Chapters 3, 6 and 16.

5. The lines of fortifications were constructed by the Russian government in the seventeenth and eighteenth centuries to control the newly acquired territories on the Volga. They are discussed in Chapter 3. This fort is part of a historical complex in the old part of Ulianovsk (called Simbirsk until 1924).

6. The majority of the inhabitants of the Volga region before the twentieth century were peasants. This is a typical hamlet on the river where the way of life changed little until the Revolution. Peasant life – of Russians and non-Russians – is discussed in Chapter 8.

7. Astrakhan is the southernmost town on the Volga. It was an emporium for trade in the seventeenth century and was home to merchants of many ethnicities, including Armenians, Persians, Bukharans and Indians. The town and its diverse population are discussed in Chapter 9.

8. Saratov was originally founded as a fortress, probably in 1590. It developed into an important port in the nineteenth century and was the location of the administrative office for the Volga Germans, who had been settled on both sides of the river in the 1760s. Volga towns in the imperial period are discussed in Chapters 9 and 10.

9. The Nizhnii Novgorod fair was very important in the late nineteenth century and it attracted merchants and goods from all over Russia, Central Asia, China and Europe. The fair is discussed in Chapter 10.

10. The Chernetsov brothers were commissioned by Nicholas I in 1838 to travel down the Volga to paint scenes which were then displayed in a special panorama. This is one of the first paintings of the river, and their voyage is discussed in Chapter 11.

11. Ivan Shishkin is best known for his paintings of forest scenes in Russia, but in this painting he is following the European Romantic tradition in depicting the ruins of Bolgar (paintings of ruins in Italy and Greece were popular at the time). His paintings are discussed in Chapter 11.

12. This is the best-known realist painting of the Volga which depicts the suffering of the barge haulers. The painting and its reception are discussed in Chapter 11.

13. Isaak Levitan painted a number of scenes on the Volga based mainly around the artist's colony of Ples (or Plyos), often featuring churches and monasteries on the river. His painting is discussed in Chapter 11.

14. The source of the river Volga is in the Valdai hills, north-west of Moscow. It has been sacralized and blessed by the patriarchs of Moscow. The development of the site from the late nineteenth century is discussed in Chapter 11.

15. Ulianovsk was the home town of Lenin (family name Ulianov), and in the 1960s the core of the administrative centre was destroyed to build new memorials to the town's most famous son. Since the collapse of the Soviet Union the town has had to re-invent itself as the home of several significant writers as well as the birthplace of Lenin. This memorial features in Chapter 11; the reconstruction of Ulianovsk is discussed in Chapter 15.

16. The Volga Germans were invited to settle in the Volga region by Catherine II in the 1760s. They suffered from famine in 1921–22 and from the excesses of collectivization in the 1930s, as discussed in Chapter 14. They were forcibly deported to the east by Stalin in 1941 during the Second Word War; very few have returned to the Volga region.

17. Workers can be seen on the shoulders and head of the memorial, illustrating its size. This statue is only one part of the memorials to the battle of Stalingrad, 1942–43, as discussed in Chapter 15.

18. This is part of a bizarre modern complex in the capital of the Mari El Republic constructed in Flemish style. It is discussed in Chapter 16.

19. Power stations now control the flow of the river Volga from Rybinsk in the north to south of Volgograd, and are discussed in Chapter 17. The Zhiguli (formerly called Kuibyshev) Hydroelectic Station was constructed in 1950–57 and is one of the largest power stations on the river.

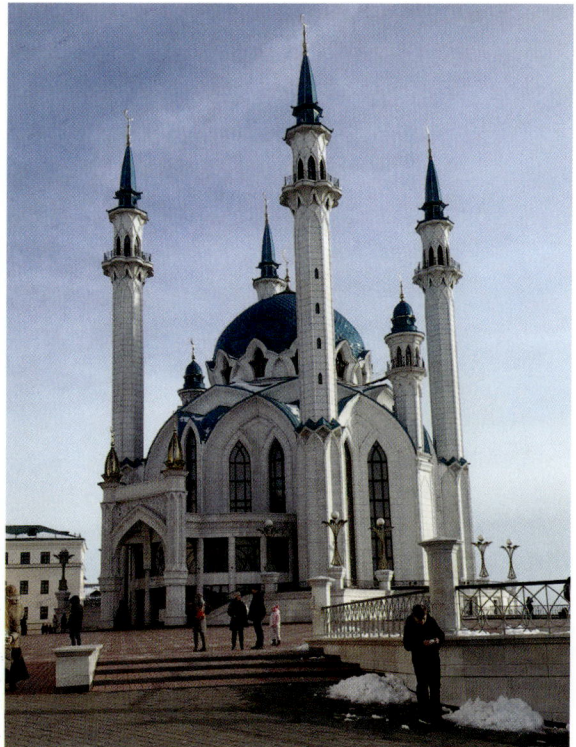

20. The kremlin was traditionally the symbol of Russian imperial power, and normally included an Orthodox cathedral along with the main provincial administrative offices. The Kul Sharif mosque was built in the sixteenth century but it was destroyed when Ivan IV conquered Kazan. It is particularly symbolic therefore that the mosque has been rebuilt since the collapse of the Soviet Union. It opened in 2005.

21. This new road bridge, called the President Bridge (named after Dmitrii Medvedev, who opened it), replaced the old bridge which had been constructed in 1913–16. It opened in 2009 as part of a modernization programme after the collapse of the Soviet Union. It is the second longest bridge in Russia and one of the longest in Europe.

22. Kazan was a Tatar town when it was conquered by Ivan IV in 1552. Tatars were then expelled and the mosques were destroyed, but Tatar quarters were re-established in the eighteenth century, at some distance from the kremlin, and a significant Tatar merchant and intellectual class evolved. Today the population of Kazan is roughly 50:50 Russian and Tatar. The restored houses and the restaurants of the Tatar quarter are popular with tourists and locals.

Separate institutions could be a curse as well as a blessing – as we saw above with the Tatar *ratusha* in Kazan. In 1799, an Armenian merchant complained about the slowness of judicial processes in the 'Asiatic' court in Astrakhan. The issue concerned an alleged concealment of silk, and the case had languished for two years in the Asiatic court, during which time, the merchant claimed, he had lost over 8,000 roubles and had been 'ruined'.[101] One of the weaknesses of the Russian system of courts was that it assumed that litigants would be of the same social and ethnic group. The reality was far more complex, as we saw above for Kazan. In Astrakhan, it was estimated that in 1798 more court cases in the Asiatic court involved Russians and non-Russians than litigants of the same ethnic group. The solution, with typical clumsy imperial logic, was to hear cases involving Russians and non-Russians in Russian courts, and Armenians with Armenians in an Armenian court, while special courts were set up for 'others', including Persians, Indians and Bukharans.[102]

The ethnic diversity of Astrakhan gave the town a richness and vitality that surpassed even other Volga towns:

> There is the stately Persian, and the lively Russian cadet; busy merchants, both European and Asiatic; Tartar nondescripts, German shopkeepers; a ponderous Bukharan, on one side of the street, as it were counterbalancing half a dozen officers on the other; soldiers of the line in their various costume, some wearing helmets, some sheepskin caps with red and gold striped crowns; Cossacks hurrying about on their nimble horses; with their lances vibrating and their pennants streaming out behind; the insinuating Armenian, and the sly-looking Jew both gliding about with a noiseless step.[103]

The attractiveness of Astrakhan and other towns to foreigners, and the basis for the diversity of the population, was trade, in which the river Volga played a fundamental part. It is to this that we now turn.

10
LIFE AND TRADE ON THE RIVER

Katherine Guthrie, an English traveller, remarked in the 1870s, during her travels through Russia to Persia, on the economic importance of the river Volga:

> . . . proud river – it bears upon its bosom cottons, machinery, and ships from England; fish, oil, and furs from the White Sea; ores and marbles from Siberia, rich bales from India and Persia; wine, silk, fruit and hemp from the Caspian – its own stream and shores yielding envious quantities of salted fish, caviar, hides, tallow, and bone manure, which is largely exported.[1]

The Volga, together with its tributaries, was the key artery for trade in Russia: it was estimated that a third of the river trade was conducted along the Volga in the nineteenth century; in 1895, over 50 per cent of the steamboats in Russia – 1,329 out of 2,539 – were to be found on the Volga (and 3,549 barges).[2] The river also stimulated exchange at towns on its course. This chapter looks at the nature and value of this trade, how it was conducted, the extent to which river trade was changed by new technology and by railways, and the people who worked on the river.

✿

'The Volga is, in one respect, the most wonderful river in the world; for it is navigated from its very source to its termination', enthused an English army officer travelling in the 1880s.[3] It is true that only the first 80km or so from the source of the Volga in the Valdai hills are non-navigable; and there are also very few rapids – and all those are located on the upper reaches. By the time the towns of Rzhev and Staritsa are reached on the upper Volga the river is already navigable; in the town of Tver, where the Volga is joined by the river Tvertsa, it is a substantial waterway.

That does not mean that the river was, or is, easily navigable: 'Of all the navigable rivers of Europe, there is probably none so uncertain and difficult of navigation as the Volga', remarked Laurence Oliphant in 1852.[4] The river is frozen for much of the year. It opens for navigation in late March or early April and closes again in late October or early November. Particularly harsh winters can mean the ice melts later or forms earlier, making planning difficult for ship-owners. The spring could also bring serious floods, with the water rising as much as 10 or 15 metres. Peter Simon Pallas took a scientific voyage down the Volga at the end of the eighteenth century and noted that floods of up to 12 metres had occurred in 1772, 1773 and 1798.[5] John Perry (employed by Peter I to investigate the possibility of constructing a canal from the Volga to the Don) witnessed that floods in Astrakhan 'towards the latter end of *April*, continue to a considerable Height above two Months'. He believed that:

> The Quantity of Waters discharged throughout the Year by the *Wolga* alone, may be at least reckon'd one sixth Part more, or 445522 Tons in a Minute. And considering the Number of other Rivers, that from all Sides fall into the said Sea [the Caspian], and that some of them may be reckon'd to run through well nigh as great an Extent of Country as the River *Wolga*; all the Waters and Floods that are discharged into the said Sea, I believe may, by a modest Computation, be reckon'd at least three times as much as discharged by the River *Wolga* alone, or 1336566 Tons of Water in a Minute . . .[6]

The Volga broadens out at various points into lakes, which can have their own weather systems, with strong winds and violent storms that can affect shipping even today.

The main problem for navigation, however, was not the depth and extent of the water, but its shallowness, particularly in July, when the level could fall as low as 1 metre. Adam Olearius travelled on the Volga in the 1630s, when his ship was grounded south of Cheboksary, on the right bank of the Volga. He remarked that:

> . . . the Volga is everywhere so shallow that we could barely get over the shoals. For this reason, in the course of this and the succeeding days, we had many difficulties, and annoyances in passing, and on the 10th we advanced only a half league. Aboard the ship was heard nothing but 'Pull! Row! Back!'[7]

Hazards in low water included shoals and sandbanks (both visible and submerged). It was estimated that there were as many as 35 shoals between Rybinsk and Saratov.

The role of pilots (discussed below) was crucial to the success of a voyage; but even experienced pilots could be deceived by shifting sandbanks, and barges could easily run aground. Barges often progressed by a system of 'kedging', whereby a smaller boat took an anchor a distance further up the river and dropped it. The barge would then be hauled by ropes up to that anchor, either pulled by men on the banks or wound round a windlass; anchors could get caught up in the weeds and undergrowth by the banks. Olearius's progress was hampered in the event described above when the anchor got stuck in this way. Low water levels can, even today, restrict shipping, despite the massive changes wrought by dams and lock systems in the twentieth century.

The role of barge haulers is discussed below, but their most gruelling work was to pull boats off the sandbanks when they ran aground. In 1826, a whole group of haulers abandoned their ship when it ran aground on a sandbank near the small town of Puchezh on the upper Volga (a town which was flooded by a reservoir in 1952). Water levels on the Volga were particularly low in 1826 and goods had to be left on the banks – some 200,000 sacks of grain were held between Nizhnii Novgorod and Rybinsk, and could only be moved when water levels rose later in the summer.[8] Sometimes goods simply had to be removed from large barges that had run aground and transferred to smaller vessels – a process that was time-consuming and onerous, and that added to the cost, and duration, of the journey.

These problems meant that the journey time transporting goods along the Volga could be painfully protracted. This was particularly true of the upward journey, against the current; and it should be pointed out that most goods were carried upstream rather than downstream. When, in 1742, the Russian imperial court demanded that (perishable) grapes and peaches should be sent to Moscow from Astrakhan, it was suggested that post horses should be used to complete the journey more quickly.[9] Large barges carrying heavy goods moved upstream very slowly, at between 4 and 11km an hour in good conditions. When the boat had to be hauled by men, the progress could be 11km *a day* or less – and that was assuming that the men would work for 10 hours a day.[10] In 1794, it was estimated that it took 53 days to get from Kazan to Rybinsk. In the 1830s, it could still take two or three months to travel from Astrakhan to Nizhnii Novgorod. In order to complete the journey, there were designated points where haulers were changed (the equivalent of post stops on major roads in the empire), spaced at roughly every 80km: there were 3 such points between Rybinsk and Iaroslavl, 13 between Kostroma and Nizhnii Novgorod, 15 between Nizhnii Novgorod and Kazan and 80 between Kazan and Astrakhan.[11] The downstream journey was often used by passengers, especially after steamships were introduced; by the second half of the nineteenth century, the whole length could be accomplished in a matter of five to seven weeks.

❖

The Volga was seen as the source of prosperity for the whole country, as this account, written in the 1830s of the town of Uglich, on the upper reaches of the river, illustrates:

> Volga-mother Russian river . . . from the Baltic Sea to the Ural mountains, from the northern ocean to the Caspian Sea – the Volga is a benefactor to spacious Russia, feeds the Russian Kingdom, and the products of its climate are handed to foreign powers. What Russian heart does not care, seeing every day the increase in industry and trade of Russia! The Volga unites almost all the seas of Russia through the means of canals; all the Volga towns are flourishing, are wealthy, are getting larger and more embellished, being indebted for their prosperity to mother Volga. The Volga – such majesty, such loveliness, crowned with a multitude of boats small and large! Such an enchanting reflection in its crystal waters of Uglich![12]

All the way up the river, the Volga towns and the Volga region played a crucial role in this prosperity. Astrakhan was the undisputed centre of the fishing industry in Russia (and it remains of significance today, despite the pollution of the river). The variety and extent of the fishing trade made an impression on foreign travellers. In the eighteenth century, John Bell, a Scottish doctor and traveller, commented on the fish in the Volga that 'no river in the world can afford greater variety, better of their kind, or in larger quantities'.[13] By the late nineteenth century, Astrakhan could be described as a 'huge and gilden fish-market'.[14] The fish-market hall in the town today is still impressive in the amount and variety of fish on sale. The importance of the trade, not only for Astrakhan but also for the empire as a whole, resulted in the state intervening to ensure that there was enough manpower to catch and salt the fish. In 1744, for example, the Senate decreed that the town was allowed to use exiles and convicts for that purpose, and could even employ garrison troops if it needed more men.[15]

Goods were predominantly transported *up* the river Volga, south to north. As well as fish, Astrakhan was a key point for the transit trade from Persia across the Caspian Sea and for goods from Central Asia. These included silk from Persia, a trade dominated by Armenian merchants who had secured rights to export in 1667;[16] and cloth, furs, leather, spices, nuts, wine, peppers and other goods from the east, a trade dominated by Indian merchants. Some Armenian and Indian traders became fabulously wealthy as a result of this trade. In the 1760s, the Russian government attempted to set up a Persia Company, in part to break this monopoly and give Russian merchants greater access to the trade; but the company never functioned as well as had been hoped, and trade with Persia and the east via Astrakhan declined later in the century.[17] Nevertheless, Astrakhan still accommodated some 300 ships a year at the beginning of the nineteenth century, by which time its trade was worth some 3.5 million roubles annually.[18]

The goods from Astrakhan were joined at Kamyshin, on the right bank of the Volga (currently in Volgograd province), south of Saratov, by barges carrying salt from Lake Elton. The value of salt was so great that Cossacks from the Astrakhan Cossack host were used as guards for the long convoys of salt brought from Lake Elton overland by oxen.[19] By the mid-eighteenth century, the Stroganov family produced over 4 million *pudy* (over 65,000 tonnes) of salt a year from Lake Elton; by 1900, some 12–15 million *pudy* (196,000 to 245,000 tonnes) were being extracted annually.[20] From Saratov to Kazan, the products carried by barges on the river were largely agricultural, comprising mainly wheat and other grains from the black-earth region that borders Saratov province. After 1860, Saratov also grew tobacco, particularly in German colonies, and that was also exported upstream. Samara developed early as a main depot for grain, as its position on the eastern bend of the river gave it access to the grain products of Orenburg province in the east, as well as grain from the south. In the 1760s, it was said of Samara that '20 *versty* [some 21km] from Samara you can find everywhere a great steppe of black earth, on which the grass grows to the height of a man'.[21] By the mid-nineteenth century, the town had many granaries and windmills, and was described by travellers as one of the 'busiest ports on the Volga',[22] and as 'one of the most flourishing places on the Volga'.[23]

The river Kama enters the Volga below Kazan. It carried a significant amount of goods from the east (ores from Siberia and silks, cloth, spices and medicinal rhubarb from China) and timber from the north of Russia. Those goods, in contrast to the products discussed so far in this chapter, went *south* down the Volga, as well as west and north to Moscow and St Petersburg. Kazan itself, as we saw in the previous chapter, was an important trading centre for Russian and Tatar merchants, and handled grain and timber products, as well as fish and other foodstuffs. In 1811, over 2,000 vessels visited Kazan, with freight worth over 25 million roubles going upstream and goods worth over 6 million roubles going downstream.[24] More agricultural products joined the river between Kazan and Nizhnii Novgorod – mainly rye, oats, tallow and timber.

The river Oka joins the Volga at Nizhnii Novgorod, bringing more commodities – in particular, timber. In the nineteenth century, the town fair became the great centre for the purchase and exchange of goods, from east to west and south to north, as we shall see. Even in 1796, before the fair was active, trade at Nizhnii Novgorod was valued at over 30 million roubles.[25] By 1861, over 3,000 ships a year docked at Nizhnii Novgorod, with a freight value of over 25 million roubles.[26] More grain products were transported from provinces further up the river, with the addition of manufactured goods from the towns of Iaroslavl and Kostroma, including paper, textiles and leather. In the early eighteenth century, 57 families in Iaroslavl had ship-building yards,[27] and ship-building also took place in the upper Volga towns of Gorodets, Uglich, Nizhnii Novgorod, Balakhna and Rybinsk.

Rybinsk received some 5,000 ships per year at the end of the eighteenth and the beginning of the nineteenth centuries, and over 7,000 by the early 1830s.[28] By the mid-nineteenth century, the town appeared as 'a complete forest of masts' and 'in summer the craft of various kinds are so numerous, that a person may pass from vessel to vessel, on planks, from one bank to another'.[29] From there, goods were offloaded onto smaller vessels, in order to reach St Petersburg, Moscow and Tver via a network of canals and rivers. There were three canal systems via which goods could be transported from Rybinsk to St Petersburg: the first was through the Vyshnii Volochok lock system, built in the reign of Peter I, and from there via the Ladoga canal and the river Neva, a journey of 1,273km; the second was through the Mariinskii canal system, constructed in 1810–14, through the Sheksna river and Lake Onega, a journey of 1,167km; and the third was through the Tikhvin system of locks, opened in 1811, the shortest journey (909km) but on narrower canals, which meant it was only suitable for small vessels.

Rybinsk and the canal system were important because they meant that the population of St Petersburg could be supplied with the grain and other agricultural products that had been shipped up the Volga. The port of St Petersburg was also used to export goods abroad – grain, timber, dried fish and silk from Astrakhan, rhubarb from China and ores from Siberia. The very existence of St Petersburg, however, as well as the need for supplies and the export facilities, resulted from the deliberate actions of a tsar, Peter I, and demonstrated how the policies of the tsars and the Russian government impacted on the trade of the river Volga. Peter founded St Petersburg in 1703, during the Great Northern War, on territory held at the time by Sweden. Peter was determined that his new, westward-looking capital on the Baltic would be a great city, and he forced nobles and merchants to relocate there (and briefly forbade export of hemp from the port of Archangel to boost exports from St Petersburg). The population of St Petersburg grew rapidly – to more than 200,000 by 1800 and 1,265,000 by 1897. The city was therefore in many respects an artificial creation; but its main problem was that the population could not be fed from the produce of the surrounding countryside, which was not grain-producing, and so relied on food imports from the rest of Russia, and particularly from the regions served by the river Volga.[30]

The whole canal system from Rybinsk to St Petersburg was a means to supply the city. The Volga was so important in this supply chain that it came under government regulations. It was vital that its trade should not be impeded, and this led to a number of regulations – from the construction of wharfs to (eventually) conditions of labour on the river. Catherine II placed the whole river network under the Admiralty College (Ministry), and divided all the rivers in the Russian Empire into four 'regions'. She instructed that all of the 'regions' had to be inspected regularly; but the river Volga was so important, and so long, that it cut across three of the four regions.[31] The importance of the Volga trade to St Petersburg can be gauged from the correspondence between the governor of Iaroslavl province, A.P. Melgunov, and Catherine II in the 1760s. The governor sent the empress regular, and personal, accounts of the number of ships and details of all their cargoes, including all the grain products, which passed through Iaroslavl bound for St Petersburg.[32]

The canal construction from Rybinsk to St Petersburg was an example of government policy to control the supply chain. There were further government plans in the eighteenth and nineteenth centuries to extend the canal system south, and to revive Peter I's plans to link the river Don and the river Volga; but physical and technical difficulties meant that this was not achieved until the 1950s. The route from the Volga to the White Sea was improved by the so-called Württemberg system in the 1820s, by which a series of rivers and lakes linked the Volga, via the river Sheksna, to Archangel in the north. Further plans to link the Volga directly with Moscow were made, but shelved.[33] The only major construction in the Volga region itself in the imperial period was the canal system, which was established within the town of Astrakhan, and which protected it from flooding.

The freight vessels on the Volga varied according to the goods carried and the stretch of the river. The waterway was used by a 'curious collection of river craft on the water, from the huge Oriental-looking Astrakhan barges to the little canoes which ply across the river', according to one traveller in the mid-nineteenth century.[34] The many tributaries that joined the Volga also had their own types of boats, which were particularly suited to their conditions and cargoes. Before the introduction of steamships – and to a large extent even afterwards – goods were carried on barges. By the mid-nineteenth century, some 21,500 such craft travelled from Astrakhan to Rybinsk every year.[35]

The largest barges, called *beliana*, were used on the lower Volga, where the water was deepest – and also on the river Kama. They could carry over 1,000 tonnes of bulky goods, such as timber or salt. Their size varied, but they could be up to 100 metres long, with a width of up to 26 metres and a draught of over 4 metres. The most common barge on the lower and middle stretches of the Volga and the Caspian Sea were the so-called *rasshiva*, also large flat-bottomed sailing barges but not as large as *beliana*. There were a thousand or so of these barges on the Volga in the 1840s. They varied in size, but could be 30–50 metres long, 12 metres wide and had a shallow draught of 2 metres or less fully laden, and a sail, and could carry almost 300 tonnes of goods.[36] The largest barges required enormous teams of haulers – up to 300 men – mainly to tow them off sandbanks; even the smaller ones required some 20–50 men and a number of pilots. Some of the barges were used for 7–10 years, but many were so poorly constructed that their effectiveness quickly diminished; they were often only used for one journey upstream, after which they were abandoned on the banks of the river to rot.[37]

Steamboats were slow to develop on the Volga (compared with the Mississippi, for instance; indeed, Russian steamships were modelled on Mississippi steamboats).[38] The first steamship in Russia was built by Charles Blair, a Scot, in 1815. The first steamship on the Volga was constructed by V.A. Vsevolozhsky in 1817; by the 1820s, D.P. Evreinov owned and ran a number of Volga steamships. The early vessels experienced a number of problems: it was difficult to power them upstream against the current; the depth of their draughts made them vulnerable to shoals and sandbanks, particularly on the river above Kazan; the engines were unreliable; and there was a shortage of mechanics capable of servicing them.

By the 1840s, steamboats had become more reliable and were constructed with shallower draughts. One such was the steamship *Sokol*, which made the journey from Nizhnii Novgorod to Astrakhan in eight to nine days. By this time, there was more encouragement from the Russian government to form private companies of steamship owners on the Volga. A consortium led by the British entrepreneur Edward Cayley formed the Volga Steam Navigation Company in 1843.[39] In 1849, the Mercury Company was formed under mainly Russian ownership – or, as a British commentator stated disparagingly, it was 'managed, or rather very much mismanaged, by Russians'.[40] By 1852, it was running seven steamboats on the river, had established a passenger service and carried traders of many different nationalities, as this description from a British traveller noted:

> There were Russians from nearly all the Eastern Governments, Tartars from Kazan, Germans from below Saratov, Persians and Armenians from Astrakhan, Bukharans and Bukharan Jews from Orenburg, and one or two people on their way to Tashkend [Tashkent].[41]

By the 1850s, steamships had established themselves successfully for both freight and passenger services, although they were still unable to function when water levels were too low. They also experienced problems with quality of construction and maintenance. In 1849, for example, the boiler of the steamship *Oka*, sailing on the Kama and then on the Volga to Rybinsk, exploded and killed four crew members. In 1856, the *Iaroslavl* hit rocks and sank.[42] Trade was disrupted by the outbreak of the Crimean War in 1854, but in 1856 the Volga Company had an annual income of 390,327 roubles and carried over 40,000 tonnes of goods.[43] The number of steamships continued to rise in the last decades of the nineteenth century, as river trade increased. By 1862, there were 15 shipping companies, owning 150 steamships, based in Nizhnii Novgorod alone,[44] and the number of ships, and the extent of their freight, continued to rise in the 1880s and 1890s.

Steamers could never fully replace barges for the transportation of all types of goods. Steamships carried more valuable goods, such as tea, wheat and metal products, while the bulkiest and heaviest cargoes – such as timber and less valuable grains – were transported by barge. In the 1870s, a British traveller remarked on the painted barges on the river and the carved 'eye' on the boats in black and white – a protection against evil.[45] There are large barges on the river Volga even today, used to transport timber and bulky goods, sometimes with the assistance of tugs.

The first railway in Russia was built in 1837 (linking St Petersburg to the summer palace at Tsarskoe Selo) and the St Petersburg–Moscow line opened in 1842. At first, railways did not impact on the Volga trade, but once Moscow was linked to St Petersburg by rail, the canal system at Rybinsk came to be of less significance, at least for the less bulky goods. More importantly, railway lines began to link towns across the empire with Moscow. The new network benefited the grain trade and enabled landowners (and peasants) to transport their grain more easily and rapidly to the towns of Moscow and St Petersburg, and to export grain abroad. Railways moved goods much faster than steamships (some four or five times as rapidly) and were far quicker than traditional barges, so potentially putting this trade at risk.[46]

In fact, river transport survived the immediate impact of railways, and the Volga region as a whole benefited from the new opportunities to export grain. The railways, however, did not lead to economic success spreading evenly across the Volga region. The most significant *geographical* impact of the railways was to divide the Volga grain trade west and east (even more so than the breadth of the river). The main railway line for grain ran from Riga to Tsaritsyn and opened in 1871. It made Tsaritsyn a key centre of the grain trade, both for the route from south to north and as a key link from the south-east to the north, by transporting goods from the Don region to Moscow. Further railway lines opened on the western side of the river Volga and resulted in Saratov also becoming a major centre for grain exports. The eastern side was cut off from this trade simply because of the difficulty of building bridges across the river.

The first bridge across the middle Volga – and for several decades the *only* bridge – was opened in 1880 and crossed the river at Syzran. This helped the development of Samara, which could now be reached by rail and road from Syzran, and the town became a key transit point for freight from the east (grain from Orenburg province and goods from Siberia) being transported to Moscow. In a guidebook to the river Volga in 1914, it was stated that in the period 1904–07 the town exported twice the amount of grain by rail as by boat (over 46 million *pudy* or 750 tonnes, as opposed to 22 million *pudy* or 360 tonnes).[47] Other bridges across the river were constructed much later: in 1912 in Kazan, which by that time had been side-lined as the main link to Moscow, but had developed railway links with the north, thanks to the opening of the Kazan–Viatka line; in 1913 in Iaroslavl and Sviiazhsk; in 1915 in Simbirsk, which by then was already in decline because it was not on the main railway line.[48] By the beginning of the twentieth century, the railways were beginning to dominate trade: by 1908–09, they were carrying three times as much grain as the waterways of Russia.[49]

❖

Who worked on the river Volga? From Rybinsk, horses were used (as many as 10 animals per barge) to pull the boats, but below Rybinsk boats were hauled by men, called *burlaki* in Russian, and immortalized in the painting by Ilia Repin (discussed in the next chapter). The number of haulers was a matter of dispute among Soviet historians. A major study was conducted by the historian Fedor

Rodin in the 1970s. He estimated that by the end of the sixteenth century there were already 54,000 haulers working on the Volga, and that this number had risen to 100,000 by the end of the seventeenth century; to 340,000 by the end of the eighteenth century; to some 600,000 by the 1840s; and to over 700,000 by the 1870s.[50] Another historian of Russian river transport, Enessa Istomina, challenged these figures, suggesting that there were under 200,000 haulers by the end of the eighteenth century and 500,000 by the mid-nineteenth century.[51]

There is no dispute, however, about the harsh nature of the work, as boats had to be hauled up against the current or off sandbanks – although it may be that the need for so many haulers resulted in slightly better conditions on the Volga than on other rivers in Russia. There was a popular saying among haulers that 'The Volga is at some times a mother; at other times a stepmother.' The famous painting by Repin shows men with rough harnesses across their chests, straining with the physical challenge of the work and, with one exception, bent and bowed down as they tow the barge. Witnesses commented on the physical strain that was put on haulers – chest pains, exhaustion, profuse sweating. Men worked for at least 10 hours a day, and could be asked to work longer to meet schedules. They towed boats irrespective of the weather, which could mean scorching heat in the summer and violent downpours in the autumn. The level of pay was low, and the work was, of course, only seasonal (roughly April to November). A former barge hauler recounted in 1924 that he and other haulers had fled from their jobs towing a *rasshiva* barge because the work was so hard. The owner then tracked down the fugitives and tried to fine them eight roubles each, which would have left them with almost nothing; they were only saved by the timely death of the owner.[52]

Living conditions were harsh, quite apart from the back-breaking work. Food often consisted only of black bread, *kasha* (a kind of buckwheat porridge) and cabbage soup. Men slept on the banks of the river in makeshift shelters when working; they lived in doss houses in major towns while they looked for work (one is still preserved in Nizhnii Novgorod, in the lower town by the wharfs). There was no provision for care, or pay, of sick haulers, who might be abandoned on the banks of the river if they fell ill. Haulers could also be subjected to corporal punishment: in 1761, for example, 53 haulers complained that they 'were often beaten with sticks for no reason'.[53] Boat haulers, as we have seen, participated in the Razin and Pugachev revolts – itself a sign of their desperation.

The writer Maksim Gorky, whose own childhood was extraordinarily brutal, was brought up in Nizhnii Novgorod and recorded this description of his grandfather's experience – during his own childhood – as a hauler on the river Volga:

So you came on a boat with steam to help it along. When I was your age I fought against the Volga, pulling a barge which was in the water, and I had to struggle on my own along the bank, barefooted, walking over sharp stones, and rocks, right from dawn to dusk. The sun scorched the back of your neck and your head seemed to boil like molten steel. And you, miserable wretch, bent double, your bones creaking, press on and on, till you don't see where you're going any more, you're blinded by sweat, and your soul weeps and a tear rolls down your cheek. That was a life! On and on you'd go, and tumble free from the hauling rope, flat on your mug, and glad of it. Work like that sapped all your strength, and all you wanted was rest! That's how we lived, with Jesus Christ smiling down on us! . . . I measured out Mother Volga three times. From Simbirsk to Rybinsk, from Saratov to here, from Astrakhan to Makarev, right to the fair . . . and that's a lot of miles.[54]

One of the reasons why haulers could be treated harshly was that they were often, at least until after 1861, fugitive serfs, or army deserters – or simply desperate for other reasons – and so had little choice but to accept the terms on offer. Few haulers left a record of their lives, but the handful of examples we have give some indication of why some ended up in this work. One, Pavel Ekimov, left his home in Smolensk and ended up in Astrakhan, where he first worked as an apprentice and then hired himself out as a hauler in 1659. An orphan, Vasilii Martynov, left his village and found work on a barge in Kazan. Maksim Piankov was a fugitive serf from one of the Stroganov estates, who had a fake passport and found work on boats in Kazan in 1767.[55] In 1722, there were 11,119 boat workers in Nizhnii Novgorod, of whom 8,593 were peasants; and of those, 3,945 were serfs and 400 state peasants (there were also 1,943 town labourers, 22 sons of clergy and – for some inexplicable reason – one hauler of noble origin!).[56]

It was often assumed that serfs were in a worse position than state peasants. Fugitive serfs could be reclaimed by their owners, and were therefore 'passport-less', which made them more vulnerable and unlikely to challenge poor conditions of work. Many haulers were serfs who had been kicked out of their villages for various reasons: laziness or failure to work the land – or simply because there were too many young male members of the household for the land available.[57] These were precisely the sorts of young men whom the commune could dispatch as military recruits; and, for all its hardships, work on the barges was preferable to conscription for life into the army. Some of these 'hirings' were tantamount to men selling themselves as indentured labourers – in effect, selling themselves, or being sold, into slavery.

Most haulers were aged between 20 and 40, but a study of men working on the upper Volga in the mid-eighteenth century found that almost 20 per cent of haulers were either under 20 or aged between 40 and 60.[58] It was assumed that haulers could only undertake this work for a limited time, but there were exceptions: in the early eighteenth century, a peasant on monastic land in Nizhnii Novgorod province called Zavarzin allegedly worked for 44 years on the boats![59] The majority of boat haulers were ethnically Russian, although Chuvash haulers were also used heavily. Cheboksary was a centre for hiring haulers, both Russian and Chuvash: in 1752, 1,763 men were hired in the town; by 1807, this had risen to 3,588, and numbers peaked at 5,958 in 1808.[60] Boat haulers in Nizhnii Novgorod included Tatars, and those recruited in Saratov included Tatars, Mari and Chuvash. In the 1860s, a Russian peasant, Vasilii Kochkin, from a village in Nizhnii Novgorod province, claimed that haulers included 'Chuvash, Mordvins, Tatars, and many Russians'.[61]

By the nineteenth century, formal contracts were drawn up between boat-owners and haulers. An analysis in Nizhnii Novgorod in the middle of the nine-teenth century showed that they were often concluded at the end of the fair for the following season, with wages to be paid monthly.[62] These contracts could be with individuals (mainly, in the cases examined, with Russians, but also with Chuvash, Mordvins or Tatars); or recruitment could be outsourced to middlemen, contracted to find a number of workers. One such contract, for example, was for two groups of 40 men, who were to be 'healthy adult persons, no younger than 22 years and not older than 45' for a rate of 28 roubles a month; they were to be given a cash advance if they gave up their passports (so that they would not run away). The second group of men, in fact, was found to be unsatisfactory, and a third of them were dismissed for being too young or too old, or simply unfit for this type of work. The contractor in this case had run off with some of the contracting boat-owners' money. The haulers recruited could also be cheated by contractors: they sometimes did not receive the money they had been promised, or were dismissed without cause. It is not surprising that dissatisfaction sometimes led to violence, or to men simply leaving (with or without their passports).

Barges and steamships also required more skilled labour, mainly pilots (*lotmany*) and *vodolivy* – barge leaders or skilled sailor supervisors. Pilots were often former peasants or serfs, were mainly Russian (although some Tatars were also employed as such) and were often recruited from particular villages on the banks of the Volga. Good pilots were much valued because of the difficulty of navigating the river and were not easy to find.[63] Pilots were better paid than haulers, earning anything from three to four times the amount that boat haulers received,[64] but they also had no job security and could be dismissed or refused new contracts if a boat ran aground on a sandbank. Work for them was also, of course, seasonal.

The boat work described so far was undertaken by men. But late-nineteenth-century photographs depict women hauling boats upstream using the same harnesses as men. Women certainly worked on the docks, carrying loads on and off boats. It was very heavy work which had an adverse effect on health: it was said that women could only perform this work for a maximum of 8–10 years. A description of women at the docks in Nizhnii Novgorod stated that they worked from 7 a.m. to 8 p.m. and that 'in the course of those thirteen hours they climbed up at the double with two-*pud* [about 33kg] sacks on their heads and then descended . . . the women performed all this work unfailingly with singing'.[65]

❖

So far, this chapter has concentrated on the goods, and the people, who travelled up and down the river Volga. But, as we have seen, the river was a meeting place for the exchange of goods along its shores. There were large numbers of local markets and fairs along the river, including in Dubovka, Tsaritsyn and Balakovo (all in Saratov province). It has been estimated that by the mid-nineteenth century, 419 fairs were held in the Volga region, most of them in villages and involving minor transactions.[66] Peasants and pedlars (Russians, Tatars and Jews) travelled along the river selling homemade clothing, hats, shoes and foodstuffs.

The Volga was also the main artery for goods reaching European Russia and western Europe, from south to north and east to west. 'The greatest quantity of raw produce comes from the East, either down the Kama or up the Volga', remarked Laurence Oliphant in 1852.[67] We saw in the previous chapter that the trading population of Astrakhan was enormously diverse, and included Armenians, Indians, Bukharans and Persians. This reflected the importance of trade from Persia, the north Caucasus and Central Asia, with goods entering Russia via Astrakhan and making their way up the Volga. Indian merchants dominated the trade with Asia, but merchants from Central Asia also brought their goods to Astrakhan. Other towns similarly attracted trade to and from the east: Samara was well positioned geographically in this respect. In 1753, a Samara merchant, D.F. Rukavkin, was trading in Khiva (now in Uzbekistan).[68] Goods came to Kazan from Siberia, and via Siberia from China, as well as from north Russia. The main emporium for the meeting of east–west trade was, however, the fair of Nizhnii Novgorod, which flourished from the 1820s until the end of the century.

The original market for east–west trade was positioned in the village of Makarevo, south of Nizhnii Novgorod, on the eastern bank of the Volga. The fair developed in the seventeenth century in the grounds of the monastery of the same name, and as early as 1622 it housed over 500 stores.[69] From 1624, the fair operated once a year, initially only for one day, but by the 1660s for two weeks. By the end of the eighteenth century, the total value of goods exchanged at the Makarevo fair reached some 30 million roubles.[70] We have seen that the Russian government played a central role in the eighteenth and early nineteenth centuries in constructing canals to supply the newly created city of St Petersburg: that is, it attempted to organize trade on the Volga to meet its political ends. The government also determined the fate of the Makarevo fair. It levied dues on trade at the fair; but more importantly, it was a government initiative to move the fair from Makarevo to Nizhnii Novgorod in 1818 (after a devastating fire had destroyed most of the stores), in the belief that that location would further encourage trading activity.[71]

The new fair at Nizhnii Novgorod was constructed, according to government plans, over an enormous area on a flat plain opposite the kremlin. It was always easier to build town bridges across the smaller rivers that joined the Volga, and in Nizhnii Novgorod a wooden pontoon bridge (at the time the longest in Russia) crossed the river Oka, giving access to the fair. A British traveller in the 1860s witnessed a crush of people and a:

> . . . scene of indescribable confusion. A swarm of passengers is constantly traversing it with carts, and waggons, and droshkies [carriages], and horses, and donkeys: and men are flogging, and swearing and crowding, and jostling along it all day.[72]

The complex included a large three-storey main building, which was used to sell mainly luxury goods, but also included an exchange and a bank. There were further central stores that were reached by three major streets, each of which was supposed to house 21 stores. In time, further shops – some permanent and others temporary – sprang up around these main streets and along the wharfs. The original fair comprised about 2,500 shops, but this grew to some 3,000–4,000 by the middle of the nineteenth century. In principle, there were separate rows of shops for particular types of goods; but in practice, the areas merged as the fair expanded, creating a chaotic, noisy, colourful and crowded arena for the exchange of goods. The main building was impressive, and the whole complex was enhanced by the construction of an Orthodox cathedral, a mosque, an Armenian church and a 'pagoda' for Chinese traders.

The fair operated for only a few weeks during the summer. It opened sometime in the period 1–5 August, and the large traders had left by the 23rd of that month, though business continued until the first week of September. Goods came from all over Russia, from Siberia, Central Asia, China, Persia and, to a lesser extent, from Europe. Russian goods dominated – cottons, wool, furs, metal, leather, grain, wood products, utensils, pottery, glass, foodstuffs and manufactured goods. Furs and ores came from Siberia; tea, silk, dried fruit, rhubarb, precious stones and dyes from Central Asia and China; and silk and cloth from Persia. Manufactured goods and fashions came from central and western Europe, although European goods were less prominent at the fair than Russian and eastern goods. The location, however, was on low-lying ground and frequently suffered flooding by the Volga and the Oka. The combination of flooding and the crush of people, with their mules and horses, meant that the fair area often became a muddy morass. Laurence Oliphant recorded two feet of mud,[73] but a later traveller claimed that the fair was 'hip and thigh deep in slushing mud'.[74]

In the first two-thirds of the nineteenth century, tea was probably the most important commodity sold at the fair. In the 1860s, some 60,000 chests of tea were brought there, which constituted almost 40 per cent of all tea imports into the Russian Empire. Moscow merchants bought about half of all their tea at the Nizhnii Novgorod fair.[75] The seriousness of the buying and selling of tea was recorded by William Forsyth, when he visited the fair in the 1860s:

> A tea-smeller, an expert, sits on a bale and before him a package of tea is brought. Into this a man darts a long sharp iron instrument with a hollow groove, and he draws out a sample of tea. The smeller smells it and if he does not approve of it, he puts it to one side as unsaleable.[76]

Tea was so important that it was the first commodity to be traded at the fair.

Vast numbers of people attended the Nizhnii Novgorod fair – not only as major traders, but also as labourers, hawkers, peasants, entertainers and even tourists. Robert Bremner commented in the 1830s on the 'countless throng from morning to night' and the 'bustle and activity unparalleled in Europe'.[77] It was estimated that 120,000 people a day attended the fair in 1858.[78] But it was not just the number of people which impressed foreign travellers, but also their extraordinary ethnic diversity. The colourful and exotic scenes are best described, albeit with a number of national prejudices, by Robert Bremner:

> ... costumes and faces more varied and more strange than ever before assembled in so small a compass ... [a] white-faced flat-nosed merchant from Archangel with his furs. He is followed by a bronzed long-eared Chinese, who has got rid of his tea ... Next came a pair of Tartars from the Five Mountain [Piatigorsk in the Caucasus], followed by a youth whose regular features speak of Circassian blood ... Cossacks who have brought hides from the Ukraine ... Those who follow, by their flowing robes and dark hair, must be from Persia; to them the Russians owe their perfume ... The wild-looking Bashkir from the Ural ... Glancing in another direction, yonder simpering Greek from Moldavia, with a rosary in his fingers, is in treaty with a Kalmuck as wild as the horses he was bred amongst ... Nogais are mingling with Kirghisians, and drapers from Paris are bargaining for the shawls of Cashmere, with a member of some Asiatic tribe of unpronounceable name. Jews from Brody [western Ukraine] are settling accounts with Turks from Trebizond ... cotton merchants from Manchester, jewellers from Augsburg, watchmakers from Neufchâtel, wine-merchants from Frankfurt, leech-buyers from Hamburgh, grocers from Königsberg, amber-dealers from Memel, pipe-makers from Dresden, and furriers from Warsaw.[79]

It is hard to estimate the overall value of the fair within the Russian economy; but it can be seen that the volume of goods sold there rose until the 1880s. It has been estimated that the average yearly value of goods *brought* to the fair rose from 48.8 million roubles in the period 1840–44, to over 100 million roubles by 1860–64 and over 160 million roubles by 1870–74, before peaking at 215 million roubles in 1880–84. Perhaps more revealing is the estimate that in the 1850s the Nizhnii Novgorod fair accounted for over half the value of goods brought to all the 4,670 fairs in Russia.[80] Bremner considered in the 1830s that goods exchanged at the fair could be worth £12 million – or, perhaps more revealingly, that this exchange was far more valuable than at the well-known fair at Leipzig.[81]

The fair was certainly an enormously impressive example of the vibrancy, variety and extent of trade in Russia, centred on the Volga, and demonstrated the geographical importance of the river for trade from east to west, Asia to Europe. However, commentators both at the time and later have characterized it as an indication of Russian backwardness – a substitute for more modern systems of transportation and goods exchange. Laurence Oliphant, a South African-born British author and traveller, writing in the mid-nineteenth century as a native of a far more industrialized nation, was disparaging. 'Fairs', he stated, were indicators of a 'primitive condition of trade', whose place should have been taken by large, industrial towns. They were, in his view, the 'remnants of barbarism'.[82]

In fact, the business of the Nizhnii Novgorod fair was not static: over the nineteenth century, different goods dominated, and facilities for monetary exchange and for access to credit developed. It is true that the fair declined in volume in the 1880s, and, although there was a minor revival in the early twentieth century, by 1913 it was in steep decline. In part, this was due to international competition and changes in the international climate. Tea, for example, which reached Nizhnii Novgorod from China, was unable to compete later in the century with so-called Canton tea, which was shipped to Europe from China via London. Routes from Persia and Central Asia changed to supply markets in central and western Europe.

Another reason for the ultimate decline of the fair was the development of railways in Russia, as discussed above. The railway reached Nizhnii Novgorod in 1862, and initially did not damage the fair; rather, it helped bring goods and people to it. But in time, the alternative and cheaper routes provided by the railways led manufacturers to avoid the fair and pursue more flexible export opportunities. What is perhaps remarkable is that the Volga, and the towns on it, retained such economic importance despite these changes in international patterns of trade and means of transportation. That could also, however, be seen as a sign of the economic backwardness of Russia and the slow rise of industrialization and urbanization – something that changed rapidly in the late nineteenth and early twentieth centuries, and that is the subject of Chapter 12.

11
THE VOLGA AND RUSSIAN CONSCIOUSNESS
Literature, art, tourism

O Volga! After many years
I bring you again a greeting
I am not as I once was but you are as bright
And magnificent as you were.
All about are the same distance and expanse
The same monastery is still there
On an island amongst the sands
And even the trembling of days gone by
As I heard the sound of the bells
All is the same, the same – but not
The crushed strength and spent years.
. . .
Oh Volga! . . . My cradle
Has anyone loved you as I have
Alone in the morning half light
When all the world is still asleep
When the scarlet gleam just glides
Over the dark blue waves
I escaped back to the river of my birth
. . .
Oh bitterly, bitterly I sobbed
When in that morning I stood
On the bank of the river of my birth
And for the first time I called it
The river of slavery and grief.

Nekrasov, 'On the Volga'

THE VOLGA AND RUSSIAN CONSCIOUSNESS

This poem is by Nikolai Nekrasov, who was brought up on an estate in Iaroslavl province on the upper Volga. Written in the 1860s, it illustrates some of the themes in this chapter concerning the development of a Russian identity: the sentimentalizing of the beauty and majesty of the Volga; the true 'Russianness' of the river; the personification of the river as a 'mother' or 'Mother Russia'; and, to an extent, the sacralization of the river. At the same time, the river was used as a critique of the social conditions in Russia in the second half of the nineteenth century: it was a river of grief, as well as glory. In the process, a Russian identity and a Russian consciousness developed, framed by the river.

The Volga had been the subject of poetic odes in the second half of the eighteenth century, as we saw in Chapter 5. Odes were written for Catherine II, especially following her visit to Kazan in 1767, to assert the 'taming' of the river by the tsars. The Volga was still considered at this date to be a dividing line between east and west, between Europe and Asia; but it was now a river that was obedient to the commands of the tsar. These formal odes were written for the benefit of the empress and for projection of Russian imperial power abroad; but by the nineteenth century, poems were accessible to a far wider, educated Russian audience. They were published separately and in literary journals, and were also read at salons, at social gatherings and in the houses of prominent nobles and merchants. The ideas they projected about the Volga therefore touched a much larger section of the Russian population.

In the first half of the nineteenth century, Romanticism was the dominant cultural movement in Europe. It emphasized the importance of the emotions, rather than what was regarded as the cold rationalism of the Enlightenment. In poetry and the arts, it also led to a discovery of the beauty of nature, especially the untamed wilderness of mountains and rivers. The movement started in German lands, but spread to the rest of the continent, including Russia. Russian poets and artists shared in this cultural movement, but at first they struggled to present positively what they saw as the bleakness and monotony of the Russian countryside, with its severe winters and with mile upon mile of almost unbroken forests – or, in the south, vast empty plains stretching into the distance. All this was so different from the beauty of the mountains and waterfalls of Switzerland, the colourful Mediterranean coastal villages of the Italian peninsula, graced by romantic Roman ruins, or the gentle glades and rural scenes of the English or German countryside. But in time, Russian writers and artists not only came to assert the difference of Russian scenery, but also arrived at the view that in some ways that scenery surpassed that of other European countries. The Volga played a special role in this process.

Even in the early nineteenth century, some Russian writers could be defensive about the Russian landscape, asserting not only that it was as beautiful as landscapes elsewhere in Europe, but that it was better and more spectacular; and the river Volga was central to this assertion. In *Letters of a Russian Officer*, published in 1816, Fedor Glinka criticized the 'opinions of foreigners about the severity of our climate, the coarseness of our people', and stressed the power of the Volga and its special qualities, in contrast to the conventional, and more mundane, pastoral imagery of other European landscapes at the time:

> . . . Almost everyone can enjoy spring days – flourishing nature, the babbling of spring brooks, but we felt it was possible to enjoy the autumn too, standing on the banks of a river (we were on the Volga) and watching as storms, with their foaming furrows, upturned the surface of the waters and howled in the mists.[1]

Russian writers (and artists, discussed below), however, often presented the Russian countryside, and the river Volga, as part of the Romantic tradition of sentimentalizing nature and as very similar to the picturesque and pastoral images of the countryside elsewhere in Europe. This extract from 'Evening on the Volga' by Petr Viazemsky, written in the 1820s, could have been written about a non-Russian landscape in western Europe:

> A dark row of forests here beneath a mantle of fog
> An airy range of bluish burial mounds
> In the great distance a village arrayed below the hills
> Meadows, paying their golden tribute to the flocks . . .[2]

In the same way, the experience of childhood in the countryside bordering on the Volga was often depicted by Russian writers as peaceful, tranquil and picturesque. In his semi-autobiographical novel *Years of Childhood*, written in the 1850s, Sergei Aksakov describes his idyllic life enjoying the countryside at his grandfather's estate in Simbirsk province (and life later on the estate in Orenburg province, after his grandfather moved), and contrasted this with the shock of having to live in Kazan, where he attended the grammar school. In Ivan Goncharov's novel *Oblomov* (published in 1859), his blissful and safe childhood in a village near the Volga is recalled by Oblomov in a dream:

The whole place, for ten or fifteen miles around, consists of a series of picturesque, smiling, gay landscapes. The sandy, sloping banks of the clear stream, the
small bushes that steal down to the water from the hills, the twisting ravine with
a brook running at the bottom, and the birch copse – all seem to have been
carefully chosen and composed with the hand of a master.[3]

Goncharov himself was a native of Simbirsk, and came from a wealthy merchant
background. There is a bizarre memorial to the indolence of Oblomov in Simbirsk
today (now called Ulianovsk): a sofa and a pair of slippers, set in concrete, with a
copy of the novel on a lectern! (Plate 15).

By the 1850s, however, the river Volga was already being presented as having a
distinctive and special beauty, which set it apart from the landscapes of western and
central Europe and in some ways surpassed their beauty. The exceptional size and
breadth of the Volga became a common feature in other nineteenth-century poems:
'Volga-mother, fast and long,/ This free, wide vision', wrote Vladimir Giliarovsky.[4]
This new portrayal coincided with the 'discovery' of the spectacular beauty of the
Caucasus; but that was an alien land to Russians, on the very edge of their empire,
whereas the river Volga, which had been a border once, was now seen as native
Russian. The playwright Aleksandr Ostrovsky had been sent down the Volga by the
Naval Ministry in 1855–56, as part of a literary expedition of writers, also including
the authors Aleksei Potekhin and Aleksei Pisemsky; each writer was given a 'stretch'
of the river to cover and Ostrovsky's was from near the source to Nizhnii Novgorod.
The Volga is central to his play *The Storm* (published in 1859), and the opening
lines have a character praising the outstanding beauty of the river:

It's an absolute marvel . . . Yes my lad, I've been feasting my eyes on the Volga
for fifty years now, and I still can't get enough . . . The incredible view! It's sheer
poetry! It rejoices the heart.[5]

The river, however, also plays a tragic role in this play. The disgraced and tormented
Katerina throws herself off the cliff into the waters, to commit suicide in the
last act.

One of the main themes which recur in Volga poetry and literature is the personification of the Volga as 'Mother' or 'Mother Russia'. 'Mother Volga' was used well
before the nineteenth century: Efrosin, an Old Believer monk, was alleged as early
as 1691 to have stated, when attacking another monk for turning the world upside
down (!), that: 'You make the Volga mother flow backwards, you send the Nile and

Danube back.'[6] The river Volga as 'mother' became strongly associated with its Russianness, and its power and ability to protect its children, the Russian people. Vasilii Pushkin (the uncle of Alexander Pushkin), in his poem written in 1812 'To the Inhabitants of Nizhnii Novgorod', calls on the Volga to protect its children at a time of great crisis, when Napoleon had occupied Moscow:

> Take us under your protection
> Children of the Volga shores!
> Take us, we are all relatives
> We are the children of Mother Moscow!
>
> . . .
>
> A gentle mother, fed us all!
> Moscow, what has become of you?
> Take us under your protection,
> Children of the Volga shores![7]

'Hail to you, Mother Volga,/ Russian river' commences one poem in a collection assembled in a guidebook to the river published in 1900.[8] Another poem quoted in the guidebook, this time by A. Lugovoi – 'Again on the Volga' – contains a sentimental depiction of the Volga in the opening line ('Volga! . . . My own! My beauty! My mother!') and begs the river to welcome back its 'weary son' after 'years of sad separation'. The poem ends with the plea: 'Volga! I am here before you, in repentance/ The confession of my heart I bring to you.'[9] In the early twentieth century, a series of stories by Vasilii Rozanov, entitled *The Russian Nile*, described the author's journey down the river, with frequent references to 'Mother Russia' and 'Grandmother Russia'. The female image of the Volga is associated with her life-giving powers. The Volga is sometimes referred to in poetry as a 'nurse' or a 'wet-nurse'. The river provides the fertile fields of Russia and gives life to her people. 'We are her children; we feed on her', wrote Rozanov. 'She is our mother and nurse . . . something immeasurable, eternal and flourishing.'[10]

The personification of the river as 'Mother Russia' entered ordinary language. An account of service on a steamship in the 1860s refers naturally to 'the banks of our Mother Volga'.[11] In the late nineteenth century, Aleksandr Naumov, a prominent Samara nobleman, reflected in his memoirs (as someone who had been born on the Volga and knew its 'disposition') that the river could be dangerous in all seasons, with its shallow water, storms, ice and snow, but he nevertheless referred to it as 'Mother Volga'.[12] In 1866, a collection of songs was published, which included the

'Song of the Volga Boatmen'. The song became immensely popular: it is a 'pulling song', as the boat haulers heave the vessel in the sun, and includes the verse:

Oh you, Volga, mother river,
Mighty stream so deep and wide.
Ay-da, da ay-da!
Ay-da, da ay-da!
Volga, Volga, mother river.

Personification of rivers – male and female – is not, of course, unique to the Volga: the Danube and the Yangtze, for example, are sometimes referred to as 'mother' rivers; meanwhile the river Thames and the Mississippi are male ('old Father Thames' and 'ol' man river'). Within Russia, the Don is sometimes referred to as 'Don Ivanovich' – probably because it flows from Lake Ivan; but this is also a characterization which signifies its importance for Russia. 'Mother Volga' became associated with the symbol of 'Mother Russia', and the river is sometimes described as a substitute for that term, so that the 'river' became indistinguishable from the country as a whole.[13] The rhetoric of 'Mother Russia' was used by both imperial Russia and the Soviet Union to rally the nation against an enemy, an 'outsider'; this in turn asserted the special status of the Volga as the embodiment of the national characteristics of the Russian nation.

The river was also, as we saw in Chapter 4, associated with pirates and popular rebellions in the seventeenth and eighteenth centuries. The Volga inspired popular songs and poems based on the exploits of Razin and Pugachev. In this characterization, the Volga is a river of freedom, and one which opposes (rather than supports) state power; a river which protects those who would revolt against the state. Indeed, it has been argued in a recent study of sacralized rivers in Russia (albeit with no etymological evidence) that the name 'Volga' can be linked with the Russian word *volia*, the meanings of which can include 'freedom' or 'liberty'.[14]

The river Volga, however, acquired other, more negative, literary characteristics in the second half of the nineteenth century. As intellectuals became more vocal in their condemnation of what they regarded as oppressive tsarist policy, the arts were used to portray the sufferings of the common people. Nekrasov's poem 'On the Volga', cited at the beginning of this chapter, illustrates this well. The theme is repeated in his poem 'Thoughts at a Vestibule' (1858), in which the pain experienced by boat haulers on the river is put into the broader context of the oppression of peasants and the common people throughout the Russian Empire. The poem starts with a doorman shutting the door in the face of 'slavish peasants', and then characterizes the pain of the 'desperate' people of Russia, and the suffering of Russian peasants, including on the Volga:

Go out to the Volga: whose groan is it
That rings out over Russia's great river?
Among us this groan is called a song.
Or is it the boat haulers hauling the tow-rope?
Volga! Volga! In the torrents of spring
You do not flood the meadows as much
As the immense grief of our people
Inundates our land.[15]

The writer Maksim Gorky (his pen name *gorkii* means 'bitter' in Russian) was brought up in Nizhnii Novgorod and worked in various lowly capacities as a young man on the banks of the Volga and in Volga towns. Although the river is rarely mentioned by name in his short stories, it is clearly the Volga that is described in various scenes on the river and by its banks. In Gorky's characterization, the river is melancholic – a reflection of the misery and hardships experienced by so many impoverished and desperate members of Russian society. In the story 'The Ice is Moving', set in the 1880s, the bluish ice on the river is melting:

The river exhaled with an arching melancholy: deserted, covered with porous scabs, it lay like a straight road without hope or promise of comfort leading to some murky region from which, weakly and cheerlessly, a cold wind blew.[16]

In the story 'One Autumn', the river 'felt the approach of winter and was running in terror from the fetters of ice which might be laid on it that very night by the north wind', while:

The upside-down vessel with its broken keel and the trees stripped by the cold wind, old and pathetic . . . Everything about me was broken-down, barren and dead, and the sky wept ceaseless tears. Desolate and dark it was around me – it seemed as though everything were dying, as though soon I alone would be left alive and that cold death was waiting for me, also.[17]

Evgenii Chirikov, a social critic and an acquaintance of Gorky, was born in Kazan and lived and worked in various Volga towns, including Samara, in the late nineteenth century (before going into exile in 1921). The river Volga plays a central role in his novel *Marka of the Pits* – in part because of its beauty, but also as a destructive force of nature, since it caused the death of Marka's father and thus condemned her to a dreadful fate (and also her mother had committed suicide by throwing herself into the river, following her husband's brief moment of infidelity). The opening paragraph of the book describes the deprivation of the dwellings in the ravines near the river, which Chirikov contrasts with the wealth of the town above: '. . . miserable little houses, huts and sheds; they were small, dirty and damp and from the distance resembled nothing more than the huts of primitive man . . .'[18] The townspeople refer to the river settlement as the 'pits' and the people who live there as 'monkeys'. In his semi-autobiographical *Otchii dom* (*Father's House*) Chirikov sarcastically referred to the river as 'Mother Volga' in 1892, when he was describing the terrible conditions during the famine; at that time the river banks, towns and villages were full of 'wandering, hungry people', looking for all the world like 'cockroaches'.[19]

❉

The Volga in art in many ways paralleled the evolution of several characteristics of the river in poetry. The first significant attempt to portray the river came from a government initiative, in the same way as the visit of Catherine II to towns on the Volga inspired odes in the second half of the eighteenth century. In 1838, Nicholas I commissioned the brothers Nikanor and Grigorii Chernetsov, both landscape painters, to travel down the Volga from Rybinsk to Astrakhan and to paint scenes on the river (just as the Naval Ministry later sent Ostrovsky and others down the Volga on a literary expedition in 1855–56). Their undertaking resulted in an enormous panorama of some 600 metres in length. It was displayed in St Petersburg, in a room that resembled a steamship cabin, to which sounds were added to give the full river 'experience'.[20] Unfortunately, the panorama has not survived, but sketches show that the artists were most at ease with their subject on the upper Volga, where they painted towns and monasteries along the banks of the river.

The lower Volga, however, proved problematic for the Chernetsov brothers, because they were unsure how to portray in picturesque fashion a landscape that to them seemed barren and uninteresting (in the same way as poets struggled with the Romantic vision of the Russian countryside). To them, the steppe on the eastern side of the lower Volga was empty: as they commented, 'Not a sapling nor a shrub, only the bitter sage waves in the desert.'[21] They spent days climbing in the Zhiguli hills, which rise above the river, north of Samara, looking for an appropriate view-point. Finally, they settled on the village of Morkvash, from where the view of the valley, the hills and the river was deemed 'remarkable for its picturesqueness'. Those hills could be depicted in Romantic style – with strange-shaped rocks, etc. – not too unlike the contemporary Romantic pictures of Italian and Greek ruins that the brothers had also painted.[22]

Russian painters were slower than poets and writers to appreciate the splendour of the Volga. In the early nineteenth century, Russian landscapists produced scenes similar in theme to paintings in central and western Europe. Most of their paintings concentrated on the details of life in an idealized countryside, in which the vast sweep of a river like the Volga had no place. In them, the Volga features only as a backdrop to the (idealized) activities of (mainly) serfs on the noble estates in the upper Volga region. Here, the foliage was rich, and the scenes were pastoral and not unlike pastoral scenes elsewhere in Europe. The artist Aleksei Venetsianov, for example, painted rural scenes from his personal experience of his estate in Tver province, where he spent every summer after 1819. His *Spring Ploughing*, dated in the 1830s, shows happy peasants at work in an idyllic landscape.[23] As one author commented drily: 'Venetsianov's peasants are well dressed and well fed; the sun is always shining, and the harvest abounds.'[24] Sentimental paintings of this nature were supported by the tsars and by the Imperial Academy of Arts. Alexander I bought *Cleaning Beetroots* by Venetsianov for 1,000 roubles in 1822; *The Threshing Floor*, set in Tver province, was purchased for the Hermitage for 3,000 roubles.[25] At the same time, provincial artists, including those based in towns such as Uglich and Iaroslavl on the upper Volga, painted traditional portraits of members of the local gentry, officials and merchants.[26]

The use of poetry to portray the sufferings of the ordinary people was reflected in art in Russia in the 1870s. This was at the time when, in the wake of the aboli-tion of serfdom in 1861 (discussed in the next chapter), young intellectuals 'discov-ered' the common people and the countryside. They became convinced that they had to 'go to the people' to reach them with a message of progress, which they assumed would be welcomed and appreciated. This task involved primarily a polit-ical movement called Populism, but it influenced artists, who considered that art played an important part in the Populist movement. In 1870, a number of artists formed a group known as the Wanderers. They set out to bring their paintings to people in the provinces, and to develop in those people a love and appreciation of art. At least some of them also challenged the status quo of the day.

The most famous painting of the Volga is *Barge Haulers on the Volga* by Ilia Repin. It was painted after Repin had travelled down the river with the group of artists and had spent several summers from 1868 to 1870 sketching the river and barge haulers at work. Repin was primarily based about 60km from Samara, in the town of Stavropol (where the river turns), because, as he noted in his memoirs, while travelling down the Volga he was told that the scenery below Saratov was 'boring and monotonous'. He, like the Chernetsov brothers, was advised to visit the Zhiguli hills upstream, near Samara. Repin, however, found the Volga at Stavropol 'beautiful' – at least on the western bank, which he contrasted with the sandy eastern shore.[27] His painting shows a group of haulers bent down by their back-breaking work. The men lean forward under the strain, their faces darkened by the effort; they are dressed almost in rags and toil under a relentless, hazy sun. The haulers seem resigned to their fate, with the exception of one young man who pulls at his harness and gazes upwards to the sky, as if to a better future for himself – and perhaps for Russia (Plate 12).

It is not surprising that this painting was seen at the time (and since) as a wonderful example of realist painting and a compelling social critique. It is hard not to agree with the contemporary comment by the art critic Vladimir Stasov:

Whoever looks at Repin's *Barge Haulers* will immediately understand that the artist was deeply impacted and shaken by the scenes that appeared before his eyes. He has touched these hands made of cast iron, with their sinews, thick and strained like rope.[28]

The sheer size of the painting is supposed to make an immediate impact, and it is certainly an immensely powerful depiction of human suffering in a stark landscape – even if the details are decidedly odd. At the time, it was pointed out by a fellow artist that a barge of that size would have to be hauled by three or four groups of men, rather than by one team alone, as depicted here.[29] Also, the shadows suggest that the boat is being hauled *downstream*, when in reality most boats had to be hauled upstream, against the current. Most oddly, the rope the men are pulling seems to be attached to the top of the mast, which would almost certainly have led to the mast snapping! Barges, as we saw in the previous chapter, were hauled up the river by kedging – that is, dropping an anchor upstream and pulling the barge up to it, using a windlass on the deck.

Factual accuracy aside, many of the Wanderers did not in fact set out to challenge the status quo. The reality in fact was that they had to earn a living, and the motivations for their travelling exhibitions were not so much to shock the viewers with their radicalism as to sell paintings.[30] The artists were pragmatic, rather than radical or romantic; and most of the people who viewed and bought the paintings, in major Volga towns and elsewhere, were wealthy middle-class Russians. The artists chose paintings for the travelling exhibition to appeal to potential purchasers; the *Barge Haulers* was never exhibited in their travelling exhibitions. Nor did the regime view the paintings as seditious at the time (it was in the Soviet period that these artists were lionized for having provided examples of the suffering of the people under the tsarist regime). At the time, paintings by the Wanderers were displayed at international exhibitions and purchased by members of the imperial family and by Sergei Tretiakov, a wealthy merchant philanthropist, whose collection was given after his death to the city of Moscow and was opened to the public in 1893 (and which forms the basis of the present Tretiakov Gallery). The *Barge Haulers* was bought by Grand Duke Vladimir Aleksandrovich, vice president of the Imperial Academy of Sciences, and displayed in his dining room.[31] It now hangs in the State Russian Museum, formerly the Alexander III Museum, in St Petersburg.

Russian artists also produced many landscapes of the Volga (and elsewhere) that did not feature downtrodden boat haulers or peasants or other examples of human misery. Some of the paintings – such as those by Isaak Levitan – projected light-filled panoramas of the river, with views across its shores of monasteries or other buildings. Levitan worked in the colony of artists and writers at Ples (usually known as Plyos) on the western bank of the river, some 50km south of Kostroma, between 1887 and 1880 (and today there is a Levitan museum in the colony). He loved the place for its beautiful church and the quality of the light.[32] While there, he painted *Evening on the Volga* (1888), *After the Rain* and *Golden Evening* (both 1889), which display not only the breadth and grandeur of the Volga, but also the great expanse of sky, with clouds that could herald rain or a storm.

Other well-known paintings of this period concentrated on the Russian forest, and reproduced every tiny detail of the forest floor and the foliage. The images of the forest and the river Volga, however, were combined in Evgenii Vishniakov's *The Volga Cluttered with Fallen and Rotting Trees* (painted in the 1890s), in which the artist depicted in detail the trees in a glade near the source of the Volga, where the river was still a stream.[33] Ivan Shishkin, whose forest paintings are exceptional in their use of detail and colour, also painted a series of pictures of the archaeological remains at Bolgar. These are in the popular style of Romantic paintings of ruins in Italy. They feature the remains of the towers and bath-houses of the city, but do not incorporate vistas of the Volga. Many other Russian painters, including the Chernetsovs, travelled to Italy and to other places favoured during the Grand Tour, at the same time as they were rediscovering the Volga and its romantic ruins.

The 'Mother Volga' theme was displayed not only in paintings, but also in early monuments of the river in female form. A statue of the river Volga as a female peasant feeding a double-headed eagle (!) was created by Aleksandr Opekushin, Mikhail Mikeshin and Dmitrii Chichagov for the All-Russian Art Exhibition of 1882.[34] The Volga was to appear in female form in several manifestations in the Soviet period, but this had its roots in imperial times.

Art and literature were not isolated from each other. Artists and writers met in the artists' colony of Ples. Indeed, the short story 'Grasshopper' by Anton Chekhov features what he calls ironically the 'not ordinary people' – that is, the aspiring artists, writers and musicians who embarked on an 'artists' tour of the Volga'. It is on the steamer in the evening that Olga (thought to be the artist Sofiia Kuvshinnikova) – who is assured by her friends that she has talent enough to be a painter, a writer or a singer – experiences a romantic moment with the artist Riabovsky (almost certainly Kuvshinnikova's lover, the artist Levitan), who reflects that 'it would be sweet to sink into forgetfulness, to die, to become a memory in the sight of that enchanted water with the fantastic glimmer'.[35] Chekhov mocks the sentiments of the fickle and vain Olga, who is captivated by the image and the words of her companion:

And Olga . . . thought of her being immortal and never dying. The turquoise colour of the water, such as she had never seen before, the sky, the river banks, the black shadows, and the unaccountable joy that flooded her soul, all told her that she would make a great artist, and that somewhere in the distance, in the infinite space beyond the moonlight, success, glory, the love of the people, lay awaiting her.

The next morning, it is raining and the Volga looks 'dingy, all of one colour without a gleam of light, cold looking'. Riabovsky regrets his contact with Olga, decides he has no talent as an artist and plunges his knife into his best sketch: 'in short he was out of humour and depressed'.[36]

A similarly doomed affair is depicted in the story 'Sunstroke', by Ivan Bunin, written in the early twentieth century and featuring a liaison between a young army officer and a lady whom he has met on a Volga steamer on a beautiful and romantic evening: 'the lights fell away; a strong, soft breeze rose from the darkness and blew into their faces as the steamer veered to one side – describing an expansive, slightly grandiose arc, it seemed to flaunt the Volga's breadth'. The next day, the young officer is bereft when it becomes clear that the lady does not wish to pursue the relationship, and he reflects that his life is left with no meaning.[37]

The artist Kuzma Petrov-Vodkin painted in a very different style from the artists already mentioned. He is best known for his iconic painting *Bathing of a Red Horse* (1912) and for later works in the Soviet period on revolutionary and modern social themes. But he was born in the small town of Khlynovsk on the Volga, and in 1927 he wrote his memoirs, which were as sentimental about the Volga as the earlier work of Aksakov and Goncharov. In them, he admitted what he called his own 'geographical patriotism'; but he was clear in his mind that spring was special on the Volga – and in Khlynovsk, in particular – because of a combination of colours and smells. He also recalled the 'wonderful taste' of the fish, including beluga sturgeon, consumed at festivals in the town.[38] He later painted the river Volga at Khlynovsk in a more traditional style than his subsequent paintings, depicting people staring out across the boats to the opposite shore, as if to an unknown future.

❖

The sentimentalism of the Volga bordered on sacralization, in literature, poetry and art. Certainly, depictions in poetry and art of the Volga as the true *Russian* river linked it inextricably with Orthodoxy. This was an early theme in Russian poetry, and was linked to Russia's unique history and culture. As early as 1793, Nikolai Karamzin characterized the Volga, in a poem of the same name, as 'The river, the holiest in the world/ The tsarina of crystal waters/ the mother!'[39] The poem by Petr Viazemsky, 'Evening on the Volga', written in the early nineteenth century, links the river with Russian history and Orthodoxy:

Utterly fascinated, I love of an evening
To listen, O great Volga!
To the poetic voice of your holy waves;
In them is heard Russia's ancient glory.[40]

Orest Somov's literary essay 'On Romantic Poetry', written in 1823, refers to the Volga 'with its distant flow and its blessed banks'.[41]

Many of the paintings of the nineteenth century depicted an Orthodox church or monastery on the banks of the river. And indeed many churches were built on the river Volga, in particular, to commemorate St Nicholas, the patron saint of sailors. The well-known painting *The Rooks Have Returned*, by Aleksei Savrasov (which was displayed at the Vienna International Exhibition in 1873), features a Russian church (although the Orthodox cupolas are not depicted) in the foreground and the river

Volga in the distance. Of course, the churches could be used simply to frame a land-scape; but they also asserted the distinct *Russianness* of the scenery, as opposed to other river views in Europe.

Some of this spiritual characterization is part of the sentimental depiction of the Volga. Nor is it unique to characterize a river as 'holy': both the Ganges and its tributary, the Yamuna, are regarded as holy, as are rivers elsewhere in the world – for example, the Osun in Nigeria and the Whanganui in New Zealand. However, the source (or supposed source) of the Volga is clearly linked with Orthodoxy.[42] A chapel was constructed on the site in the mid-seventeenth century, but this burned down in 1724. A small, crumbling chapel remained, ignored by most people. The source is positioned between St Petersburg and Moscow, but it was not visited by travellers in the eighteenth century (it is not easy to reach, even today). The chapel was rebuilt in the 1870s, just at the time when the Volga was being 'discovered' in art, poetry and, as we shall see below, tourism. In 1870, the municipal administration of Tver (the town nearest to the source) determined that a new church should be built on the site, in addition to the small chapel, and advertised for donations beyond the province to aid in the construction. In other words, in this case the move to have a church at the source stemmed from educated society, and was not an initiative from above on the part of the Orthodox Church or the tsar. The Church of the Transfiguration opened in 1910 near the source.

In post-Soviet Russia, the source has become not only a protected monument, but a sacralized site. A new wooden chapel has been built in the style of the old one (which can be seen in early twentieth-century photographs) and has been blessed (two plaques on the chapel wall note that the waters were blessed by His Holiness the Patriarch of Moscow and All Russia Aleksei II, on 9 July 1995, and by His Holiness the Patriarch of Moscow and All Russia Kirill, on 9 June 2017). The Church of the Transfiguration has also been restored. Religion and magic are often close companions: a mile or so away from the source is the sign 'command the waters to be clean'!

❖

The 'discovery' of the river Volga in poetry and art in the nineteenth century complemented the discovery of the Volga as a tourist attraction. The advent of steamships meant that passenger travel increased, and special steamships were built for tourists – by the mid-century, steamships could carry some 250 tourists, and accommodation ranged from luxurious first-class cabins through to third-class accommodation on deck.[43] In Repin's *The Barge Haulers*, a steamship appears in the background, showing the contrast between the world of steam and the traditional use of human labour; but his original sketch also featured a group of picnickers on the bank. This would have provided an even starker contrast with the suffering experienced by the haulers, but it is also recognition that the Volga was being used for leisure purposes by this time.

Between 1850 and 1900, over 40 books and many articles and pamphlets were written to introduce Russians to the Volga.[44] In the process, these guides created an image of the Volga as a special natural feature in which Russians could and should take pride, and which helped them to define their understanding of nature, of their country and of their own Russianness. The intentions of the authors are made clear. In *The Volga from Tver to Astrakhan*, by N.P. Bogoliubov, published in 1862, the author sets out to provide 'a clear and true understanding of the remarkable Volga region'. In the same year, P.P. Neidgart's *Guide to the Volga* stated that all Russians had an obligation to familiarize themselves with the river. The three-volume detailed guide to the Volga by Viktor Ragozin urged readers to take their children on a trip down the river, so that they would become acquainted with the river from an early age. The writer A.N. Molchanov proposed hiring a steamer that could hold 500 passengers to familiarize urban, educated Russians with the countryside and the life of the Volga peasants.

It took time, however, to move from presenting the Volga as something which the writers considered Russians had an *obligation* to see, to something which they would *enjoy* seeing as tourists. In this respect, the perceptions of the Volga by the educated elite mirrored the developments in poetry and art, as the Volga gradually took on a new significance. Molchanov even called the river 'dirty', 'grey' and 'monotonous'. By the 1870s, as the number of steamships increased, so the guides placed more stress on the scenic beauty of the river, the prospect of relaxation and the pleasures to be found on the trip. The Volga became special – special within Russia, but also something unique to Russia that deserved to be celebrated. The travel writer Vasilii Nemirovich-Danchenko, in his 1877 guide *Along the Volga*, wrote eloquently that 'The Volga is an endless poetic song, an endless epic poem' and also that 'No matter what kind of sorrow you may feel, come to the Volga and you will forget it!'[45]

By the turn of the century, the Volga had become a fashionable destination for Russian tourists, and was being presented as not only beautiful, but also unique to Russia and superior to anything that central and western Europe could offer. The guidebook by E.P. Tsimmerman, *Down the Volga*, published in 1896, stated that it was impossible to find 'such strikingly and picturesquely wild views on any river in western Europe'. An 1895 guide, *Volga*, by A.S. Razmadze, criticized what he regarded as the unnatural preservation of ruined castles on the banks of the Rhine and contrasted this with the Volga, where 'everything is natural and everything is beautiful in its natural state'.[46]

Guidebooks in the early twentieth century also drew attention to the advances in technology that could be witnessed along the river, and in which the Russian tourist could take pride. The guidebook by G.P. Demianov, *A Guide to the Volga from Tver to Astrakhan*, published in 1900, pointed to the bridge at Saratov as 'one of the longest in the world', built at a cost of over 7 million roubles, and drew attention to the fact that the railway was developing the town of Tsaritsyn (it also reproduced a number of poems about the Volga for the reader).[47] A guide by N. Andreev, *An Illustrated Guide to the Volga*, also drew attention to bridges across the river, including the new (both 1913) crossings at Iaroslavl and Sviiazhsk. The bridge at Syzran (opened in 1880) was noted as the tallest in Europe (it had cost over 7 million roubles to construct). That guide also noted that the fair at Nizhnii Novgorod was declining, with the growing importance of railways, and gave statistics for the amount of grain exported from Samara by rail (it stated that twice as much was now being exported by rail as by boat).[48]

In the late nineteenth century, tourists could buy postcards to remind them of their holiday.[49] Many of these views were of sites along the way – towns, monasteries, churches, the Zhiguli hills; but there were also postcards of the steamships on which they travelled. Other cards were sentimental depictions of people enjoying the trip with the river in the background, and were very typical of the postcards produced elsewhere in Europe at the time: beautiful women lounging on deck; lovers entwined, with the sun setting on the river; small, cute, beautifully dressed, angelic-looking children playing on the shore. Some postcards depicted the exotic dress of Tatars, but for the most part they confined themselves to the experiences of the *Russian* tourist, and the buildings featured were predominantly Orthodox, *Russian* churches and monasteries.

It has to be said, however, that neither the home nor the foreign tourist always appreciated the special characteristics of the Volga. Anton Chekhov travelled down the upper Volga and the river Kama in 1890. He was underwhelmed by the experience, writing to his sister:

My first impression of the Volga was poisoned by the rain, by the tear-stained windows of the cabin, and the wet nose of G., who came to meet me at the station. In the rain Yaroslavl looks like Zvenigorod . . . it's muddy, jackdaws with big heads strut about the pavement. In the steamer I made it my first duty to indulge in my talent – that is, to sleep. When I woke I beheld the sun. The Volga is not bad; water meadows, monasteries bathed in sunshine, white churches; the wide expanse is marvellous, wherever one looks it would be a nice place to sit down and begin fishing . . . The steamer is not up to much . . . Kostroma is a nice town . . . It's rather cold and rather dull, but interesting on the whole . . . the sun is hiding behind the clouds, the sky is overcast, and the broad Volga looks gloomy. Levitan ought not to live on the Volga. It lays a weight of gloom on the soul. Though it would not be bad to have an estate on its banks.[50]

The 1914 *Baedeker* remarked rather disparagingly:

'Matushka Volga' or 'Little Mother Volga', is spoken of in Russia so often and with such affection that it is easy to cherish too high hopes of the attractions of a voyage upon it. The scenery is nowhere of an imposing character, and the length of the voyage is very fatiguing.

The guide goes on to warn that 'in July and August the steamers are often very late', and 'the waiting-rooms on the piers are by no means as comfortable as they might be', although the cuisine on the larger steamer was 'excellent', especially caviar, fish dishes and crab.[51] This did not put off foreign travellers, however; many of the contemporary accounts of the river and the Volga towns in this book come from foreign tourists who enjoyed a voyage down the river.

Despite the reservations expressed by some, in general by the late nineteenth and the early twentieth centuries the Volga was being presented as uniquely and emotively *Russian*, something which could unite all classes of society in their Russianness. Anna Petrovna Valueva-Munt expressed this clearly in her 1895 *Along the Great Russian River*:

The Volga has meaning for us [Russians] not only as one of the largest and most interesting rivers on the globe: it is a purely Russian river and on that basis endlessly admired by all Russian people. Notice with what love the third-class passenger looks at it – the peasant, the artisan, the poorer merchants. To them the Volga is fine not only in those places where it strikes one with its pictur- esqueness, like at Nizhnii-Novgorod, Zhiguli, or Vasilsursk. For the majority of the Russian people it is beloved even in those places where there are only bushy willows and white sand banks with snipes running along them. Something native, heartfelt, and poetic breathes out from these places.[52]

12
REFORMS, REVOLT AND RUSSIFICATION ON THE VOLGA

This chapter illustrates the ways in which the Russian Empire and the towns and villages of the Volga were changing in the late nineteenth and early twentieth centuries. Railways improved transportation of vital products and transit goods, and led to rapid population growth in some Volga towns, which put additional pressures on urban resources. New economic forces brought opportunities for some, but for many also led to social disruption and hardships in the countryside and the towns. The dissatisfaction culminated in mass political and social unrest in 1905, in the wake of the humiliating defeat of Russia in the Russo-Japanese War. The 1905 Revolution was brought to an end by a combination of repression and concessions, the most significant being the October Manifesto, which established for the first time an elected Russian duma. In the second half of the nineteenth and the early twentieth centuries, the Russian government passed legislation of immense importance for peasants – the emancipation of the serfs in 1861 and the Stolypin reforms of 1906–07. The Volga is an ideal testing ground for examining the impact of these initiatives, not least because the region included such a variety of agricultural practices – from very large noble estates to subsistence farming. In contrast, little was done to alleviate the conditions that gave rise to urban social unrest in 1905.

At the same time, rural and urban society became more conscious of their separate ethnic and religious identities, not least in reaction to the policies of Russification adopted by the Russian government in the late nineteenth century. The previous chapter discussed the evolution of a sense of 'Russianness' through the medium of poetry and art, and the 'discovery' of the Volga as a unique *Russian* natural feature. This chapter focuses on the experience and evolving identity of the non-Russian peoples of the Volga as they experienced economic and social change. We have seen that conflicts arose in the second half of the nineteenth century between 'newly baptized' Christian Tatars and Muslim Tatars. A greater awareness of Muslim identity, new initiatives in education and the rapid increase in publications in the Tatar language intensified these conflicts from the late nineteenth century, but also created a new, more culturally and politically aware, educated Tatar elite, who by 1905 identified themselves as *Volga* Tatars. People in the multi-ethnic and multi-confessional Volga region faced particular challenges as the Russian Empire modernized, and this chapter reflects on the extent to which those challenges were being met at the beginning of the twentieth century.

❖

In 1861, Tsar Alexander II (reigned 1855–81) emancipated the serfs. There were several good reasons for doing so: the perception that serfdom was holding Russia back economically; a fear of serf revolt; and the moral objections to human bondage that were commonly expressed by writers and senior bureaucrats alike. By the mid-nineteenth century, there was a strong feeling in government circles that serfdom was anachronistic – a view that was strengthened by the humiliating defeat of Russia in the Crimean War, which exposed the weakness of Russia's economic infrastructure and raised questions about the effectiveness of a predominantly serf/peasant conscript army. The Emancipation Act made the serfs legally free and allotted them a proportion of their nobleman's estate; it also abolished their labour and fiscal obligations to the noble landowners and the judicial authority of the nobleman on the estate.

The significance of the Emancipation Act cannot be overstated. For all that the institution of serfdom had come in for much criticism, it was still a radical – and brave – step on the part of the Russian government to expropriate almost half the land of the nobility, and probably it was only the shock of defeat in the Crimean War that made it possible to countenance that risk. The Russian Empire was not wealthy, which meant that compensation for the landowners had to be in the form of long-term government bonds. Serfs were allocated their own land for the first time. They had, however, to pay for the loans made to the landowners at a level determined by the government – initially over a period of 49 years (at an inflated price in 1861, although the land increased in value over the next four decades). State peasants were allocated their land in a separate act (in 1866), but also had to pay the state over a period of 45 years.

Implementing the Emancipation Act was an enormous task, and almost inevitably led to peasant resentments that they had been badly treated. Overall, serfs lost about a fifth of the land that they had formerly cultivated. The allocation to the former serfs of plots on noble estates was largely determined by the landowners, and this led, almost inevitably, to variations across the empire. Not unnaturally, the better the quality of the land, the more the landowner wished to retain it (or at least to be fully compensated for its loss), and so the smaller the plots allocated to the peasants; conversely, the worse the land, the more the peasants received. Those peasants who could not afford to pay for their land were granted very small plots, which were known as 'beggarly allotments'. It was recognized that this land was very often inadequate for subsistence. Although the peasants were now legally free, some restrictions remained. Of these, the most significant was that peasants were not allowed to leave the village without the permission of the peasant commune, which continued to determine the agricultural cycle of the village (although its role in conscription was abolished, with the introduction of compulsory, short-term military service in 1874). In effect, this meant that many of the economic weak-nesses of serfdom were retained, despite emancipation, including the difficulty of freedom of movement, strip farming, the potential repartitioning of lands within the village by the commune to equalize land holding (and tax burdens) among the peasants, and the reluctance of many communes to risk innovation in agriculture.

The Volga offers a good case study of the results of emancipation, because of the variety of land holding and peasant status in the region. The largest noble estates were in the excellent black-earth territory on the western bank, and predominantly in Saratov province, where the majority of serfs were ethnically Russian. But even within a province, or a district, the outcome of emancipation varied according to the quality of the land, the size of the noble estates, the availability of water (rivers and ponds) and simply the attitudes of individual landowners and peasants. The peasants had a conviction that the only 'fair' outcome was for them to be given all the land they had worked. As a result of emancipation, the peasants in Saratov province lost about a quarter of the land they had previously worked, though some lost up to half (that is, the peasants lost more land than the average for European Russia), reflecting the good quality of at least some of the land in the province. The small land allocations, combined with population increase, made it hard for the peasants to subsist, let alone take advantage of the railway-driven expansion of the grain trade in that province in the late nineteenth century.

Throughout the Russian Empire, emancipation led, perhaps unsurprisingly, to disturbances which lasted several years. Some of the initial disruption was due to wild rumours about the terms of the act; but when the peasants were made aware of the outcome, many felt resentful on account of the loss of lands, including forest and common pasture land, and the continuing tax and other obligations due to landlords and the state. In black-earth provinces, like part of Saratov province, these disputes were bitter and often violent. Conflicts in some villages lasted for years: in the village of Romanovka, for example, which was owned by the immensely wealthy Vorontsov family but run by a steward, there were disturbances and outbreaks of mass disobedience in 1861, 1862 and 1863. Soldiers were brought into the village and brutal punishments inflicted on the perpetrators, including flogging and running the gauntlet; but this did not prevent further revolts from breaking out.[1]

In the period up to 1905, peasants and non-peasants (merchants and artisans) in Saratov province acquired more noble land, either through purchase or by leasing the land. In the period 1877 to 1905, noble landowners in Saratov province sold 238,900 *desiatiny* of land (over 2,600 square kilometres) on over 500 estates, and the percentage of land owned by peasants increased a little in this period – from 48 to 49 per cent.[2] It was, however, more common for peasants to lease, rather than purchase, noble land, especially from the largest estates. By the end of the nineteenth century, peasants in three districts in Saratov leased about a third of their land (higher than the average for European Russia).[3] Arson attacks (known as the 'red cock') on noble estates intimidated nobles and forced some to sell up both before and after the 1905 Revolution. Many noble landowners diversified into other occupations – in the towns, rather than the countryside – although some also innovated in agriculture in this period. Countess Elena Shuvalova, for example, who owned estates in Saratov and Samara provinces, hired an innovative estate manager called Morits Roland in 1884. He experimented with irrigation techniques on one of her Saratov estates, and was awarded the gold medal at the Saratov agricultural exhibition of 1889. The yield on the lands he irrigated was over double that on unirrigated lands.[4]

The middle and lower Volga regions were susceptible to extreme climatic conditions, including harsh frosts in the winter and droughts in the spring and summer. The winter of 1891 was particularly harsh, with temperatures plummeting to minus 31 degrees Celsius, followed by flooding, with the result that seedlings died. A windy spring in 1892 removed the topsoil; and this was followed by a hot, dry summer. The result was a devastating famine in the central black-earth region; indeed, it extended from the western provinces to east and north of the Urals. The Volga, however, was at its core. The writer Aleksei Tolstoy witnessed the famine on his estates near Syzran on the middle Volga:

> Great cracks appeared in the earth, the trees turned colour and shed their leaves and the crops stood brown and scorched. A misty wave of heat quivered low

over the horizon, burning up every vestige of plant life. The roofs of the houses in the villages lay bare and exposed, the straw having been used to feed the cattle, and the emaciated beasts which survived had to be tied to crossbeams, to keep them on their feet.[5]

The central government failed to halt exports of grain quickly enough, and there were inefficiencies and inconsistencies at the local level in supplying aid. The situation was particularly fraught south of Nizhnii Novgorod, because the railway network was not comprehensive enough to supply stricken areas, and supplies could not be shipped along the river Volga later in the year when it froze. Despite these problems, some 13 million peasants received state aid in 1891–92 – a considerable achievement, given the difficulties of reaching those most in need.[6] That stands in striking contrast to the famines of 1920–21, when the Soviet government relied heavily on foreign aid, and the famine of 1930–31, the very existence of which the Soviet government denied. Indeed, it could also be said to compare favourably with famine relief in other European countries, not least with that of the British government in Ireland in the mid-nineteenth century. The 1891–92 famine was particularly devastating in Samara and Saratov provinces, and allegedly left half the inhabitants of Saratov province destitute.[7] It was estimated that two-thirds of horses and almost 90 per cent of cattle had perished, either slaughtered for food or dead for lack of fodder. The peasants had not recovered by the end of the century, when the harvest failed again. Famine was followed by cholera and typhus, all the more lethal when sufferers were malnourished. In all, perhaps 300,000 died from the 1891–92 famine, followed by many more from cholera.[8]

The often poor allocation of good-quality land and the repayment costs, combined with a significant increase in the peasant population in the second half of the nineteenth century (despite the famine), led to heightened resentment among former serfs. This anger was expressed in 1905 across the Russian Empire, and manifested itself in a wave of peasant violence in the countryside. Peasants took their revenge on the former noble landowners by seizing timber and breaking down fences, and in some cases by grabbing land. Access to woods was crucial for the peasants – not only for timber, but also because they grazed their livestock there and collected berries and mushrooms. Peasants could be violent and would seize lands and burn down estate buildings, including manor houses. In Simbirsk province, the seizure of timber was described as taking on a 'mass character', but the peasants also stole grain and goods and burnt down barns and noble houses. In November 1905, a nobleman from Kurmysh district (now in Nizhnii Novgorod province) reported that over 20 estates had been plundered and ruined, and that the noble landowners had fled; a year later, it was reported that 30 estates had been attacked.[9] In Saratov, some 300 estates were destroyed in the year 1905 alone.[10] Former serfs were particularly active on the black-earth lands (as, for example, in part of Saratov province), but it was not just the poorest peasants, or those with least land, who protested in and after 1905. In Samara province, Aleksandr Naumov, a prominent local nobleman who was a marshal of the nobility and chairman of the Samara *zemstvo* (a local, primarily liberal, tsarist administrative organ), was horrified when he returned to his estate of Golovkino (on the banks of the Volga) in 1906 to find the peasants disrespectful and their mood 'excitable'. He found this particularly true of young peasants and of those with little land, who had been incited by agitators and made 'demands' for 'freedom'.[11]

The events of 1905–06 exposed the fact that emancipation had not solved the 'peasant problem'. An acknowledgement of this came with the cancellation of further redemption payments by peasants to the state for their land. In 1906, a new set of significant reforms concerning the peasantry was enacted by the prime minister, Petr Stolypin. Stolypin knew the Volga region: his family had estates in Kazan and Nizhnii Novgorod provinces, and he had been governor of Saratov in 1903–06, during which time he had put down peasant disturbances. The 1906 law allowed, and encouraged, peasants to choose by a two-thirds vote to abolish their peasant commune altogether and set up as separate farmsteads; individual peasants were also allowed to leave the commune and set up their own farmsteads by consolidating their strips. The aim was to create over time a new class of prosperous peasants, or independent farmers, who would not only increase the productivity of the countryside, but also provide some stability in the country and be loyal to the tsarist regime. New rural banks were established to enable peasants to invest better in their land.

The reaction in the Volga region to the Stolypin reforms varied according to the quality of the land, not only within provinces, but also from district to district and even village to village. In general, the peasants living on good-quality land, including in the provinces of Saratov and Samara, responded more positively to the Stolypin reforms, and some whole villages chose to dissolve the communal strips. Those on poorer land were less likely to 'separate' and more likely to be influenced by immediate factors within the village, such as the extent of pasture land and the availability of water, in rivers, streams or ponds. In Iaroslavl province on the upper Volga, there was a significant increase in the number of peasants who left the commune, collectively or separately – some 15,000 in all by 1914 – but this varied greatly by district (between 7 and 13 per cent, depending on the district); on the whole, it was the poorest peasants who left the commune altogether to look for work or land elsewhere (in the towns or in Siberia).[12] In Kazan province, only 8.5 per cent of households left the commune, although – as in Iaroslavl province – the proportion varied considerably, according to district.[13]

Some peasant reluctance could be ascribed to fear of the risks of going it alone. In Saratov province, one peasant, when questioned, explained: 'At present, all my land is in six places. If we have a hailstorm or a fog some grain would be destroyed, but look, you can collect it from another strip.'[14] Peasants, not without cause, feared the impact of natural disasters (like storms or fire) were they to cut themselves off from the village. For the most part, only wealthy peasants could take advantage of the right to set up on their own, against the will of the commune (such a farm was known as a *khutor*, and the peasants who set them up were known as 'separators'), and only they had the confidence to use the new peasant loan banks. As one peasant, an M.V. Savinov, commented in Kazan province: 'a *khutor* holding of course is fine, but only when you have a sufficient holding [of land]'.[15]

'Separating' could cause disputes within the village, if individuals, rather than the whole commune, took this course. In Kazan province, it was reported that:

If the peasants move onto *khutora* [the plural of *khutor*], the others become disgruntled and use every means to make life difficult for them. They deprive them of pasture to which they are entitled . . . they do not allow them to use roads in the commune, and do not give them access to water on the village's territory.[16]

The resentment against 'separators' was to be exploited by the Soviet state post-1917, when it expropriated grain from the peasants – and then again during collectivization.

The experience of the Volga countryside raises the question of the extent of success of the emancipation and of the Stolypin reforms. In the Russian Empire as a whole, some 10 per cent of peasants had claimed 'title' to their lands by the end of 1915; but this did not mean they had necessarily consolidated their plots and separated themselves from the rest of the village. Statistics on the peasants' reactions to the reforms are difficult to assess accurately, not least because both those who supported Stolypin and those who opposed him (including Marxists and later Soviet historians) have used the figures to prove that the peasants were (or were not) ripe for revolution, or that a wealthy class of peasants had emerged (or had not) as a result of the reforms. A study of Tver province, where land was generally poor, found that there was some increase in the variety of crops in the late nine-teenth century, and that some further modest changes were made after 1906 to respond to the market, but that fundamentally agricultural practices remained the same despite emancipation and the Stolypin reforms.[17] A survey undertaken by the government in 1908 found that even in Saratov province, where at least some of the land was good and where the peasants had responded more favourably to the reforms, the new, separate farms and communities had not changed agriculture significantly: 'In the majority of cases we do not observe any changes in these farms compared with communal farms.'[18] This was, however, only a couple of years after the Stolypin reforms, and change would inevitably take time.

The emancipation had less impact on non-Russian peasants in the Volga region, because the vast majority of them were state peasants and not serfs. It has been estimated, for example, that only 8.6 per cent of Chuvash peasants were serfs. However, those who were had much the same experience as the Russian peasants: Chuvash former serfs received less land than they had expected or (in their view) needed and deserved, as a result of emancipation. The same pattern could be observed for Chuvash as for Russian peasants: a gradual loss of land owned by nobles, and an increase in land purchases and leases by peasants. In Simbirsk prov-ince as a whole, where the majority of the Chuvash peasants lived, land owned by nobles dropped from 12 per cent in 1877 to 8.9 per cent in 1905, while land owned by peasants increased over the same period from 45 per cent to almost 48 per cent. The process of transfer of land away from nobles accelerated after the Stolypin reforms of 1906.[19] Disturbances also took place among non-Russians after 1861. Particularly savage reprisals were taken against Mari peasants who revolted against the tax burden in the southern region of Viatka province in 1889. Peasants were repressed by a battalion from Kazan, and punishments of 60 to 80 birch blows were inflicted on numerous peasants.[20]

The reaction to the Stolypin reforms was muted among non-Russian peasants. Many of them lived on poor-quality land, and so were less able to respond to the fresh opportunities to market grain and other agricultural products offered by the new railway system. Very few Tatars, Mari, Chuvash or Mordvins became 'separa-tors' after the Stolypin reforms. Indeed, the primitive conditions in many Mari and Chuvash villages meant that the peasant communes rarely, if ever, repartitioned lands at all. There was some increase in credit and cooperative activity after 1906, but the non-Russian villages changed little.

Nor did Volga German colonists change markedly in their agricultural organization in this period. Although the perception of contemporaries was that German colonists were richer than Russian peasants, in fact a combination of the persistence of traditional farming methods and a high birth rate meant that this was not the case.[21] Perhaps because of this, however, Volga Germans did take advantage of the greater ease of movement after the 1906 reforms, and some Volga Germans moved to Siberia in the hope of finding more land and a better life.[22] This was in part because many colonies had not recovered from the devastating effects of the famine in the 1890s. Many more Volga Germans also emigrated to the United States in this period – 'in large numbers', as one British traveller noted from Saratov.[23]

There is the question of what 'success' meant for peasant agriculture. One historian has argued that Stolypin was not setting out to destroy the peasant commune, or to transform rural society, but was intending to change peasant economic activity over time.[24] Other historians have seen the reforms as designed completely to transform peasant communities, in which case the reforms failed. The truth is that the task of reforming the countryside was of 'such magnitude', as one historian has put it,[25] that it could not be solved without more fundamental political and social reform, and could certainly not be solved overnight. If the Russian Empire had had more time to consolidate these reforms before the outbreak of the First World War, would the economic and political situation in the countryside have become more stable? We shall never know.

The fundamental problem was one of *perception*. Whether they lived in Iaroslavl, Saratov or Kazan, and whether they were Russian or non-Russian, the peasants did not feel that they had enough land. And for that they blamed both the landowners and the state. To the peasants' way of thinking, all the land should be theirs, because they were the people who worked it; in their eyes, emancipation had only served to remove some of what they anyway regarded as 'their' land and oblige them to pay for the rest. The Russian government had clearly failed to satisfy the majority of the peasants on the eve of war and revolution. However, it is unlikely that this could ever have been achieved: only a revolutionary government could contemplate complete expropriation of all the property of landowners.

❖

The event that sparked the disturbances in the 1905 Revolution was Bloody Sunday, 9 January, when troops massacred hundreds of strikers who were attempting to present a petition to the tsar in the Winter Palace for a constituent assembly. The brutal handling of this event set off a wave of strikes in factories and protest marches in towns across the Russian Empire.

Volga towns were volatile because in the late nineteenth and early twentieth centuries they had experienced significant industrial expansion and population growth, with the related social pressures (albeit not on the same scale as St Petersburg). To an extent, Volga towns were the victims of success – the Russian Empire was expanding its industry and export trade faster than any other European country. This was admittedly from a lower economic base, but the expansion of roughly 8 per cent in the 1890s was impressive nonetheless. New opportunities for the grain trade brought by the railways had also led to an increase in the factory population of Volga towns. In Saratov, for example, industry developed after the town became a centre for the grain trade, following the construction of a railway link to Moscow. Major leather, brick, tallow, potash and metal works were established – by 1910 there were 136 factories in the town.[26] If Saratov grew as an industrial town, then Tsaritsyn (at the time merely a district town in the province of Saratov) was something of a boom town. Known as the 'Russian Chicago', its prime position on the railway – as well as on the river – had led to a rapid increase in workers and factories. The population had soared from 6,700 in 1861 to 67,650 by 1900 (134,683 by 1915). Its factories also included large concentrations of workers – the Ural Volga metal works (French–Belgian owned), for example, employed over 3,000 workers.[27]

There were strikes and demonstrations during 1905 in almost all the major towns in the Russian Empire. The Volga towns were no exception, and even those that had seen little in the way of disorder in the period 1895–1904 – such as Saratov and Samara – experienced strikes in 1905.[28] Dockers went on strike in Rybinsk, as did 10,000 textile workers in Kostroma. Factory workers were often seasonal workers, but both the seasonal and the permanent labour force suffered from poor housing (often barracks for single men), low pay and poor working conditions, and were attracted to the programmes offered by socialist and revolutionary parties, which addressed their practical concerns. In 1905, factory workers were at the forefront of the protest. They were influenced by socialist propaganda, but their demands often called for improved pay and conditions, rather than change in the political and social order. Ivan Gavrilov, a worker at the French–Belgian-owned metal factory in Tsaritsyn, recalled in his memoirs in 1940 that they worked for up to 12 hours a day, with poor pay and conditions. Their demand was for an eight-hour day and a lunch break. Soldiers refused to fire on the workers, but the strike collapsed when the strike committee was arrested.[29] In Astrakhan, workers asked for a nine-and-a-half-hour working day.[30] Workers at railway junctions in

218

the Volga region went on strike for an eight-hour day, longer meal breaks and fewer fines.[31]

In many cases, individual circumstances – the extent and nature of the factories, the presence of a military garrison, the activities of local socialist parties, and the personality of individual governors – determined the form and the degree of violence on both sides. To give one example, in the town of Tver, radicalization could mainly be found among textile workers, of whom there were some 22,000 in the early twentieth century. These workers were poorly paid (their wages considerably lower than in St Petersburg) and were often employed in large factories (of which the largest was the Morozov textile works), where it was easier for socialist parties to make inroads and for workers' leaders to mobilize support. Tver textile workers had been involved in major strikes in 1902 and 1903. It was no surprise, then, that the day after Bloody Sunday a meeting of some 600 workers called for an immediate strike. Attempts to down tools were thwarted, however, not only by the actions of the factory owners (who threatened to close the factories), but also by the combined actions of skilled workers and a militia of apprentices, who fought with the unskilled workers and turned the ringleaders of the strike action over to the police. Sporadic demonstrations continued during the summer, but matters came to a head when a right-wing mob, led by policemen, attacked the provincial *zemstvo* building in October, allegedly because of its support for workers. The mob rioted, set fire to the building and brutally attacked local inhabitants. When the riot finally died down, 64 people were left dead or severely injured. This incident was followed by a workers' demonstration, which was brutally suppressed by sabre-wielding Cossacks. In December 1905, workers barricaded themselves in the Morozov and other factories, but had to call off their strikes when faced with the prospect of being attacked by armed soldiers. In the aftermath of these events, many workers were arrested and the level of strikes and political activity among the Tver workers dropped; but the memory of their brutal treatment remained. After the fall of the tsar in March 1917, Tver workers went on a two-day violent rampage – a clear indication that the resentments which had caused the 1905 Revolution had not been addressed either by the state or by individual factory owners.[32]

The events in Tver showed that Volga towns were not immune to the right-wing backlash that occurred in the later stages of the 1905 Revolution. The so-called Black Hundreds were ultra-nationalistic, anti-socialist and anti-Semitic groups. Originally founded around 1900, they flourished in the wake of the disturbances of 1905 and were particularly active in upper Volga towns. In Iaroslavl, for example, a Black Hundred march on 20–21 October 1905 involved at least a thousand people and resulted in attacks on Jewish shops and on workers.[33] A pogrom took place in Kostroma on 19 October, when a mob of over a hundred people, many of them young men, attacked Jewish shops and set out to 'beat the Jews'. One person was killed and 40 were badly injured. The perpetrators were traders and shop assistants, but also included seminarists and peasants who had come to the town from the country.[34] As much as the demonstrations, this back-lash in 1905 illustrated the volatility of social relations in towns – a volatility that had not eased by the time the First World War began.

❖

The social and economic changes in the late nineteenth century also had an impact on the evolution of a specifically *Volga* Tatar identity in this period. There had always been a sophisticated group of Tatars in the Volga region, and by the end of the nineteenth century a number of Tatars had risen to economic prominence in Volga towns. This increased the confidence of at least urban Muslims about their status in the empire. For example, among the wealthy Tatars in Kazan (where the Tatar merchant and industrial elite was concentrated) was Ismail Apakov, who in the late nineteenth century established himself not only as a wealthy merchant, but also as a member of the urban elite – he served on the district court and was a member of the Kazan Imperial Economic Society.[35] There were also wealthy Tatar industrialists in other Volga towns. In 1892, Tatars had 10 trading houses in Saratov, and by 1914 that number had risen to 154.[36] A number of wealthy Tatar merchants formed philanthropic societies to aid poor urban Tatars in towns with a significant Muslim minority; in Kazan, for example, a Society for Impoverished Muslims was set up in 1901.[37]

The new confidence of the urban Volga Tatar elite was demonstrated by a deter-mination to assert their independence through their own urban institutions. In the eighteenth century, government policy had been to establish separate legal and administrative institutions for non-Russian people in the empire (not only for Tatars, but also for non-Russians in the Caucasus and the western borderlands). The question of whether non-Russians could be urban representatives was poten-tially controversial. In 1868, there were disorders in Mamadysh district of Kazan province, when Muslims found that they could no longer elect a Muslim represen-tative as an elder.[38] This was shortly after the emancipation of the serfs, when there were several disturbances in the countryside. In 1885, there was an attempt by the local authorities to merge the Russian and Tatar artisan 'societies'. Opposition to this came from the Tatars in Kazan and Chistopol, who considered it to be an attack on their autonomy. The merger went ahead, but with separate elections; and it was reported that at the first meeting, the Russian and Tatar representatives in Kazan sat on separate sides of the hall.[39]

For Tatars, however, the most obvious way to assert their separateness and new-found confidence was through their religion, Islam. And the most common way of doing so was to construct new mosques in the towns and the countryside. In 1892, for example, a wealthy Tatar petitioned the authorities for permission to open a new mosque in Kazan, and provided testimonies from other wealthy town citizens.[40] The other way of asserting separateness was to establish more Muslim schools attached to these mosques, to teach (mainly) boys the tenets of Islam and the Arabic language. It was in these two areas that conflict broke out between the Russian government and the Volga Tatars – namely over the attempted conversion (from the Orthodox perspective) or reconversion (from the Muslim perspective) to Islam of Tatars who had previously converted to Christianity, and over the syllabus and use of Russian in Tatar schools.

We have seen that attempts were made at a local level in the Volga region to persuade Christian Tatars (strictly speaking, 'newly baptized' Tatars who had converted to Orthodoxy in the eighteenth century) to become Muslim. This movement became more active and more aggressive in the late nineteenth and early twentieth centuries. At village level, this was in part a reflection of the revival of popular Islam, and was often led by the Muslim women in the village.[41] Mosques were regularly and illegally constructed in Tatar villages where Tatars had reconverted to Islam.

The movement against Christian Tatars was intensified after the 1905 Revolution, when the government issued an edict on 'freedom of conscience'. In the words of Prime Minister Sergei Witte, 'after two hundred years of the policy of religious restrictions Russia has embarked on the path to religious toleration'.[42] In the Volga region, the practical result of this decree was that the Russian government permitted Christian Tatars to petition to convert back to Islam. It has been estimated that as many as 32,000 Christian Tatars in Kazan province became Muslim between 1905 and 1910 – allegedly leaving only 3,000 Christian Tatars in the province. In the same period, 4,360 Christian Tatars became Muslim in Simbirsk province (plus small numbers in Samara and Saratov provinces).[43] In fact, so many Christian Tatars petitioned to become Muslim in the middle Volga region that the Russian government became nervous and started to refuse petitions. Local authorities were also alert to forced conversions. In 1909, for example, a Kazan peasant named Aleksei Stepanov was found guilty of forcing Christian Tatars to become Muslim against their will.[44]

After 1905, Muslim Volga Tatars asserted their rights to practise their religion more openly. One poignant example involved the determination of Tatars in Kazan to celebrate Muslim religious holidays, and not to observe Russian ones. In part, this indicated a confidence in their Muslim Tatar identity; but equally, it enabled pragmatic Muslim traders to open their shops on Russian holidays. The campaign was conducted through columns in Russian and Tatar local newspapers in Kazan, and culminated in a rally in the town of 2,000 Muslims in 1910 in support of their own holidays.[45]

The education of Tatars – Muslim and Christian – also became a sensitive issue in this period. Education had always been an area where the Russian government had interfered directly with the syllabus of state schools (in a country where there were no religious teaching orders, so that education was in the hands primarily of the state or private tutors). Until the late nineteenth century, Muslims were permitted to organize the basic education of their own youth (mainly boys). Tatar education was traditionally linked with the mosque: basic schools (*mektebs*) and higher-level schools (*madrassas*) were attached to the mosque; mostly their activity was limited to reciting texts from the Koran and learning Arabic by rote. By this time, however, the Russian government was worried that these schools served to encourage apostasy and were failing in what it regarded as the main purpose of education: to produce not only literate, but also loyal citizens of the Russian Empire. A report received by the State Council noted in 1872 that it was impossible to establish secular education for Muslim Tatars and to 'unite them' with Russians 'through Russian schools, or by means of missionary activities'.[46] A new policy was implemented under the guidance of Nikolai Ilminsky, professor of Turkic languages at Kazan Theological Academy and Kazan University.

Ilminsky recognized that the Russian language and secular subjects had to be taught, in order to counter the influence of Islam and to inculcate loyalty to the Russian Empire. But at the same time, he realized that it was not practical to eliminate the teaching of Islam altogether or to curb teaching in local languages. The syllabus of the new, so-called Russo-Tatar schools was dominated by the study of the Russian language. The Russo-Tatar school in Astrakhan, for example, taught 11 hours of Russian per week in 1896–97, alongside 3 hours of the Tatar language in the lower and middle classes and 14 hours in the upper class – far more than any other subject.[47]

The schools recruited slowly, mainly because Tatars were suspicious of the new approach, fearing (not without reason) that it would undermine their separate educational system and lead to the assimilation of Tatar youth into Russian society. These fears were not allayed by the statutes which established the schools, and which stated clearly: 'The Russification of the Muslim-Tatars can be achieved only by disseminating the Russian language and education.' Traditional village schools also had to teach more Russian. Some villages in the Volga region petitioned against the introduction of the new syllabus, some simply stating, 'We do not want to learn Russian.' But others voiced genuine and more perceptive concerns:

> . . . we do not have enough money to hire a Russian teacher . . . the reason for its introduction [Russian language] is that not knowing Russian brings harm to Muslims. Well we never had any problems not knowing Russian. We are working people and peasants and with our work we pay the taxes.[48]

There was, however, a recognition among more educated Tatars that the old educational system was backward. They realized that a more secular education was helpful, and that a knowledge of Russian was essential for young Muslim men to become successful and to embrace the opportunities that were then opened up to them. By 1874, there were 29 Russo-Tatar schools in Kazan; by 1913–14, the figure had risen to 155.[49] By the twentieth century there were 19 Russo-Tatar schools in Astrakhan and a further 137 in the province.[50] At the same time, teachers' seminaries in Kazan trained a large number of non-Christian and Christian Tatar teachers to implement the new syllabuses in the villages. The new schools also, however, provided a basis for the development of an educated cohort of Tatars who were receptive to new intellectual approaches to national identity and religion. The potential danger of this was seen by the director of the national schools in Samara, D. Bogdanov, who wrote in 1914: 'For part of the Islamic intelligentsia a Russian education gave them the armour for life's struggle, a mission that was not the mission of the Russian government.'[51]

In the Volga region, it was not only Tatars who were affected by the education policy. The Chuvash, Mari and Mordvins were, for the most part, Orthodox (although we have seen that Christianity existed in fusion with older, pagan beliefs). Russian-language teaching in their schools also increased and was often unpopular. Many non-Russian peoples saw the increase in Russian teaching as an unnecessary intrusion into their way of life; meanwhile the Russian authorities, as we have seen, saw it as a means of integrating them more firmly into the empire. The practical problems on the ground that ensued from this difference of perception were described by a missionary, Evgenii Bolshakov, in the 1860s:

... as soon as I entered the room, the people started shouting in Mari and Russian that they were against the school, but later they agreed to it ... I am quite sure school should be the main means by which to influence non-Russians (*inorodtsy*), but its impact on the masses is slow, almost invisible, while the Cheremis [Mari] remain in the dark, which is a deplorable fact.[52]

The Volga Germans had their own schools attached to churches, both Lutheran and Catholic. By 1913, there were 327 German schools in 192 colonies in the Volga region, with some 68,000 pupils.[53] The state largely left these schools to devise their own syllabuses until 1897, when the teaching of the Russian language became compulsory. The attitude of colonists towards Russian-language teaching was ambivalent. On the one hand, they wanted to preserve their cultural distinctiveness; in 1905, for example, German colonists were not revolutionary, but they did demand that they should retain autonomy over the education of their children.[54] On the other hand, the colonists recognized that their children needed Russian in order to prosper economically, particularly as the grain trade expanded in Saratov province, where many of the colonies were located. Added to this, the colonists had lost the privilege of exemption from military service in 1874, and the practical result of this was that German conscripts needed to speak at least a little Russian.

It was during this period that Volga Tatars became conscious of their own, separate and special identity among Muslims in the Russian Empire as *Volga Tatars*. The intellectual roots of this consciousness emerged from a new, reformist, Islamic scholarship, much of which centred on Islamic thinkers in Kazan.[55] In the mid-nineteenth century, the scholar Shihabeddin Merjani, a teacher in Kazan at a *madrassa* and then at the Teachers' Seminary, and a reformist religious scholar, wrote the first serious history of the Volga Tatars in Arabic (he was the first Muslim writer to refer to the Volga Muslims as *Tatars*). By the late nineteenth and early twentieth centuries, a number of liberal Islamic religious thinkers from the middle Volga region were challenging the status quo, upheld by more conservative religious thinkers in Central Asia. Many of these discussions were conducted through Tatar newspapers in Kazan – itself an indication of the vigour of debate over religion and education among the Tatar educated elite by this time. At the same time, social issues concerning women within Islam were being discussed more openly, including their education (schools for girls opened in Kazan in the early twentieth century), use of the veil, and polygamy.

Alongside this religious debate, there was a movement in the late nineteenth and early twentieth centuries to establish a Tatar literary language and to encourage the printing of books in Arabic. A leading figure in this movement was Kayyum Nasiri, born into a merchant family near Sviiazhsk, a teacher of the Tatar language at Kazan Theological Seminary and the Russo-Tatar Teachers' Seminary, who developed a literary Tatar language based on the Tatar vernacular of the Volga region. Later Volga Tatar scholars based in Kazan built on his ideas and produced grammars, as well as books and pamphlets on Tatar history and current affairs. Until the nineteenth century, all Islamic religious books used in the Volga region were produced in Bukhara or Constantinople; but from 1802, printing presses operated in Kazan. The activity of these presses grew significantly in the first years of the twentieth century. The Kharitonov publishing house, for example, was founded in 1896, but only published its first Tatar-language book in 1902; by 1917, it had published 666 titles in over 3 million copies. Between 1900 and 1917, some 20 Tatar presses operated in Kazan alone, publishing over 5,000 titles in nearly 39 million copies.[56] Many of these books were secular in content, covering science, history and geography. Some were of concern to the Russian government, particularly those directed at Christian Tatars, which seemed to encourage them to petition to become Muslim. In 1907, the government prohibited the Tatar village book trade, a ban which the local police vigorously upheld by confiscating books at Volga fairs.

After the 1905 Revolution, and in particular after the decree on 'freedom of conscience', Volga Tatars involved themselves more in national politics. Informal student political circles had existed before 1905, and a political party called Union was set up by students from the Russo-Tatar Teachers' Seminary in 1906. Most political activity by Volga Tatars in and after 1905 was channelled through Muslim congresses, rather than through political parties (something which damaged Tatar interests after 1917). The first All-Russian Muslim Congress was set to open in Nizhnii Novgorod in August 1905, but in circumstances approaching farce the governor refused to allow it to go ahead, and the congress instead assembled for a riverboat trip up the river Oka – because river transport was not covered by legislation on the rights of assembly! The impromptu congress approved a number of aims, including the unification of Russian Muslims for the purpose of achieving reforms, equal legal rights for Muslims and Russians, and the freedom to develop Muslim schools and publishing.[57] The Volga Tatar representatives adhered to the Kadet party in the first duma set up after the 1905 Revolution; but as later dumas became more conservative, so the number and the influence of Muslim deputies waned. The Muslim faction of the fourth, and last, duma of tsarist Russia (1912–17) numbered only six deputies, five of whom were from the Volga-Ural region.

An important part of this new consciousness was that Volga Tatars began to trace their origins back to the Bolgars: that is, to Turkic or even European origins of the Bolgar state on the Volga, as opposed to what they regarded as the less civilized Mongols from the east. At the same time, this established that *Volga* Tatars were different from Astrakhan Tatars or Siberian Tatars.[58] The following extracts are from the mid-nineteenth-century poem 'The Sacking of Bolgar', by Gali Chokri, an Islamic scholar. What is significant is that, in bemoaning the loss of the great Islamic city, he blames the Mongols (who in fact let the city survive), rather than the Russians (who left it in ruins):

Bolgar, the holy refuge of Islam,
Now lies in ruins; nothing else is there.
. . .
Its books and letters in the past were famed;
Its scholars were renowned for skill and grace.
. . .
Then Timur came, his black and villain horde
Slew children, and the old were beaten down.
To crush the Muslims was his evil plan;
He longed to bring destruction to the town.
Bolgar was laid to waste, its buildings razed.
How many men and women died in shame.[59]

By the early twentieth century, the Volga region had undergone significant economic and intellectual development. The process had, however, led to tensions within the countryside and the towns, and within communities, both Russian and non-Russian. It was in this potentially unstable situation that the region faced two enormous challenges: world war and revolution.

PART 4

SOVIET AND POST-SOVIET VOLGA

CONFLICT, IDENTITY AND MANAGING THE RIVER

13
THE VOLGA IN WAR, REVOLUTION AND CIVIL WAR

'To the Volga the lot, to the Volga!'[1]

The Russian Empire was transformed and traumatized in the period 1914–21. The First World War caused strains that it could not sustain, and the tsarist regime fell in February 1917; but the new, provisional government structure could not maintain authority over the country. The Bolshevik Revolution in October 1917 did not immediately herald a new era, and led almost immediately to a bitter civil war. By 1921, the Bolsheviks had established a new state, but in the process they had waged war not only on the Whites and on perceived enemies of the regime, but also on the peasants in the countryside.

The main developments between February and October 1917 took place in St Petersburg, but the Volga towns were crucial to the outcome of the revolutions. Events in these towns demonstrated the strength of the social forces that made the revolutions, but also showed how the outcome could be affected by local circumstances – whether the local political parties, the presence of a garrison, or simply the role played by particular individuals. The ethnic and religious make-up on the Volga was also significant, including the participation of Tatars, German colonists and other non-Russians in events. Furthermore, the Volga was crucial to the outcome of the Civil War. For a period, an alternative government to the Reds and the Whites existed on the Volga, in Samara. Most importantly, the river was a vital strategic concern in the Civil War. The fact that the Whites could not sustain a foothold in Volga towns and that White forces from the east and south were not able to converge on the lower Volga largely determined the outcome of the Civil War. And it was on the Volga, in particular, that a second war took place between the Bolshevik state and the peasants, with tragic consequences. In short, the Volga played a crucial role in the outcome of the Civil War and helped to shape the future Soviet state.

❧

All towns in the Russian Empire suffered from severe shortages during the First World War. The Volga towns were particularly vulnerable, because they depended so much on the river to transport goods, but many steamers had been diverted to the river Dnieper to support the front against Austria-Hungary.[2] Prices increased dramatically during the war. In Tver, it was said that 'prices for food have increased ten times',[3] and in Syzran 'everything increased two, three or four times'. In Saratov, it was estimated that in the two years from October 1915, the price of buckwheat increased by over 2,000 per cent and the cost of butter by over 600 per cent, while some items, such as sunflower oil, were not available at all. Rationing was introduced in most Volga towns from 1916 for goods such as sugar and bread, and those on a low income inevitably suffered most. The situation was exacerbated by seemingly arbitrary shortages: Iaroslavl had no sugar in September 1915; Syzran only suffered from lack of sugar in 1916; Samara had shortages of milk and butter in January 1916.[4]

The initial patriotic reaction at the outbreak of the war in 1914 was soon replaced by discontent and defeatism, and trouble broke out at recruiting posts in Kazan and other Volga towns. The presence of garrisons and of deserters led to widespread disorder, and attempts to control the sale of alcohol were ineffective. Soldiers, including the wounded, could spread defeatism; for example, a soldier in Astrakhan was reported to have said in October 1916 that the war would destroy the people and would not end.[5] The situation in factories was made more acute, in part because of the increase in wartime industries (in Saratov, for example, there was a decrease in work in food and clothing factories and an increase in chemical, oil and metal works),[6] and in part because workers were most affected by shortages and increases in prices for basic foodstuffs. Volga towns, especially those on the upper and middle Volga, also received an influx of refugees in the first years of the war, including Poles, Lithuanians and Jews, which added a further element of social instability. Over 400 refugees from Courland (western Latvia) arrived in Rybinsk in July 1915, and there were over 28,000 refugees in Saratov by February 1916.[7]

The Volga towns, as we have seen, were exceptional in the complexity of their ethnic and religious diversity. During the war, the presence of Muslim Tatars and German colonists in the towns and countryside was perceived as a potential source of instability by the authorities. The government was concerned about the loyalty of non-Russian conscripts, and feared in particular that the German colonists would be disloyal and that the Tatars would not fight against Turkey. In the event, these fears were largely groundless, although German colonists (of whom some 40,000 were conscripted) were mostly deployed in the Caucasus. Tatars fought as officers and ordinary soldiers, and their main complaint was that their limited knowledge of Russian held back their promotion. Patriotic mullahs prayed for victory and urban Tatars made donations to the war effort.[8] Most Tatars in Volga towns were more concerned about the shortages of food than about government policies towards Turkey.

Such actions did not allay the suspicions of the authorities. In Kazan, 'undesirables' were put under surveillance, including Germans (both German nationals and colonists), Tatars and Jewish refugees. The police reported the 'hostile attitudes' of the population of Samara and Saratov towards the German population in 1915, exacerbated by the perceived pro-German sentiments of the unpopular Empress Alexandra, who was a German princess from Hesse. Anonymous denunciations were made of Germans in Saratov and elsewhere who supposedly supported the war. The police investigated allegedly treasonous comments to the point of absurdity: in the German colony at Sarepta, claims were investigated that a colonist had allegedly said 'our tsar is a fool'.[9] A German was arrested at a railway station in Nizhnii Novgorod province for allegedly expressing 'hostile' views about Russians. A 13-year-old boy was reported in February 1915 for being disrespectful to the tsar while out skiing with a friend at Kamyshin (on the Volga, some 190km by road north of Tsaritsyn): he supposedly said the tsar did not care about the poor, and was said to have added 'to hell with the rich!'[10]

Germans who had settled on the Volga and made a life and career there had their lives disrupted or ruined by the actions of the police. In Iaroslavl, the police deported a German professor, the latter protesting that he had lived peacefully in the town for five years.[11] Some Germans in Volga towns were sacked and even arrested as spies. The situation was exacerbated by the presence of German prisoners of war in several Volga towns. In 1916, there were 23,000 Germans in Astrakhan, who were mainly prisoners of war.[12] On the eve of the February Revolution, Nicholas II extended the right of the state to expropriate the lands of Germans – from territory on the western frontier to all German land in the Russian Empire. This would have led to the confiscation of the land of the German colonists on the Volga, had not the regime changed in February 1917.[13]

In Saratov, a Muslim called Sattar Manafov was arrested and accused (among other things) of supplying fruit from his store to aid the health of a Turkish officer and, more seriously, of helping Turkish prisoners of war to escape to Siberia. Manafov was arrested and imprisoned, but was freed after a trial.[14] The heavy-handed reaction by the state authorities to non-Russians on the Volga demonstrated their nervousness about the popular mood.

By 1917, food shortages and a general feeling that the war would never be over made the situation in Russia volatile. After a series of mass strikes and army mutinies in Petrograd (St Petersburg; renamed Petrograd in 1914, after the outbreak of war, to remove the German-sounding name), Tsar Nicholas II abdicated in March 1917. The abdication was greeted with enthusiasm by most of the population: a letter from Aleksandr Markov to a member of the Simbirsk archive commission (that is, a member of the tsarist administration and not a worker) stated, 'I congratulate you, dear Petr Aleksandrovich, on the triumph of the Revolution and on the new free Russia!'[15]

Kazan being a university town, students enthusiastically participated in marches alongside workers: 'Kazan then was a time of meetings. In the streets, one demonstration replaced another. Orchestras rang out. People sang the "Marseillaise". Red bows flamed in the sun.'[16] In Nizhnii Novgorod, some 20,000 people marched to celebrate the downfall of the tsar, and prisoners were freed from jail.[17] Saratov was witness to similar marches and to the singing of the 'Marseillaise', and again prisoners were freed.[18]

It was slower for news to reach small towns and villages on the Volga. There were local marches, particularly in places where there were factories. One participant later recalled that workers at his factory in Melekess – a small town (now called Dimitrovgrad, after the Bulgarian revolutionary) in Simbirsk province on the river Melekesska, a tributary of the river Volga – decided to demonstrate, and 300 of them marched through the town waving red flags, before holding a meeting in the local theatre.[19] The celebrations of the downfall of the tsar were, however, primarily *Russian* celebrations in the countryside. It was reported by an activist from the village of Kumor in Kazan province (a predominantly Russian village) that the Tatars, Mari and Udmurts on the outskirts of the village were less enthusiastic about the Revolution and displayed 'a general lack of trust towards the Russian population, as a result of dark forces and the adherents of reaction'.[20]

The February Revolution was bloodless in most towns. But the Russian Empire was large and diverse, and local circumstances could lead to events taking a different course, in the Volga region and elsewhere. We saw in the previous chapter that the 1905 Revolution in Tver was particularly violent, and that workers' strikes had been brutally crushed. The workers particularly hated the governor, Nikolai von

Bunting, who had banned meetings and had tried unsuccessfully to suppress the news from Petrograd. During the war, prices in the town had risen sharply, but wages had not kept pace and there were shortages of basic foodstuffs and oil. On 2 March, there was a major demonstration of some 20,000 workers and other inhabitants. Things soon got out of hand: wine cellars were broken into and one reserve regiment mutinied. In this volatile atmosphere, Governor von Bunting chose to make a courageous, if reckless and arrogant, stand. He strode out to meet the crowd, dressed in a splendid black greatcoat with red epaulets and, according to one eye witness: 'Stood as if of stone; not one muscle of the body moved. Finally, the crowd lost all patience and in the doorway of the guardhouse, he fell from two bullets and numerous bayonet wounds.' The crowd, drunken and now enraged, dragged his body into the main street and trampled on it; the greatcoat, the symbol of his authority, was flung into the upper branches of a tree. The mob then destroyed the market, raided several shops and looted the police station. It was an event that shocked everyone, on the left as well as the right (it was 'bad business; the masses degenerated into drunkenness' wrote the leader of the Bolsheviks in Tver), and can probably only be explained by a combination of drunkenness and the provocative behaviour of the governor. October 1917 passed off peacefully in Tver; it was as if the bloodletting had already occurred.[21]

After February 1917, an uneasy power-sharing arrangement existed in Petrograd between the Provisional Government (formed from the fourth duma) and the Petrograd soviet (representing workers, soldiers and sailors) – an arrangement that was only supposed to remain in place until a constituent assembly could be elected to determine the future government of the country. This provisional dual-power structure was replicated in large and small towns throughout the Russian Empire, with power shared between an organ of the former administration, normally called the Provisional Executive Committee (or PEC), and a soviet (or soviets) of workers and soldiers. These structures were fundamentally unstable, and they began to break down in the course of 1917, especially after August, when General Lavr Kornilov, commander in chief of the forces of the Provisional Government, moved his forces against Petrograd and was repelled by workers.

What is striking about the situation on the Volga (and elsewhere outside Petrograd) is that, at least at first, there was considerable cooperation between these organs – something that was quite at odds with the situation in Petrograd. On the one hand, this might suggest that in the provinces political divisions were not as sharply defined as in the capital. On the other hand, it could simply be an indication that in the chaos of 1917 no body was capable of keeping order – either in the towns or in the countryside, where peasants had started to take matters into their own hands by seizing land (see below).

The PECs and the soviets worked together in Kazan, Nizhnii Novgorod and Saratov, for example, in the immediate aftermath of the February Revolution, and in all those towns representatives from the soviet attended the PECs. There were distinct local features in each town: the Nizhnii Novgorod PEC was dominated by members of the former town duma; the Kazan PEC, by contrast, included representatives of so many organizations that its membership grew to 260 (it is said there was even a representative of the beekeeping society!).[22] The ethnic make-up of the Volga towns could also be a significant element in the composition of the provisional administration. In Kazan, Muslims and Jews had separate political parties which wanted to be represented, although the ethnic parties were also split – there were 15 Muslim and 3 Jewish parties.

In the countryside it was difficult to overcome the suspicions of Tatar and Chuvash peasants and to encourage them to take part in elections at all. This was partly because the parties lacked the expertise to translate their propaganda into non-Russian languages, but also because of the lack of interest shown in events. There was particular resistance in Muslim villages to allowing women to vote for the Constituent Assembly. One peasant refused to let his wife vote separately: 'Are you saying I am not master of my woman, are you soft in the head? What could my woman keep secret from me?'[23]

The situation in 1917 in Volga towns was affected above all by the presence of soldiers. Many towns on the Volga had garrisons of reserve soldiers, because the region was a safe distance behind the battle lines, but at the same time the extensive railway system could transport soldiers back to the front quite rapidly, if necessary. Reserve soldiers tended to be the most radical, because of their fear of being sent back to the front (there was a major, though unsuccessful, offensive in June 1917). As a result, during 1917 they became more attracted to the Bolsheviks – the only party to offer an immediate end to the war. One soldier elected to the Constituent Assembly from Saratov summed up their feelings: 'We are dark [uneducated] people. What sort of Bolsheviks do we make? For us it's all the same if only they would end the accursed war and go home as quickly as possible, otherwise it is impossible.'[24]

Nizhnii Novgorod and Kazan had garrisons of 40,000 and 50,000 men, respectively. These garrison soldiers supported Bolshevik demonstrations, and also created disturbances in the towns through their drunk and disorderly behaviour. Volga towns also housed a large number of wounded soldiers in military hospitals. In Saratov, the soldiers formed their own military committee, which put pressure on the soviet and the PEC. In Tsaritsyn, there was a garrison of only 15,000–20,000 men, but, unlike in Saratov, soldiers were billeted on the local population rather than housed in barracks; this meant they became more directly involved in events and helped to radicalize workers in the town.

In this volatile situation, the role of individuals could be crucial. In Cheboksary, at the time a small, largely non-industrialized port with a considerable number of Chuvash and Mari inhabitants, a young Latvian Bolshevik hothead, Karl Grasis, almost single-handedly incited soldiers, sailors and workers to break off cooperation with other socialist parties and the PEC in the town. The soldiers – some 500 of them – had been sent to the town to carry out work in the fields, and were easily brought over. When the Provisional Government deputy sent from Petrograd to calm the situation denounced Grasis at a public meeting, soldiers beat him up and imprisoned him. Grasis went on to lead the Kazan Bolsheviks.[25]

The radicalism of Tsaritsyn after February 1917 was in part due to the large concentration in the town of workers (including some 12,000 in metal and armaments factories) and soldiers, and possibly to the presence of some Polish prisoners of war. But a Bolshevik party was initially slow to form (Mensheviks dominated the soviet), and only became prominent because of the role played by Semen Minin. The son of a priest and himself a former seminarist, Minin had been arrested in 1905 and exiled for revolutionary activity in the town; but he returned in February to organize and radicalize the local Bolsheviks. The town authorities responded in a heavy-handed manner by bringing in 500 Cossacks from Saratov to arrest local Bolsheviks and ban meetings; but this only provoked sympathy for the Bolsheviks and increased their popularity, especially among garrison soldiers. After the Cossacks departed, the Bolsheviks made electoral gains and were, in effect, in control of the town even before the October Revolution.[26]

The Bolsheviks took power in Petrograd on 25/26 October 1917. By this time, Bolshevik parties were dominant in many Volga towns (not just large, industrial centres), with support primarily coming from garrison soldiers, sailors and workers. For the most part, the October Revolution passed off peacefully in Volga towns – not least because many people thought that a new, broader-based political organization would be put in place once the Constituent Assembly met in January 1918. There were, however, local disturbances. Saratov was one of the few major Volga towns to experience a violent seizure of power in October. Some members of the PEC barricaded themselves in the town hall and, helped by the presence of armed officers and military cadets, tried to resist the Bolsheviks by force. The presence of an officer training school in the town may have played a part in their decision; or it may simply have reflected a lack of realism among some PEC members. Whatever the case, their resistance was met with a sporadic bombardment of the building by the Bolsheviks, during which one officer was killed, one cadet was fatally wounded and seven others were badly hurt. The councillors were forced to surrender and were led out, hands above their heads. At bayonet point, they were marched along the main street through a hostile crowd which cried out: 'To the Volga the lot, to the Volga!'[27] In the end, the cadets and the councillors in Saratov were released without harm, but the anger and hostility shown by the crowd towards the representatives of the 'old order' was a precursor to the onset of the Civil War.

❖

Armed resistance to Bolshevik rule broke out almost immediately, but the Civil War really dates from early 1918, after the Constituent Assembly was forcibly closed down by the Bolsheviks in January. This was followed by the Treaty of Brest-Litovsk in March, as a result of which the new state ended its participation in the war, with the loss of huge parts of the Russian Empire in the west. A loose coalition of monarchists, conservatives, imperial army officers, Cossacks, liberals and non-Bolshevik socialists moved against the Bolshevik state from the east, the north and the south, with support from Allied governments. The river Volga and the middle and lower Volga regions played a crucial role in the outcome of the Civil War, and in the first instance a Volga town was the location of an ill-fated alternative government to both the Reds and the Whites. The failure of this government, in Samara, demonstrated the virtual impossibility of a more moderate outcome to the Revolution.

The background to the Samara experiment was that elections to the Constituent Assembly had produced a majority for the Socialist Revolutionary party (SRs), who won roughly 40 per cent of the vote, compared with 24 per cent for the Bolsheviks. No other party was able to muster significant support (the Kadets, a liberal party, received just under 5 per cent of the vote). The SRs were a revolutionary socialist party, but also had roots in the Russian late-nineteenth-century Populist movement, which attempted to attract peasants to the cause of overthrowing the tsar. Support for the SRs was concentrated in the countryside rather than the towns, because the party promised to give the remaining noble land to the peasants. The Bolsheviks, on the other hand, had the support of the majority of workers in the towns, and crucially had the support of soldiers. Towns on the Volga where there were garrisons voted heavily for the Bolsheviks in the Constituent Assembly elections. In Saratov, for example, some 20,000 residents (about a third of them soldiers) voted for the Bolsheviks, compared to 8,000 for the SRs, 2,000 for the Mensheviks and 12,000 for the liberal Kadets.[28] Many people at the time assumed that the Constituent Assembly would form some sort of socialist coalition; but when the Bolsheviks dissolved the assembly, their opponents were forced to take violent action to overthrow them.

The situation created a dilemma for other socialist parties, which rejected single-party rule by the Bolsheviks, but were loath to join the reactionary forces of the 'Whites'. The SR solution was to seek to reconvene the Constituent Assembly as a basis for a new government; and because they could not safely do so in Petrograd or Moscow, they chose Samara. The town was considered appropriate not only because of its strategic importance on the river Volga, but also because it was already experiencing armed unrest, and the SRs had heard that the peasants in the region were resisting the Bolsheviks. The SRs planned an uprising against the Bolsheviks in September, and hoped to encourage the Czech legion (which was fighting to liberate the Czech lands from the Austro-Hungarian Empire) to join them. This legion had been formed originally from Czech and Slovak volunteers, and had been attached to the Russian 3rd Army during the war. In 1917, it had been allowed to supplement its numbers from among prisoners of war in Russia, and by 1918 had reached a total strength of some 40,000 men. Some 7,000–8,000 members of the Czech legion entered Samara on 25 May and captured the city. The new government called itself the Committee of the Members of the Constituent Assembly, usually known by its acronym as the *Komuch*.

The Komuch formed an army and, in combination with part of the Czech legion, scored a couple of early successes on the Volga, capturing the towns of Syzran and Simbirsk in July 1918. At this point, there was some dispute about the strategic direction of the army: namely, whether it should head south, down the river, in an attempt to take Saratov (which would have consolidated its southern flank), or aim for Kazan, which would potentially open the way to Nizhnii Novgorod and Moscow. The decision was taken to attack Kazan, which fell to the combined forces; but by this point, the Komuch forces were already too stretched to hold on to the Volga towns they had taken. The Bolsheviks retook Kazan in September 1918; and two days later, they recaptured Simbirsk, forcing the Komuch army to retreat eastwards, away from the river Volga, to Ufa, where eventually it joined the Siberian forces of Admiral Aleksandr Kolchak. 'The taking of Kazan is the beginning of the death agony for the bourgeois scoundrels on the Volga, in the Urals and Siberia', declared Leon Trotsky – correctly as it turned out.[29] On 24 September, Samara was taken by

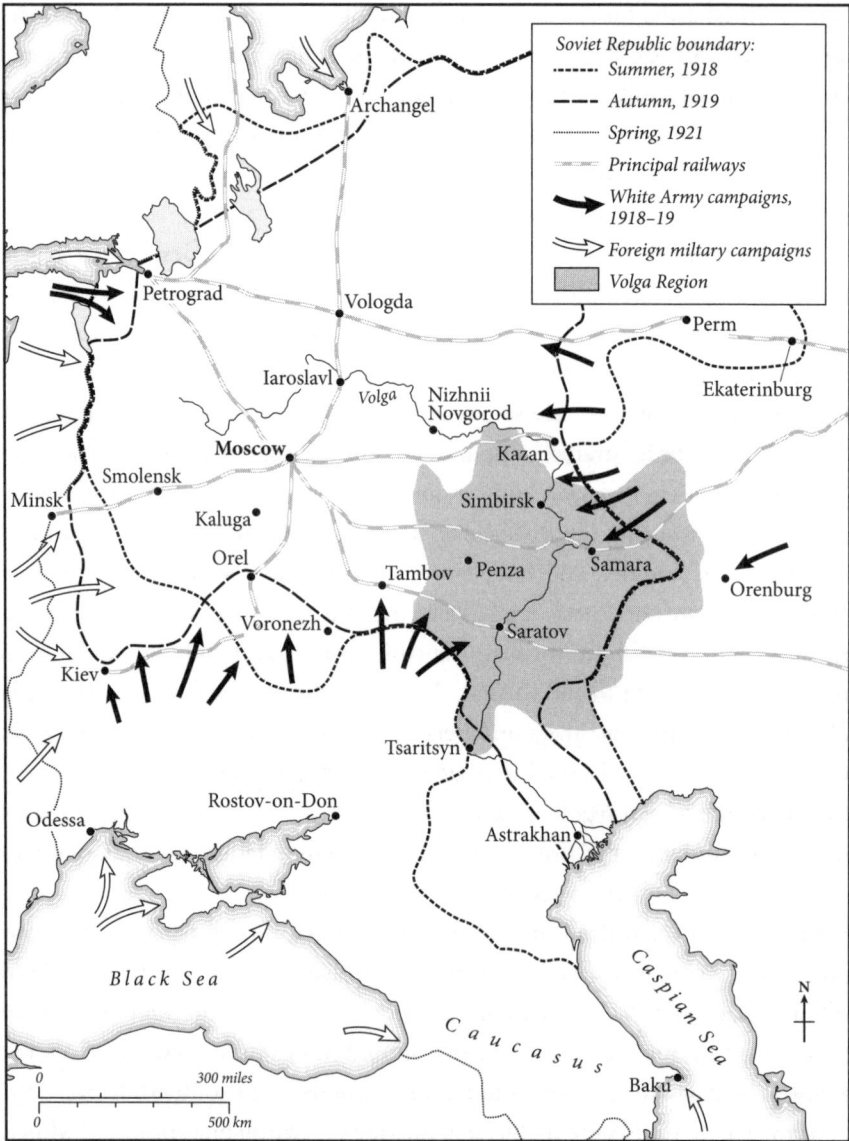

Sovjet Republic boundary:
- - - - - Summer, 1918
— — — Autumn, 1919
·········· Spring, 1921
Principal railways
White Army campaigns, 1918–19
Foreign miltary campaigns
Volga Region

Archangel

Petrograd

Vologda

Perm

Iaroslavl   Volga   Nizhnii Novgorod   Ekaterinburg

Moscow   Kazan

Smolensk   Simbirsk

Minsk   Kaluga   Penza   Samara

Orel   Orenburg

Tambov

Voronezh   Saratov

Kiev

Tsaritsyn

Rostov-on-Don

Odessa   Astrakhan

Black Sea

Caspian Sea

Caucasus

Baku

N

0   300 miles
0   500 km

7. The Russian Civil War

the Red Army, and the members of the Komuch had to flee. In November 1918, the Komuch was dissolved and the short-lived experiment of an SR government ended.

Why did the Komuch fail? In principle, it had a legitimacy that the Bolsheviks lacked, in that it claimed to represent the Constituent Assembly, an elected body. The SRs would have been the largest party in the assembly, and that fact also gave its government more credibility than either the Reds or the Whites. It had a 'people's army', mainly comprising ex-imperial army officers; in General Vladimir Kappel it had an experienced military leader; and it enjoyed support from the well-armed Czech legion. Furthermore, the Bolshevik behaviour in the first months of power was not only unpopular among industrialists and liberals, but also alienated some workers (because factory production collapsed) and some peasants (who resented their grain being requisitioned – see below). The Komuch, however, never commanded key areas of support. It was always an exclusively SR government and failed to gain the participation of liberals or other socialists; this limited its claim to represent the country, or even the aborted Constituent Assembly. It could not address the land question in a way that satisfied both landowners and peasants, and at the same time it could not get enough grain from the peasants to supply the cities under its control or to feed its army. The Komuch enacted a number of industrial reforms, including confirming an eight-hour day; but this did not bring immediate relief to beleaguered factories, and workers were alienated by its policy of curbing the power of factory committees. Trying as it did to balance the needs of factory owners and workers, of landowners and peasants, the Komuch never had genuine support from any group.[30]

Most importantly, in order to overthrow the Bolsheviks, the Komuch had to have an army, and that could only be formed from peasant conscripts. Tired of the war and wanting only to work their newly seized land, the peasants resisted conscription, and many of those who were conscripted deserted *en masse*. An army that should have numbered some 50,000–75,000 men was in practice never able to muster more than 10,000 at any given time. The Bolsheviks, on the other hand, were able to mobilize peasants more easily, because they promised that the landlords would not get their land back, whereas the Komuch could not give such clear reassurances. Peasants also feared the Red Terror more than reprisals from the 'people's army'.[31] Nor was there enough support for the 'people's army' from Kolchak's Siberian armies, because they distrusted what they perceived to be the 'red' nature of the Komuch, and because the Komuch land policy alienated White army officers who were landowners. This was a civil war, and ultimately could only be won by military means.

The river Volga played a key role in this stage of the Civil War. A new fleet, the Volga fleet, was founded by the Bolsheviks in June 1918 in Nizhnii Novgorod (at the time, the Bolsheviks had no control over the river below Nizhnii Novgorod). The fleet was involved in a number of skirmishes with White ships on the Volga and the Kama.[32] The Volga, however, was more important for transporting industrial goods, foodstuffs and men to the various fronts than as a naval battle ground. The middle and lower stretches of the river remained a crucial border between the Reds and the Whites after the defeat of the Komuch. Had the Siberian forces of Kolchak and the southern forces of Denikin been able to meet on the Volga, the outcome of the Civil War could have been very different. But after those Volga towns that had initially been captured by the Komuch fell to the Bolsheviks, the Siberian forces were never able to penetrate further west than Ufa and Perm. By November 1919, the Bolsheviks had seized Kolchak's main base in Omsk, western Siberia.

In the south, the forces of Generals Anton Denikin and Petr Wrangel (Vrangel) largely comprised Cossack forces from the Don and the Kuban (in the north Caucasus). Denikin was wary of moving too far north from the Cossack home-lands, and would have preferred to concentrate his forces in the south; but he was persuaded to move towards the Volga at Tsaritsyn and Astrakhan. The southern forces recorded a number of successes in 1919, of which the most significant was the taking of Tsaritsyn in June, with the capture of over 40,000 prisoners and a large amount of supplies and munitions.[33] Buoyed by this success, it seemed for a time that the White forces could move on Moscow up the Volga from the south, and there was a real possibility that the White forces, east and south, could meet on the Volga at Saratov. Kolchak's forces were, however, pushed back east, which enabled the Bolsheviks to concentrate their troops on attacking Denikin's forces in Ukraine, and to cut the southern armies in two by crossing the Don. At the same time, the Bolsheviks stabilized their forces north of Tsaritsyn and prevented further advances up the Volga. The White southern front was, in any event, greatly overextended, and Denikin's troops were forced to retreat. Tsaritsyn was retaken by Bolshevik forces in January 1920 (the city was renamed Stalingrad in 1925, to honour the role that Stalin allegedly played in the recapture of the town). By this time, the Whites were in retreat on all fronts: in the south, Denikin's troops were evacuated, with heavy losses, from Novorossiisk to the Crimea in February 1920; and the final evacuation of Wrangel's troops from the Crimea took place in November 1920. In the east, Kolchak's army disintegrated – he was captured and shot in February 1920. By October 1922, the last resistance ended as Vladivostok was taken.

❖

The Volga towns had been at the forefront of the conflict in the Civil War. Several of them on the middle and lower Volga were taken and retaken – Tsaritsyn, Samara,

Simbirsk, Syzran, Kazan – and others, such as Saratov and Astrakhan, had been under the threat of attack. The war, however, did not take place only on the battle-field: a second, internal, war was waged in these years – first by the Bolsheviks against the urban bourgeoisie, and second, by all armies against the peasants. In this, the Volga villages were a crucial arena.

The shortages that had plagued Volga towns during the war increased during 1917, and were further exacerbated by the Civil War. In particular, the Volga was not able to act effectively as a conduit for trade or passenger traffic, because almost all the bridges over the river were destroyed in the course of the Civil War. At the same time, railway lines were damaged; and those lines left in operation were often commandeered by troops from one side or the other. The few steamers that remained were poorly maintained and often also used for troop transport. In effect, the transportation links of the Volga reverted to the situation in the eighteenth century.

Towns had been at the forefront of revolutionary movements, and within the towns the factory workers had been the main instigators of strikes and demonstrations. It was ironic, then, that industry collapsed under the strain of the Civil War, and that life in the towns deteriorated so badly that many workers abandoned them and fled to the countryside. In Astrakhan, for example, the fishing industry, which had employed 120,000 workers in 1914, contracted sharply: by 1918, there were only 40,000 workers, and by 1919 only 24,327.[34] There was a crisis of production in Volga towns. By 1920, half of the factories in Tsaritsyn, and two-thirds of those in Saratov, were no longer operating, and over 75,000 workers in Saratov were left unemployed.[35] In Tver, where we have seen that the workers had welcomed the Revolution enthusiastically, the period from late 1917 through to 1921 was marked by a succession of bitter (albeit poorly organized) strikes, as the living standards of workers fell. There were shortages of basic foodstuffs, and rations were constantly being cut. At one point, in May 1918, workers at a Tver textile factory, concerned about the supply of bread, threatened to hang the members of the supply organization in the town and throw them into the Volga, if their demands were not met![36]

Class war was waged on the 'bourgeoisie', in part to find scapegoats for the economic situation. This was despite the fact that some small towns had few residents who could be categorized as bourgeois. In Kuznetsk, it was reported almost wistfully that 'we have almost no bourgeoisie'. In the town of Kamyshin, 125 people regarded as members of the bourgeoisie were rounded up by the police and forced onto a barge in the river, under threat of execution. In Saratov, some 300 members of the bourgeoisie were forced onto a barge in the river Volga and kept there for several months, again under threat of execution. One of those held was A.A. Minkh, a prominent member of Saratov society who had been the chair of a commission on adult education and a member of the city duma. He had been one of the town duma members who had been bombarded and forced to surrender to the Bolsheviks in October 1917. After keeping his head down for a while by working for a soviet organization, he was exposed, arrested and imprisoned in August 1918. Minkh and other prisoners were removed from their cells and lined up against a wall; they expected to be shot, but were instead taken to the barge. Twenty-five prisoners were released from the vessel in October, but Minkh and the others were kept behind. Finally, the remaining prisoners were taken out onto the deck; again, they expected to be shot, but instead were released. His ordeal almost certainly contributed to Minkh's subsequent nervous breakdown.[37]

Saratov experienced a further influx of refugees and prisoners of war in the early stages of the Civil War, and by March 1918 there were over 150,000 refugees in the town. With insufficient housing available (one contemporary described the damaged town as 'a heap of garbage and bricks'), refugees built makeshift, unhygienic camps on the banks of the Volga. The number of refugees rose steadily throughout the Civil War as a result of war and famine, and by mid-1919 the town was said to be plagued by 'all sorts of vagrants, pick-pockets, fortune tellers, singers, people with talking parrots promising happiness, petty tradesmen, Chinese etc'. Disease, including cholera, accompanied the refugees, spreading down the banks of the Volga and along the few railway lines still in operation. By 1920, the city was a place of 'hunger, cold and typhus'.[38]

We know what daily life was like for a member of the bourgeoisie in Saratov from the diary of Alexis Babine, who lived in the town from 1917 to 1922. Babine was born in a Mordvin settlement and had attended the local grammar school. In 1889, he had emigrated to the United States, where he became a librarian. He returned to Russia in 1910 to become an inspector of schools. In 1917, he was in Saratov, working as an English-language instructor at the university. Most staff at the university were liberals and had opposed the Bolsheviks in October 1917 (the same was true of Kazan University, where a number of professors joined the Whites – their photographs now appear in the Kazan University Museum). Babine could, therefore, be classified as a 'bourgeois enemy', albeit a minor one. He was scathing about the inability of liberals in the town to resist the Bolsheviks in October, because in his view – they were 'incapable of resisting force by their own exertions, discredited in the eyes of a deceived people, are utterly helpless, and merely shudder at the prospect of the impending doom'. This 'impending doom' arrived soon enough. In January 1918, Babine recalled that:

> The well-to-do people are being expelled from their residences to unsanitary basements and to hovels on the outskirts of the city . . . Landowners, priests, physicians, rich merchants, and businessmen are daily reported shot in cold blood and without even a semblance of a trial.

He noted that '135 conservatives are being kept on the Volga barge as hostages' (described above). His own life was dominated by the search for basic foodstuffs – or 'hunting for provisions', as he put it. There were shortages of meat, eggs, tea and sugar, and of cooking and heating oil. The problems were exacerbated by the arrival of refugees: already in 1918, on a rare train journey, he encountered along the tracks towards Kozlov (renamed Michurinsk in 1932) 'large crowds of tattered men and women going south on foot in quest of food'. Refugees crowded onto the few boats still on the Volga, so that 'the smell and dirt are indescribable among third-class passengers, especially the Tatars'. By 1921, 'famished Germans' (see below) were fleeing their homes and camping on the banks of the river. In March 1921, he recalled a dreadful tale:

> Two small boys [German colonists] came to the Mitrofan market this morning. A kind-hearted milkwoman gave them a good feed, after which the boys dropped to the ground, and died.[39]

Spirits were only kept up by constant rumours that the Bolshevik leaders had died, or that the Whites or their western allies had captured key towns.

War, followed by civil war, had left armed and sometimes desperate soldiers roaming the countryside and often causing violent disorders. In Rtishchevo, a railway junction in Saratov province, a particularly gruesome incident occurred. The mill, owned by a local German called Teilman, had burned down in mysterious circumstances. Rumour had it that Teilman had started the fire deliberately, to get his own back on local peasants with whom he was in dispute. Teilman tried to slip away unnoticed, but he was spotted at the railway station and subjected to an impromptu 'trial' by soldiers. Found guilty, he was shot and bayoneted. His body was then laid on the railway tracks and a train driver was forced to drive back and forth over it several times. The soldiers then stuffed cigarettes in the corpse's mouth, set the body on fire and danced around it, to the abiding horror of the train passengers looking on.[40]

Peasants in the Volga countryside had not reacted with instant enthusiasm to the Revolution of February 1917; indeed, it has been noted above that it was hard to get the news to the villages, and even harder in non-Russian villages because of the lack of translators. Peasants had, however, been quick to seize land owned by nobles in 1917 and to force 'separators' back into the village communes. This was a reflection of the traditional peasant belief that the land they tilled belonged to them individually, and to the commune. An indication of this is that peasants were often prepared to allocate some land to 'their' nobleman, but were not prepared to do the same for 'outsiders' from other villages. Peasants also seized woodland in noble ownership (and there was mass woodcutting in case the forests were reclaimed at a later date). Most of this activity was conducted by *Russian* peasants; the German, Mordvin and Tatar villages were less active, but that may have been because those peasants were primarily former state peasants, and not serfs, so that there were fewer noble estates to seize.

Some land seizures took place without violence, but the whole process could be anarchic. One peasant, Ia. Bergishev, recalled that in Kazan province:

They started to destroy . . . teams of some hundred men and women began to steal and carry off livestock, grain, etc. The livestock was driven into the village, the noise, the cries were such that the animals ran through the village, were lashed, and not getting away, fell against each other . . . blood flowed.[41]

Many noble landowners left the country, if they could; those who remained behind – usually on account of necessity, infirmity or old age – faced an uncertain fate. Olga Aksakova, a descendant of the writer Sergei Aksakov, was a well-known philanthropist and had donated 25,000 roubles to Samara University. She died from starvation on her estate in 1921, aged 73 (a plaque has recently been erected in her memory at the site of her former estate).[42]

The main concern of the peasants during the Revolution and the Civil War was about the use of the land and woods they had seized, and not the future political organization of the country. A village teacher in Nizhnii Novgorod province complained that the peasants 'are silent about the schools, about the state structure and about *zemstvo*, but they talk and talk about who is to get the most land'.[43] There was particular resistance to the presence of any *Russian* officials by non-Russian peasants, who were no less wary of representatives of the Provisional Government in 1917 or the Bolsheviks thereafter than they had been of tsarist officials. Peasants feared census-takers, because they were convinced (not without cause) that any register would lead to the seizure of the little grain they had. When the census-takers came to a Mari village in Kazan province, 'they found empty houses and stores'. In the Chuvash village of Bolshoi Sundir, the president of the provisions administration was murdered. In particular, peasants, and non-Russians above all, suspected that any 'norms' of supply with fixed prices would disadvantage them.[44]

It was the food supply crisis in the towns during the Civil War that led to violent conflict between the Bolsheviks and the peasants. The Bolsheviks – often from the city and with little understanding of local conditions, or of rural social relationships generally – attempted to stimulate a 'class war' in the village by inciting poor peasants against so-called middle and wealthy peasants, and by blaming shortages on rich, exploitative peasants, speculators and deserters (and on Muslim clergy in Tatar villages). Instead, they found themselves in effect fighting a civil war against the peasants. The conflict was particularly acute in the middle Volga region, in part because of its importance as a grain-producing area, and in part because the region was located just behind the front line. In the process, the Bolsheviks set the scene for later conflict, and ultimately for the complete defeat of the peasants.

In the summer of 1918, the Bolsheviks set up committees of poor peasants (known by the acronym *kombedy*) to expropriate grain from rich peasants (called *kulaks*, from the Russian word for 'fist') to feed the towns. Over 800 *kombedy* were set up in Saratov province alone. The official picture was that poor peasants willingly undertook this class war: 'we, the poor, will always be oppressed by the kulaks, and see the only rescue in Soviet power', stated poor peasants in the village of Bobrovka, in Samara province. The reality was that grain shortages occurred because less land was sown during the Civil War, and this forced expropriation only exacerbated the situation. Furthermore, the 'class' element was completely at odds with the reality of peasant relations in the village. Some villages reported that they had no kulaks; others that they were all poor peasants. A village in Melekess district suggested that the middle peasants should decide such things, and proposed that *kombedy* should be replaced by a 'middle peasant soviet'.[45] The reality was that *kombedy* were often made up of non-peasants, including hooligan elements, and were at odds with all, or almost all, the peasants in the village. Bolsheviks themselves blamed the failure of the *kombedy* on their non-peasant members: one called them 'criminal elements, hooligans, drunks, robbers, and loafers'.[46] In Astrakhan, the *kombedy* included poor Kalmyks, Kirgiz and Mordvins, presumably to take revenge on perceived richer Russian peasants.[47]

Peasants complained vigorously about grain requisitioning generally, and about the work of the *kombedy* in particular, as these extracts from peasant delegates to the Syzran district soviet assembly demonstrate:

245

They always blame everything on the peasant, even if he has just delivered his last sack of grain to the delivery centre . . . The peasants gave their best in both the imperial war and the civil one: they gave both their sons and their products. But who cares about the peasant? . . . [The commissars] take away our meat, but then let it rot . . . Also it is said that workers in the factories are given jam and sausage while the peasants see none of these . . . The authorities take nothing into account, neither the needs of the peasants, nor anything else . . . Nobody trusts the peasants – neither the workers nor the speculators; they are treated like the lowest of the low.[48]

Resentment led inevitably to violence. Armed detachments accompanied the *kombedy*, but this often provoked more violence. In the village of Akseevka, a village on the river Volga in Saratov province, enraged peasants killed and injured 12 soldiers and a Bolshevik official. They then disbanded the soviet, destroyed documents and replaced the portraits of Lenin and Trotsky with icons! In response, punitive detachments captured over 500 villagers and shot 34 of them, leaving behind a detachment of 50 men to keep order.[49] At the same time, peasants were terrorized by bandits who roamed the countryside, often led by army deserters from Reds or Whites. By 1921, the countryside was in a state of lawlessness, chaos and, worst of all, hunger.

The combination of the forced requisitioning of grain (which left the peasants with no reserves) and a harsh winter in 1920–21, followed by a dry spring, led to famine on the Volga in 1921. In August and September of that year, Carl Eric Bechhofer travelled down the river Volga and was faced with a human nightmare. In Samara, he found 'people yellow with under-nourishment' and the children 'listless and moody'. Worst of all were refugees from villages – in 'absolute despair' in a 'terrible, slow, public waiting for death' – desperately trying to buy a ticket for a Volga steamer to take them somewhere – anywhere – where there might be food. Thousands of refugees lined the banks of the Volga, in 'the filth, the stench, and the flies of the banks of the Volga, which is this year many feet below its normal level'. When a steamer came, there were fights to board, and on the boats the 'lower decks were chockablock with fleeing peasants', while the upper decks, 'which had once been the promenades of holiday-seeking Russian aristocrats, were covered with miscellaneous rags and bundles'.[50]

The middle and lower Volga regions were affected by grain shortages in 1921. There were some 36,000 homeless children in Simbirsk alone, and children's homes throughout the Volga region were overcrowded. Food kitchens in Kazan, Samara,

Simbirsk and Ufa fed over 6,000 people a day.[51] In Kazan province, it was estimated that peasants had lost 65 per cent of their livestock and 96 per cent of their sheep, compared with their holdings in 1917.[52] A contemporary in Samara province commented that 'In 1921 there was a Great Hunger in Samara province . . . which had never been experienced since the beginning of time.' In one village in Samara province, there was a case of cannibalism involving a family of six – a mother, father and four children – where only one child remained alive.[53]

The Volga Germans were particularly badly affected by the famine; indeed, their colonies were referred to as the 'centre of the disaster' (and note the comment above by Babine on German refugees in Saratov). The colonists had been regarded by the Bolsheviks as potentially disloyal, and grain requisitioning had been exceptionally brutal and extensive in their villages. A common practice was to position a machine gun in the centre of the village and fire wildly before searching every possible hiding place, events which were referred to as 'brooms' by the colonists, because they were left with so little. Atrocities were committed by both sides. Peasants caught hiding grain were beaten and thousands were shot, and so much was taken that 'village after village was left with nothing to plant in the spring and nothing to eat during the winter'. Some colonists in desperation joined local anti-Bolshevik bandit gangs. When in 1921 a bandit group captured some Bolsheviks, 'a mob of people attacked the [Bolshevik] prisoners, cut off their ears and noses, gouged out their eyes and killed them with pitchforks'. In another incident it was reported that 'almost all the communists and Komsomol [Young Communist League] members in the meadow-side settlements were either butchered or shot, or bound and shoved under the ice [of the Volga].' The rebels were then 'gunned down' by the Red Army.[54]

The result of the savage requisitioning was that colonists had no seed to sow. Distrust on both sides meant that it was late on before it became clear that the colonists, especially those on the eastern 'meadow' side of the river, were suffering from famine. People ate 'various grasses, cabbages, onions, wild garlic, dogs, cats, rats, frogs, field mice, hedgehogs, and dead fish gathered from the Volga'. Famine victims were swollen, too weak to speak and resigned to death. An Extraordinary All-Russian Commission Investigating the Famine Area began to report back at the beginning of August 1921. The scenes it recounted were shocking. In one village:

> In one house was found a young woman dead from hunger who had just given birth; in the next room lay another woman who had also starved to death. In another house were found children ranging in age from seven to seventeen

gnawing on the bones of one of their slaughtered dogs. The condition of these children, four in number, was wretched; all were swollen, emaciated, and debilitated, incapable of moving on their own . . . In the fourth house the Commission found a couple so badly swollen that they were wholly incapable of moving; they lay on the floor with their two fourteen- and sixteen-year-old daughters and when asked what they still hoped for they replied, 'We have begged God's forgiveness and are prepared to die.'[55]

It is estimated that 48,000 colonists died (10 per cent of the population) and 70,000 fled. Many of those who survived were left weak and in poor health.

By 1921, the Bolsheviks had won the Civil War. The cost of the Great War and then the Civil War had, however, been enormous. Industry had collapsed and workers were either living in appalling conditions or had fled the towns and returned to villages. In the countryside there was almost open warfare between Bolsheviks and peasants. In the novel *Zuleikha* by the contemporary Tatar writer Guzel Yakhina, which is set in a Tatar village not far from Kazan, Zuleikha reflects that:

Her father had told her a lot about the Golden Horde, whose harsh, narrow-eyed emissaries collected tribute for several centuries in this part of the world and took it to their merciless leader Genghis Khan, his children, grandchildren, and great-grandchildren. The Red Hordesmen collected tribute, too, but Zuleikha didn't know who received it.[56]

To the simple peasant, it seemed that nothing ever changed: the Mongols, the tsar, the noble landowner, the Bolsheviks – all took his grain and his money and abused him. In the novel, Zuleikha's husband is shot dead for hoarding grain, she is arrested and her blind and elderly mother-in-law is left behind alone to face almost certain death. This brutal resolution to the 'peasant problem' is the subject of the next chapter.

14
COLLECTIVIZATION AND REPRESSION ON THE VOLGA

In 1929, the rumour spread on the middle Volga that the reign of the Antichrist had begun. 'Soviet power is not of God, but of Antichrist' was the saying at the time, and the focus of evil was the collective farm (abbreviated to *kolkhoz* in Russian). Peasants were warned not to join the collective farm for fear of being stamped on the forehead with the mark of the Antichrist, to identify them for damnation at the time of the Second Coming. On the lower Volga at the same time, a Cossack warned: 'the collective farm – this is the devil's branding, from which [you] need to save yourself in order to enter the kingdom of God'.[1] In a deeply troubled countryside, apocalyptic rumours were rife in the late 1920s and early 1930s, as were warnings that Cossacks would massacre peasants who had joined the collective farms. To understand the depth of these fears, and the peasants' conviction that their world was being turned upside down, we need to examine the process and impact of collectivization in the countryside.

Collectivization was implemented in all regions of the Soviet Union and affected all forms of activity (even reindeer herds were collectivized in the far north) and ethnic Russians and non-Russians alike; but the process was particularly intense in the Volga region, which was such an important source of grain. Rich peasants could be deported, imprisoned or simply starved to death after Stalin ordered 'the liquidation of the kulaks as a class'. The numbers who died as a result of this policy are hard to estimate, because the devastating economic consequences of collectivization lasted longer than the immediate shock. It could be that up to 5 million peasants who were defined as kulaks died in the villages or as a consequence of being deported; but as many peasants again may have died in the famine of 1930–32, which has largely been attributed to the policies of collectivization.

'Dekulakization' was followed by the purge of a wider section of society, where anyone who could be defined as an 'enemy of people' – or was simply associated with such enemies – could be sent to a corrective labour camp (*gulag*). At the peak, in the early 1940s, some 4 million people were in labour camps and other prisons in the Soviet Union.[2] Far more people passed through the camps – perhaps as many as 18 million from the 1920s to the 1950s. The young Soviet state had declared war on its people.

❖

The introduction of the New Economic Policy (NEP) in 1921 allowed peasants to be in control of what they grew and to sell their products at market prices. In the short term, this led to an increase in productivity and a recovery from the famine years of 1920–21, in the Volga region and elsewhere in the country. NEP created, however, both an economic and an ideological problem for the Soviet state. By giving control back to the peasants, the state could not manage the market forces that determined production. The government could not force the peasants to sow more grain at a time when other agricultural products, such as milk, butter or cheese, might be more lucrative; nor could it override the market and determine the price at which grain was sold. By the late 1920s, the situation had become serious because there was not enough grain to feed the workers in the rapidly growing new industrial cities – a situation that was exacerbated by a poor harvest in 1927. The shortage of grain was also due to the unwillingness of the peasants to sell their grain at artificially low prices, particularly when the manufactured goods they could purchase in return were priced very high, were of poor quality or were simply not available at all. Under these circumstances, the peasants preferred to consume their grain rather than sell it; in other words, they turned their backs on the market just at the point when their grain was needed most. This led to the reintroduction of grain requisitioning in 1928, presented at the time as an 'extraordinary measure' just for that year. The peasants resisted the expropriation of their grain, and the use of force by the state only served to discourage them from sowing more grain for the following year.

A second, and more fundamental, problem was that the Communist Party and its general secretary, Joseph Stalin, regarded NEP as a purely temporary measure and ideologically unsound. Private peasant farming was regarded by the new Soviet state as unacceptable capitalism. Collective farms were, in its view, the proper socialist solution; they were also supposed to be more efficient and to provide the economic conditions for social progress in the countryside. More efficient agriculture would lead to a smaller labour force in the countryside, which would in turn increase the number of peasants who could join the new industrial workforce and lead to greater industrialization. NEP was also seen to have been most advantageous to those social groups that had been characterized as 'enemies of the people' during the Revolution and the Civil War – speculators and wealthy peasants.

During the Civil War, the Bolsheviks had attempted to introduce collective farms on the Volga and in other parts of the country which they controlled; but this was a voluntary movement and had little success. There were two requirements for mass collectivization: first, there had to be an absolute determination on the part of the state to implement the policy from above; second, an enemy had to be identified and scapegoated. Those two conditions were met in 1929. In November of that year, the Central Committee of the Communist Party took the decision to go ahead with mass collectivization; the process, in other words, was determined and implemented from above (as so many earlier initiatives concerning peasants had been). At the same time, the enemy was declared to be the wealthy peasants – the kulaks – who were blamed for creating shortages by hoarding their cereals and sabotaging grain collections. Local party officials and village soviets were ordered to identify kulaks and confiscate their land and belongings. Collectivization proceeded at an extraordinary pace throughout the Soviet Union: by 1931, some 50 per cent of farms were in collectives; by 1940, the figure was over 96 per cent.

The Volga was one of the main areas in which collectivization took place rapidly, on account of its importance in the production and export of grain. By 1930, the lower Volga was already heavily collectivized (70 per cent in some regions), and the process continued relentlessly over the next decade. When mass collectivization occurred, it could happen very quickly, following just a few meetings (or even only one). For example, in the village of Lom (in what is now the Mari El Republic) four meetings were held and then the livestock of eight kulak families was simply seized.[3] The official story of collectivization presented at the time – and then in Soviet publications from the 1950s to the 1980s – was that the poorest peasants welcomed collectivization, and easily and willingly identified the kulaks as their enemies. Poor peasants in a Samara village, for example, were recorded as denouncing the kulaks in their village for 'hiring labour', and blamed kulaks for their poverty and lack of horses.[4] In Chistopol district (Kazan province), the peasants in the village of Buldyra allegedly declared that 'we, the poor' saw the kulaks 'as our age-old enemy'.[5]

Of course, peasants had always been aware of disparities in wealth and good fortune (such as having several healthy sons), and everyone knew the wealthiest peasants in their village. Nor was it against human nature for peasants to want to see richer neighbours put in their place, or want to settle old scores. As we have seen, before the emancipation of the serfs in 1861, peasant communes deliberately conscripted boys from poor households because those households were less able to contribute to collective taxes. After the revolutions of 1917, peasants had forced 'separators' back into the commune – a clear indication that they resented those wealthier peasants who had the ability, or the courage, to set up separate farms outside the commune. However, during collectivization the definition used for rich peasants – kulaks – was almost entirely artificially created by Communist Party officials in Moscow. 'Exploited labour' was the defining communist characteristic of kulaks; but this was a difficult concept for rural society, given that peasants regularly hired other peasants at harvest time as a normal part of agricultural relationships. Wealth, as perceived by peasants, was usually related to the ownership of land and, above all, of horses (hence the comment above from peasants in Samara about poor peasants lacking horses).

It was not surprising that peasants, on the Volga and elsewhere, often found it hard to identify 'the enemy' in their midst, and it was common for villages to claim that 'we have no kulaks here'. A 'quota' of kulaks was required to satisfy local Communist Party officials, which meant that the peasants often had to penalize other peasants who were not wealthy and who seemed, on the surface anyway, to have been loyal to the Soviet regime. For example, a peasant in Tver province called Aleksei Marov, who had served in the Red Army during the Civil War, owned 3.4 hectares of land. Despite having only one horse and one cow, he was identified and condemned as a kulak. Another Tver peasant, Vasilii Paskin, had six young children and an invalid wife, yet his was one of 22 families expelled as kulaks to Siberia.[6] In Samara province, one N.F. Lykov appealed against his identification as a kulak, arguing that he was not wealthy and had served the regime loyally in the Red Army. He offered to give his home to the collective farm, and pointed out that not only was he now unable to live as a peasant in the village, but the same applied to his wife, who had been a teacher and had also lost her rights.[7]

Among the non-Russians on the Volga, it proved particularly difficult to identify kulaks using the Soviet definition. Collectivization was particularly savage in the German colonies: the prosperity of those villages, and the neatness of the farmhouses, meant that *all* farmers could look like kulaks to local officials. Collectivization there was swift: by January 1930, 68 per cent of the German colonists were in collective farms, and by 1931 that figure had risen to over 85 per cent.[8] By contrast, most Mari and Chuvash peasants on the Volga were poor, and on the whole poorer than Russian peasants; these communities often found it very difficult to identify *any* rich peasants at all. One delegate from a Mari village in 1928 stated that he could not forge shared loyalties between poor and middle peasants against the wealthy peasants, because all of the peasants were so poor that they had nothing to lose from collectivization. In the Mari village of Gornomari in 1930, the women of the village stated, 'we have no kulaks, so a *kolkhoz* is not needed'. Some peasants in the Mari villages who had to be designated as kulaks were clearly not wealthy: one A.N. Solovev, in the village of Kuguner, was said not even to own a cow or basic household utensils.[9] The same was true for Tatars. A report on dekulakization from the Tatar Republic noted that a certain Mutiashin was condemned as a kulak because he had purchased a feather bed during a famine year – he had one horse, one cow and had served in the Red Army, in the service of which he had been wounded. The same report noted the petty items taken from kulaks – clothes, bast shoes, children's skates, bric-a-brac.[10]

Resistance to collectivization often came from many peasants, irrespective of wealth and ethnic background, and sometimes from *all* peasants in a village. Peasants facing collectivization would sometimes slaughter all their livestock and gorge themselves in great feasts, rather than hand over their beasts to the collective farm. Resistance could also be open and violent, if ultimately futile, given the absolute determination of the state to collectivize. Peasants would sometimes break up the meetings of the village soviet when it was about to collectivize the land. Their feelings were clearly expressed. In January 1930, in the village of Arkhangelskoe in Kuznetsk region, a poor peasant called Surkov stood up and declared: 'You are pillaging the peasants and have pillaged the kulaks . . . under the Tsar we lived better, the collective farm is a noose. Down with slavery, long live freedom.' His speech was greeted with shouts of 'hurrah', 'that's our Surkov' and 'Down with the collective farm'. In another Volga village, called Aleksandro-Bogdenovka, it was the middle peasants who broke up several meetings: one woman banged her fist on the desk and yelled 'To hell with your collective farm!'[11]

A recent fascinating publication attempted to give the 'peasant voice' by using memoirs, reports from peasant delegates and the oral histories of peasants who had experienced collectivization. Many peasants were unable to remember clearly the expulsion of kulak families, perhaps in a sort of collective amnesia; but some claimed to recall it with sorrow, or stated that they had been unable to oppose the actions out of fear – not an unrealistic concern, given that expressing support for the 'enemy' was tantamount to being an enemy oneself.[12]

There were mass disturbances of peasants in the 1930s across the country, but they were particularly numerous on the middle and lower Volga, perhaps because of the speed and intensity of collectivization in that area, or perhaps because the traditions of peasant violence – from Razin and Pugachev through to the nineteenth century – lingered on in the popular memory. Over 1,000 disturbances were recorded on the lower Volga, involving over 100,000 peasants; and 777 disturbances were recorded on the middle Volga, involving over 140,000 peasants. Some of these disturbances could involve over a thousand peasants. In the village of Nachalova, Astrakhan province, a crowd of some 700 people marched on the village soviet and stormed the building. In the ensuing chaos, six or eight party members and villagers were killed and another eight people injured. The police arrested almost 150 peasants, including kulaks, middle and poor peasants (and one collective farmer).[13] There were also major disturbances in the German colonies on the Volga, with riots in over 20 villages. The target of peasant anger could be anyone associated with the regime. A female member of the rural soviet in the Volga region, Anastasiia Semkova, was murdered and her body burned on a bonfire.[14] Village teachers, including women, were beaten up and murdered, as were party officials and grain requisitioners.

Religious belief could be a factor in opposition to collectivization, because the process often went hand in hand with the closure of at least some village churches and with the use of church buildings by the collective farm – as schools, granaries and clubhouses, for example. It was symbolic that attacks on the building of the village soviet frequently involved taking the bell from the office and returning it to the church. In the village of Novinka in Samara province, women returned the bell to the church and then attacked a workman who was dismantling the altar with the cry of 'let's kill all the communists'.[15] The rumours concerning the coming of the Antichrist, quoted at the beginning of this chapter, were linked to the view of peasants, often led by priests, that communists and those behind the collective farms were 'godless'. An activist reported from the middle Volga that:

> Everywhere the priests are spreading the legend that in Penza at the Maiden's Convent a light issuing from the cross is burning day and night, and it is necessary to say that the people go there, the devil knows how many, to look at those miracles. Besides this [the priests] say that soon the Roman pope will come, the government will fall, and all the communists and collective farmers will be crushed.[16]

When a priest in Tver province told a crowd of female peasants that he was no longer going to be able to hold religious services, the group marched to the soviet building, declaring 'down with Soviet power and communists, long live the priest and the Church'. Protests were not the sole preserve of Christians. In the Muslim village of Enganaevo in Ulianovsk (former Simbirsk) province, preparations were under way to collectivize several villages when a crowd of some 1,500 peasants marched on the soviet, shouting, 'We will not let them exile the wealthy and the mullahs.' Troops had to be sent in to restore order, and 13 people were arrested.[17]

The process of the expulsion of kulaks – 'dekulakization' – could be as arbitrary as their identification. In some cases, richer peasants fled before collectivization could take place. It has been claimed that up to a fifth of peasants fled from the middle Volga area, rather than face 'dekulakization'.[18] Many wealthy peasants from Tver province left for work in Leningrad (the former St Petersburg), rather than wait to have their land and possessions confiscated. In other villages, peasants designated as kulaks lost their lands, but were allowed to remain. That could be seen as traditional 'equalizing', rather like a 'separator' being forced back into the commune, or nobles after 1917 having most of their estates seized but being given some land. For example, a peasant called V.E. Kokorin, again from the village of Lom, was described as 'clearly a kulak' and had both his reaping and his threshing machine confiscated, but was left in his home.[19] In the villages of Bulaevo and Novoe Nikitino in Kashirin region of Astrakhan province, poor peasants stated that they did not wish to expel kulaks and declared that 'we dekulakized them, now let them stay and work with us'.[20] In the village of Shakino on the lower Volga, the peasants refused to allow the deportation of kulaks: the women surrounded the carts holding the kulaks and refused to let them leave.[21]

Such examples were, however, the exception; and kulaks who promised to hand over their goods if they could join a collective farm were often rebuffed as 'enemies'. Richer peasants were frequently thrown out of their houses with their families – often with almost no possessions or warm clothing, whatever the weather – and were sometimes left to die in the cold. One Tatar peasant recalled that 'all was taken: the house, the cattle, the samovar from the table, and weight clocks from the wall'. Her grandmother was wearing a new dress: 'the village activists demanded that she take it off and put on an old one; they were pulling her dress'. Another recalled that everything was taken, including clothing, mattresses and towels. She was a baby and had been wrapped in a blanket and put to bed; but she was thrown on the bed boards, as the village soviet members tried to seize the bedsheets and the blanket. Many alleged kulaks were deported under appalling conditions, with almost no food, water or shelter. A Tatar recorded her experience as a child: 'In August 1932, they put our mother and her five children on a cart and took her to an unknown destination.' They were sent north of Irkutsk, in eastern Siberia,

255

where 'Our clothes were no good for such cold weather . . . And so we suffered in the freezing cold, without knowing why we were being punished.' Another peasant recalled that they had a two-storey house, 2 horses, 2 cows, 15 sheep, tools and a Singer sewing-machine. The family was sent to Krasnoiarsk, in western Siberia, where 'We worked in the mines and built roads in the taiga' and 'we lived in dugout shelters and tents'.[22]

One German colonist wrote the following in a letter dated 31 March 1930, without any expressed emotion:

> I was taken in the middle of the night by the president of the village soviet and the militia . . . I was sent with all my family, comprising nine people: my wife who was 42 years old, my son Georg – 18 years old, Gottlieb – 14 years old, Heinrich – 11 years old, daughter Anna – 9 years old, son Ivan – 8 years old, Jacob – 6 years old, Philip – 4 years old. Three of them have already died from impossibly harsh conditions.[23]

Some 2 million peasants defined as kulaks were deported from European Russia to Siberia and Kazakhstan; of those, some 500,000 died, either on the journey or once they reached their destination, where there were no means of support. Kulaks could be imprisoned; many were simply shot.

The immediate economic consequences of collectivization were disastrous. The amount of grain sown dropped dramatically, and the number of livestock plummeted. By 1932, there was a full-scale famine, which affected much of the middle and lower Volga, the north Caucasus and, in particular, Ukraine. It is impossible to know how many people perished in the famine – not just because of the difficulty of interpreting statistics, but because, for political reasons, the numbers may have been deliberately distorted. Stalin denied that there was a famine at all; some peasants were accused of faking it (it was claimed that peasants in Volga towns were 'passing themselves off as ruined kolkhozniks [collective farm members]'),[24] and in other Volga villages a new round of anti-kulak dispossessions took place, as scapegoats were sought for the famine. The number of deaths in the Soviet Union may have been over 12 million, with some 7 million of those in Ukraine.

Peasants on the Volga who survived to record their experiences many years later recalled that this was a 'terrible' time. Some of them could recall the famine of 1921 and noted that – unlike then – this time there was no international help available (not least because of the official denial of the famine). A peasant called I. Merkulov, from the village of Teplovka, in Saratov province, recalled his family selling off their livestock and then their household articles in order to buy bread. In 1933, a peasant woman from the village of Atamanovka (Volgograd province) recalled trying to steal a little grain from the collective farm by pouring some into her gloves and pocket. She was searched and the grain fell out: 'the hunger was terrible!' Two elderly residents of the same village recalled that, while out collecting nettles, they had come across a stray calf, which they stole and slaughtered: 'that was what the hunger was like then!' They added that in those days, the struggle for survival led simply to such dishonest actions.[25] Many peasants saw the famine as a deliberate policy by the Soviet state to starve them into submission: in Saratov and Penza provinces, the story was told that the government was treating them like circus animals that were starved to get them to obey.[26]

The famine was particularly acute in the Volga German colonies. Letters from colonists recorded the disaster: 'Four of Brother Martin's children have died of hunger and the rest are not far from it'; 'We have had no bread, meat or fat for five months . . . many are dying'; 'dogs are no longer to be seen, nor cats'. It is estimated that some 140,000 colonists may have starved to death.[27] Andrew Cairns, a Canadian, witnessed the distress on the river Volga at first hand while on a steamer. On the lower deck were 'hungry' people, who were trying to escape starvation: 'thousands of miserable people packed like sardines on a wet filthy floor, on top of barrels of fish'.[28] Soviet reports revealed the appalling consequences: in Cheboksary district in March 1933, there were allegations of murder and cannibalism; and the following month, a similar allegation was made in the lower Volga region.[29]

Life on the collective farm was hard, even after the immediate threat of famine had receded. Peasants resented having to work so hard when they received so little in return from the state. People in what is now the territory of the Mari El Republic complained that they had been middle peasants (that is, not the poorest, but not kulaks either) but now 'had nothing', for which they had to 'work day and night'. Other peasants in the same region declared 'This is not life but hard labour' and 'Soviet power robs us.'[30] There was an instinctive dislike of the forced 'equalization' of peasants, irrespective of their contribution to the community. A 'middle' peasant in Samara province complained: 'I work day and night, and on the *kolkhoz* others do not work or work any old how, but grain is given equally. Why should I work? Better that I do it alone for myself.'[31]

Nobody was fooled either by the collective farm names, which reflected Soviet power but lacked any hint of irony – 'Path to Socialism', 'The Red Dawn', 'The Paris Commune', 'Lenin' or 'Stalin' – or by the establishment of 'socialist' holidays on former religious feast days. Most resistance was passive: the new collective farmers worked only as much as was absolutely necessary. There was no point in being innovative or independently minded, because there was no reward for it – and conversely, there was always the danger that anyone who stood out could be denounced for some form of anti-Soviet activity, or condemned out of jealousy or fear. Theft from the collective farm came to be seen as not a real crime, because everything was owned in common. Life was dull and monotonous in the villages, and the spirit of the peasants had been broken by collectivization. Those who could escape left the village for the town.

❖

Kulaks had been identified as 'enemies of the people'. In the novel *Zuleikha*, Guzel Yakhina puts the following justification into the mind of the Red commissar, Ignatov, after he has delivered his 'good crop' of kulaks to prison in Kazan:

> One more useful thing had been accomplished, one more grain of sand had been tossed on the scales of history. This is how a people shapes its country's future, one grain at a time, one after another. A future that will certainly become a world victory, an unavoidable triumph of revolution both personally for him, Ignatov, and for millions of his Soviet brothers.[32]

That definition, however, spread to encompass many other members of Soviet society, and the grain of sand became a mountain. At first, these enemies were clearly associated with opposition to the Soviet state: former White officers; police and other officials of the tsarist regime; imperial army officers; leaders of alternative socialist parties. Intellectuals, writers and ministers of religion (Christian and non-Christian) posed a particular threat to the regime and were persecuted from the 1920s. It has been estimated that 3.7 million people – mostly former officials, traders, 'exploiters of labour' and clergy had been disenfranchised by 1929 (that is, not allowed to vote, but this label also in practice deprived them of work and housing).

Stalin used the assassination of Sergei Kirov, first secretary of the Leningrad Communist Party, in December 1934 as an excuse to purge what he regarded as opposition within the party in a number of high-profile show trials. The People's Commissariat for Internal Affairs (NKVD – that is, the secret police) had quotas to fulfil of 'enemies', in the same way that local officials and *kombedy* had to find quotas of kulaks to condemn. The purges mushroomed to include anyone who could be accused of disloyalty to the regime, or who sheltered someone who was so defined. A careless word, an expression of support for someone arrested, a casual comment that was deemed ideologically incorrect, a foreign name, a letter from abroad – anything could result in arrest. This created an atmosphere of hysteria throughout the Soviet Union. Neighbours denounced neighbours; students denounced teachers and professors; children denounced parents. Over a million people died in the Great Purge, and many more had their lives destroyed by arbitrary arrest and imprisonment in labour camps. This was a reign of terror, waged by the state against the people.

How the individual had behaved was often irrelevant, and simply a negative label was enough to be condemned out of hand. For example, Kirill Zarubo had been a police officer in pre-revolutionary Nizhnii Novgorod and, according to him, had been spared by the crowd in February 1917 because, as he recorded, 'the citizens had nothing against me'. He then served in the Red Army, and after he had been invalided out had worked with the river police. Despite this service, his former post led to his disenfranchisement; he was arrested in 1937 and shot. Arbitrariness was endemic in the Soviet system: V.A. Chernogubov, for example, had been a police officer and then assistant to the head of prisons in Nizhnii Novgorod before the Revolution, and then became supervisor of the market from 1913 to 1918, but continued to work for Soviet institutions and retired on a pension in 1925.[33] He was one of the few lucky ones.

Arrests took place in the Volga region in the 1930s, as they did throughout the Soviet Union. An example of one region gives an idea of the extent of this repression. Balakhna was an average-sized district town in Nizhnii Novgorod province (population: 33,000 in 1968), on the western bank on the Volga. It was known for ship-building and industry, including a paper factory. One thousand people were arrested in the Balakhna district; of those, 218 were shot and 109 died in prison or in the camps; 220 were sentenced to 10 years or more in the camps, 162 to 5–9 years, 162 to 1–4 years and 129 were exiled or imprisoned for under a year. The victims were ordinary people, mostly factory workers or employees in local institutions; but they also included 76 peasants, 40 clergy, 21 teachers and students, and 12 doctors or hospital employees. Six victims were categorized as 'pensioners, invalids and children'.[34]

When the NKVD set up its headquarters in Black Lake Street in Kazan, the nature of the whole neighbourhood changed. Eugenia Ginzburg, who wrote one of the most harrowing and insightful accounts of life in the labour camps, lived in Kazan and recalled that the Black Lake park had been a 'favourite haunt of shopkeepers out on a spree and there had been an expensive restaurant and a vaudeville theatre in the grounds', but after the NKVD moved in:

. . . the name lost its original association and took on the same meaning as 'Lubianka' in Moscow. People said: 'Watch your tongue, or you'll find yourself at Black Lake', or 'Have you heard? They took him off to Black Lake last night'. The very words 'cellars at Black Lake' aroused terror.

Ginzburg and her husband were both loyal Communist Party members. In 1935, she was interrogated and then reprimanded by the local party for not denouncing a friend, Professor N. Elvov at Kazan University, who had been arrested because of his perceived errors on the theory of permanent revolution, which he had expressed in a chapter on the 1905 Revolution for a history of the Communist Party. Already under suspicion of disloyalty, Eugenia was arrested again in 1937 for not having denounced Elvov before, and this time was expelled from the party. At this point, she and her husband realized that her arrest was imminent. Eight days later, she was taken to the NKVD headquarters in Black Lake Street, where she met up with several former Communist Party acquaintances:

How many steps down – a hundred, a thousand? I don't remember . . . And here was hell itself. A second iron door led into a narrow passage dimly lit by a single bulb close to the ceiling. The bulb gave off a special prison light – a sort of dull-red glow . . . On the left was a long row of bolted and padlocked doors . . . Behind these doors were friends of mine, Communists who had been cast down into hell before me; Professor Aksiantsev, Biktagirov of the municipal committee, Vekslin, the University Rector, and many others.[35]

Ginzburg's experiences as a Communist Party member were not unusual. A recent publication of documents relating to Simbirsk (renamed Ulianovsk in 1924, in honour of the birthplace of Lenin) during the 1917 Revolution records in the footnotes the biographies of many of those who participated in soviet committees during 1917–18, and many of whom perished in the 1930s. They included Vladimir Ksandrov, who was president of the executive of the Simbirsk soviet from December 1917 to February 1918. After the Revolution, Ksandrov moved to Moscow, where he held another post with the Soviet government. Returning to Simbirsk in 1919, he presided over the soviet of labour and defence and was in charge of the export of grain from Simbirsk to the industrial centres of Russia (that is, he loyally carried out requisitioning for the Soviet state). He was arrested in 1938 and died in prison in Kirov province (formerly Viatka) in 1942. Aleksandr Shver was a Bolshevik and editor of the *Simbirsk Soviet News* and other Bolshevik publications in 1917 and 1918. He was arrested for

Trotskyism in 1937 and shot in 1938 in Khabarovsk (Siberia). Nikolai Cheboksarov was a Menshevik, but participated in the PEC of Simbirsk in 1917. He was arrested in 1937 for spying, allegedly for Japan, and shot. After the death of Stalin in 1953, Nikita Khrushchev, the new leader of the Soviet Union, implemented a policy of 'de-Stalinization' which, among other things, 'rehabilitated' or exonerated many of those who had been executed, imprisoned in the gulags or internally exiled. Ksandrov was rehabilitated in 1955; Shver the following year; and Cheboksarov in 1957.[36]

Intellectuals in Volga towns were regarded with deep suspicion by the Soviet state. Over 40 members of the faculty of Kazan University were arrested; 10 were shot and 16 died in camps. The denunciations, often from students at the university, became more and more surreal, as more writers and academics were arrested. The biologist Professor Slepkov was arrested because he allegedly stated in a lecture that 'Leninism has become obsolete' and 'it is impossible to use sociology in biology'. He was shot in 1937. Other professors were denounced as Trotskyists and 'anti-party historians', and for 'slandering Stalin'. Members of the Tatar publishing house were arrested and independent publications ceased.[37] The worst period was 1937–38, when it is estimated over 11,000 people were arrested across the Tatar Autonomous Soviet Socialist Republic (ASSR), including over 5,000 alleged kulaks. Newspaper correspondents, Soviet officials, factory managers and the director of the Tatar theatre – all were arrested. Over 3,000 were shot.[38]

Other non-Russian intellectuals were also targeted. In the Mari ASSR, 26 Mari writers and intellectuals were shot in November 1937.[39] The Chuvash poet Ille Takhti (Ille Tăhti) avoided the camps but was, as the recent anthology of Chuvash poetry states, 'subjected to fierce literary harassment'. Some of his pain may be seen reflected in the verses of his poem 'Song of Praise to the Chuvash Poets' (1925), and could have been echoed by many others:

Rather than be a Chuvash poet,
Better to be a rotting log,
A dried up stump of a perished tree
That the wind has tumbled into the bog;
Better by far to crash to earth
And be carried off by the waters of spring
To float downstream on the Volga's waves,
Tossed and rocked by the Volga's swirl,
And then at last to be borne away
To the edge of the world.[40]

Artists were as vulnerable as writers, Russian and non-Russian. Iakov Veber (Weber) was a Volga German who was much influenced by the portrayal of the suffering of ordinary people in Repin's paintings, including *Barge Haulers on the Volga* (see Chapter 11). In the 1920s and 1930s, he painted a series of landscapes of the Volga, including industrial and commercial activity, in the Soviet realist style. *The Evening Post*, painted in 1932, shows workers ending their shift and eagerly reading the latest Soviet newspapers . . . He was 'repressed' in 1937.[41]

Anyone could be arrested in this unreal world, and anyone with a foreign name was suspect. Boris Gofman (Hofman), a Volga German colonist, worked as a teacher in the town of Tver (renamed Kalinin in 1931). He was shot in 1938 as an 'enemy of the people' and rehabilitated in 1956.[42] Kalinin province was also home to a small number of ethnic Karelians, who had lived there since the middle of the seventeenth century. They were also targeted as potentially disloyal to the Soviet state. A number of Soviet officials with a Karelian background were arrested in the 1930s, in a similar pattern to Russian officials in Simbirsk, noted above. Viacheslav Dombrovsky had held senior positions in the Karelian Autonomous Republic in the 1920s; he moved to a government post in Kursk, but by 1936 had returned to Kalinin as head of the local NKVD. He was arrested in September 1937 and shot; he was rehabilitated in 1956. Petr Rabov had held a number of posts in what we would now call communications ('agitation – propaganda' in the language of the time) in Kalinin province in the 1920s. By 1934, he was the first secretary of the Vyshnii Volochok (also in Kalinin province) regional committee, and in June 1937 he became first secretary of the Kalinin *oblast* (province) committee of the Communist Party. He was arrested in March 1938, convicted in 1940 and it is assumed he was shot. In 1938, some 140 Karelians were arrested as members of a Karelian organization that was allegedly colluding with Finland against the Soviet Union and they were condemned as 'traitors to their country and enemies of the people'.[43]

The attack on individuals was accompanied by an attack on religion. We saw above that the attack on churches and mosques played a significant part in the peasant resistance to collectivization. Religion was never officially banned in the Soviet Union, but believers were carefully monitored, and religious communities – Christian and non-Christian – had to be officially registered (and the names of members of the community recorded). The 1930s witnessed the destruction of many churches in the Soviet Union. The Volga was the location of major Orthodox monasteries and churches – partly because of the role played by the Church in settling the lands of the middle and lower Volga, and partly because prosperous merchants in Volga towns had built private chapels. Iaroslavl, for example, was well known for the number and beauty of its many such churches. In the 1930s, 27 churches in the town were destroyed, including the Cathedral of the Assumption.[44]

In Kazan province, the number of active churches fell from 802 to 495 by 1930.[45] Monasteries and convents became schools and hospitals, or were left to decay. The St Paraskeva convent, some 30km from Saratov, housed 600 nuns and novices in 1915. It became the headquarters of the Red Army in 1918, then a military hospital in which the nuns worked as nurses, and then the main building of a collective farm (it reverted to being a church in 1991).[46] Mosques were also closed and used for other purposes. In Kazan province by the 1940s, there were only 16 officially approved mosques and 25 non-official ones; in 1956, only 11 official mosques remained. In Samara, there were only 14 functioning mosques in 1963.[47] In Astrakhan, the archives still contain the inventories made at the monasteries and churches before they were closed down. Some toleration was shown, however, in Astrakhan towards non-Russian believers. The Persians in the town were allowed in 1924 to hold a religious procession, although the 'religious ritual on the street' had to be under supervision.[48]

Church leaders – Christian and non-Christian, Orthodox and non-Orthodox – were also targeted by the state. Ten Orthodox priests from Karelian villages in Kalinin province were arrested: six were shot and the others were sentenced to 10 years in the camps.[49] In Gorkii (Nizhnii Novgorod) province, two brothers, sacristans Sergei Borisov (born 1887) and Ivan Borisov (born 1896), were arrested in 1937; the former was shot in January 1938 and the latter was sentenced to 10 years in the camps. He died in 1948.[50] On the territory of the Tatar ASSR, V.I. Nesmelov and N.V. Petrov, former professors at the Kazan Theological College, and Bishops Iosif and Nektarius were arrested. The Raifa monastery housed 44 monks; six were arrested for anti-Soviet activities. It has been estimated that between 1917 and 1923, the Soviet state executed 28 Orthodox bishops and some 1,200 Orthodox priests; church records show that over 100,000 priests were shot in the years 1937 and 1938. Muslim clerics were equally persecuted: in Kazan, mullahs allegedly serving illegal mosques were arrested and some were shot. In total, it has been estimated that 234 ministers of Christian and non-Christian religions were repressed in the Tatar ASSR.[51]

Those sentenced to forced labour were sent far away, often to camps in Siberia. But prisoners were also incarcerated on the Volga. Sviiazhsk, at the confluence of the Volga and the Sviiaga rivers, was a community based around a fort and the Cathedral of the Assumption and Sviiazhsk monastery. During the 1930s, the administrative buildings housed a psychiatric prison. The Raifa monastery, a short distance from Kazan, was a labour camp during the 1930s. Prisoners were used as forced labour in the construction of hydroelectric plants and canals on the Volga (discussed in Chapter 17) both before and after the Second World War.

The families and relations of those declared 'enemies of the people' were also penalized. 'The walls have ears' – Nikolai Semenov recalled his mother telling him that when he was a child in Saratov, after her two brothers had been arrested. She removed their photographs from her albums.[52] One personal account from Kazan, taken from oral interviews, illustrates both the suffering of the children of those arrested and the absurdity of their treatment by the local authorities. Giuzel Ibragimova was the daughter of the Tatar writer Gumer Galiev. Her father was arrested twice: in 1937 he was sentenced to 10 years in the notorious Norilsk labour camp in the far north; he returned to Kazan in 1947, but was rearrested in 1949 and sent into internal exile in Krasnoiarsk, in western Siberia, where he died. Giuzel's mother, Sufkhari Biktogirova, was arrested in 1937 as a 'member of a family of enemies' and spent five years in the Siblag camp in Siberia, only returning to Kazan in 1943. Giuzel, aged two when her mother was arrested, and her older sister were initially sent to an orphanage in Irbit in the Urals. Her grandfather spent three years searching for the girls – in part because the official records were inaccurate – before he could bring them back to Kazan. She recalled from her childhood that the experience of her parents bred secrecy and fear. Life was very difficult for her mother even after she was released, because as a former prisoner it was hard to get work and she was frequently sacked. She told her daughter: 'Don't keep letters'; 'Don't talk in public places'; 'If there are more than two of you, don't discuss any political themes.' Giuzel referred to this as 'a panicked terror that this might happen again'.

Giuzel was 'innocent', but the difficulty she experienced in enrolling at university sums up the absurdity of the official suspicion of the children of enemies of the people:

> The first time it hit me [that I was the daughter of 'enemies of the people'] was when I applied to university. Well, when I went to the university to apply, to Kazan University. I was applying to the Chemistry Department. I really loved chemistry. They told me they could not admit children of 'enemies of the people' to the Chemistry Department. Because, they said, 'It's like this: we send students in their third year on an internship to chemical factories, to large enterprises, to closed factories. And how could we send you? You, after all, might organize an explosion! . . . We [she and her sister] were standing there crying in the hall, a

man came up to us, very handsome, Boris Lukich Latin. He was the Dean of the Physics and Math Department at the University, he embraced us. We told him what was the matter. He said, 'Don't cry, children.' And they had arrested his brother, an artist. He understood and he said, 'Don't cry. Come to be admitted to my department, to Physics and Math. Mathematics is a science outside of politics. You will always need it. You will never get into trouble – it is an indifferent science. Not ideological.' And he convinced us to go to Physics and Math. I graduated from there.[53]

By 1940, the Soviet state had crushed opposition. Almost all peasants were collectivized, kulaks had been repressed and opposition to the state had gone underground, or had taken the form of passive resistance. Repression had, however, come at a terrible price. Millions of people had died, been displaced or been sent to camps, and famine had taken hold in the Volga and other parts of south Russia. Furthermore, the crushing of peasants and the political repression had bred a cowed people – downtrodden, deadened and unwilling to take any risks that might bring more oppression. Eugenia Ginzburg expressed this eloquently after her release from the camps:

Hardly anyone, except the doctors, worked at or even wanted to work at his old profession. The animal hatred of the authorities for the intelligentsia was all too familiar from our experience of our years in camp. The thing to do was to be a tailor, a cobbler, a cabinet maker, a laundress – to crawl into some quiet, warm nook, so that it would never occur to anyone that once upon a time you read seditious literature.[54]

Nevertheless, the people of the Volga, and of the Soviet Union in general, had to face another great challenge – one that would change many lives forever: the outbreak of the Second World War, and the 1941 invasion of the Soviet Union.

15
THE VOLGA IN THE SECOND WORLD WAR
Conflict, reconstruction and identity

Stalingrad means the Volga. The meaning of the Volga for Russia cannot be overemphasized. There is no river like it in Europe. It cuts through Russia. The people have composed hundreds of songs devoted to mother Volga . . . Would despicable Germans bathe their horses in the Volga, the great Russian river?

An essay on 'Stalingrad', written in September 1942
by the publicist Ilia Ehrenburg[1]

The young Soviet state perceived that it faced internal threats in the late 1920s and 1930s – albeit most of the so-called 'enemies of the people' were of the state's own making and not infrequently were a product of its own imagination. The Soviet Union, however, faced a far bigger external challenge in 1941, when Nazi Germany invaded and at first inflicted a series of crushing defeats on the Red Army. The river Volga was of vital strategic importance for the Soviet Union in the Second World War: first, the town of Kuibyshev (Samara) was earmarked as the possible venue for the Soviet government, should Moscow be taken; second, towns on the upper Volga were on the front line as the German troops advanced towards Moscow; third, and most importantly, the battle of Stalingrad was the bloodiest battle of the Second World War, in which some 700,000 soldiers on both sides lost their lives, and was a major turning point not only in the fortunes of the Soviet Union and Germany, but in the war as a whole. The soldiers of Nazi Germany reached the Volga just north of the city, but could not control the river and never crossed it (except as prisoners of war; a mural in the memorial complex in Volgograd shows German soldiers crossing the Volga in captivity). Not since before the construction of railways and bridges had the Volga been such an important barrier between east and west; but this time it lay between the occupied (or at least threatened) west and the relative safety of the lands to the east.

A major reason, if not the main reason, why the Soviet Union was able to turn the tide against the German armies was its ability to relocate industry from west to east – which included relocating 200 factories to towns on the river Volga (and even more to the Urals and Siberia) – and then massively step up the production of armaments.[2] After the war, those towns damaged by the fighting (in particular Stalingrad) were reconstructed, other towns were built on the Volga and yet others were vastly expanded to create a new socialist urban way of life. In effect, the state would control who lived where and how they lived. And, as we shall see in the final chapter, it even set out to control nature itself. Given the prominence of the Volga in determining the destiny of the Soviet Union in the war and its role in modernizing the country, it is not surprising that the cultural importance of the river to the Soviet people should have been reasserted and redefined in this period. 'Mother Volga' became the patriotic symbol and the protector of the Soviet Union, as much as it had been in imperial Russia.

❖

Operation Barbarossa commenced on 22 June 1941, when some 3 million men – mainly German units, but supported in the south by Romanian forces, and by over 2,000 aircraft and 3,350 tanks (plus between 600,000 and 700,000 horses) – invaded the Soviet Union at points along a front stretching over 3,000km. It was the largest invasion force in the history of warfare. The German troops advanced rapidly through eastern Poland, Belarus and Ukraine, helped by the almost total unpreparedness of the Soviet forces. The first few weeks were a disaster for the Soviet Union: many of the units that faced the Germans were under strength; at least 1,200 Soviet planes were destroyed on the ground at undefended and un-camouflaged bases within just a few hours of the invasion; hundreds of military supply dumps fell into German hands. By mid-July, the Germans had taken Smolensk, a key town on the route to Moscow. By the end of September, Leningrad was encircled, Kiev had fallen and the German army had taken control of Ukraine. By early October, German troops were only 64km from Moscow and were seemingly unstoppable: 'the name Moscow will disappear forever', predicted Hitler confidently.[3]

In the ensuing panic in Moscow, the Soviet government had to develop plans to evacuate if Moscow were taken. In one ghoulish episode, Lenin's embalmed body was put in a specially refrigerated railway carriage and taken to Tiumen in western Siberia, where it lay in a former school, guarded by soldiers and scientists; the guards at the Lenin mausoleum continued to stand to attention as if nothing had happened. Plans also had to be made to move Stalin and his government. On 1 October, Stalin ordered the evacuation of government staff and papers to Kuibyshev (Samara), on the Volga. Kuibyshev was a good choice geographically. It is just over 1,000km by road east of Moscow – far enough away to force the German army much further east than they had planned if they wanted to attack the new 'capital', but not so far away as to prevent it from acting as an effective base for air attacks and as a location for reserve troops. Kuibyshev was highly industrialized and could support further factories that moved east in the wake of the German invasion. It was also, of course, protected by the river Volga. The town is situated on the eastern (left) bank of the river, which acted as an additional barrier to invading troops; and it is also protected by the large bend in the river, which turns east by the town of Syzran. Any invading force attempting to take Kuibyshev would have to cross the Volga at Syzran before reaching Kuibyshev, or else approach the town from the north, from Kazan. The government in Moscow was conscious, however, that Kuibyshev could potentially be within reach of German bombers if they advanced further east. A special deep bunker was constructed for Stalin; this was never used and is now a tourist attraction in the town. Foreign embassies also moved to Kuibyshev: the British embassy was housed in the fine mansion of Aleksandr Naumov, whose comments on peasants in 1906 were quoted in Chapter 12. The irony that Samara had previously been the capital of the anti-Soviet Komuch in the Civil War was clearly lost on Stalin. Nor were comparisons made with the French invasion of 1812, when Alexander I had spoken about fighting on, whatever the fate of Moscow – even if it meant retreating to Kazan ('I will not make peace until I have driven the enemy back across our frontiers, even if I must . . . withdraw beyond Kazan').[4]

As the German troops tightened their grip on Leningrad and moved towards Moscow, several towns on the upper Volga became strategically important. Volga towns were bombed, along with Moscow and Leningrad. Galina Kosterina recalled her childhood in Iaroslavl during the war:

> When the war started the entire city literally went dark. It was bombed every night so lights were prohibited. Our dark, cold, hungry days started and the most difficult thing was the darkness because not much could be done during the long evenings. We told fairy tales to each other.[5]

Her father was sent to the front on the first day of the war, because he was a driver; he was killed in 1942 on the road between Tula and Moscow.

The line of the river Volga from Rzhev to Kalinin (former Tver) became part of the German front line. Rzhev was taken by the German army on 14 October 1941 and Kalinin was captured on 18 October. By late November, German troops had crossed the Volga–Moscow canal and were only 20km or so from the city. By this stage of the operation, the Germans had lost some 160,000 men, but the Red Army losses were horrendous: over 2,600,000 killed and a further 3,350,000 taken prisoner. For every German killed, 20 Soviet soldiers had died.[6] The German forces had, however, been severely weakened by their losses of men and equipment. Furthermore, they lacked the necessary supplies for a winter engagement: there was a shortage of warm uniforms, proper footwear and winter camouflage, and vehicles and planes lacked effective antifreeze. On 5 December 1941, the Soviet army launched a counter-offensive, in deep snow, with the early-morning temperatures at minus 10 degrees centigrade. By the end of December, Kalinin had been retaken and German troops had been pushed back from Moscow. The immediate danger to Moscow was over, but the Germans remained in control of swathes of territory to the west, and the town of Rzhev was still in German hands.

The northern sector of the war centred on Rzhev, a town of some 50,000 people which was an important rail junction on the Volga. It has received far less attention than, for example, the devastating battle of Stalingrad or the dramatic tank battle at Kursk. The northern front had been formed after the Germans had been pushed back from Moscow in December 1941. On 30 July 1942, a Soviet offensive was launched with almost 400 tanks and over 1,300 guns and 80 rocket launchers. The weather was unusually wet, and the tanks and men got bogged down in the marshy terrain. The Volga was a contributory factor in the slow progress of Soviet troops: it is some 130 metres wide at this point, and acted as the border between German troops on one side and the Red Army on the other. For two months, the Red Army advanced slowly towards Rzhev, pushing the Germans back; but it was unable to defeat them completely. Russian losses in the main assault on the Rzhev line in the summer of 1942 were extremely high: some 300,000 men (the campaign was known by Red Army soldiers as the 'meat grinder'). German losses were far lower – some 53,000 men – but the operation did significantly weaken the German position in this sector.

This campaign has been referred to as the 'forgotten battle', as one Russian told me recently – probably because, compared with the battle of Stalingrad, it did not end in a resounding victory, despite the human cost (General Georgii Zhukov, who commanded the assault before he was transferred to Stalingrad, hardly mentioned it in his memoirs). The surrounding woods are allegedly full of skeletons that have never been recovered. The poem 'I Was Killed Near Rzhev', by Aleksandr

Tvardovsky, describes the death of an anonymous ordinary soldier ('I was killed near Rzhev/ In a nameless bog'). The dead soldier, however, imagines that the Red Army will triumph, and hopes that he will have played a small role in its victory:

And in the steppes beyond the Volga
You dug your trenches in haste
And in battles you marched
To the limits of Europe.

In February 1943, another push was made in the northern sector, after the victory at Stalingrad. Rzhev was retaken in this campaign, but only because the German army fell back in good order to a new defensive line.

German forces had been pushed back from Leningrad and Moscow in the winter of 1941–42, but they still controlled Ukraine. In May 1942, they took Kharkov (Kharkiv) in eastern Ukraine, and then occupied the Crimean peninsula. German commanders wanted to concentrate their efforts on taking Moscow after the winter, but Hitler, convinced that the Red Army was near to collapse, had grandiose plans for southern Russia: he intended to sweep east to Stalingrad and then south to Astrakhan and the Caspian, while German forces would also advance south, through the Caucasus, to cut the Soviet Union off from its source of oil in Azerbaijan. The German Operation Blue started on 28 June 1942. By the end of July, the German army had taken Rostov-on-Don, where it had met with very little resistance. Confident of victory, the German armies then split: one army advanced south to the oil fields of Azerbaijan; the other force, the German 6th Army, advanced on Stalingrad. On 28 July, Stalin issued the famous, or infamous, Order 227 to the armies: 'Not a Step Back'. Although this was an order to fight to the death, it was less a patriotic statement than a desperate attempt to shore up demoralized forces by threatening summary execution for desertion or for disorganized retreat (some 13,000 Red Army soldiers were executed for cowardice during the Stalingrad campaign).

By 19 August 1942, General Friedrich Paulus, commander of the German 6th Army, was ready for an assault on Stalingrad. The city stretches in a thin strip for miles along the western bank of the Volga. On 23 August, the Germans reached the Volga for the first time just north of Stalingrad, at the river settlement of Rynok, and 'there the soldiers of the 16th Panzer Division gazed on the Volga, flowing past right before their eyes'. As one of the soldiers recalled: 'We had started early in the morning on the Don and then we were on the Volga.' They took photographs of each other staring at the distant shore on the other side through their binoculars. It was intended to be a memento of the moment when the Third Reich reached its furthest point east. The photographs were displayed at the headquarters of the 6th Army with the caption: 'The Volga is reached!'[7] 'There stands a soldier on the Volga bank' were the words to a popular German song composed at the time.

That same day, the German 6th Army reached the outskirts of Stalingrad to the west and south. That day, 600 bombers attacked the city, killing some 40,000 citizens, according to Soviet sources (to put this in context, a similar number of civilian deaths occurred in Britain during the Blitz, about half of which were in London). General Vasilii Chuikov, the commander of the Red Army in Stalingrad, surveyed the scene:

> The streets of the city are dead. There is not a single green twig on the trees: everything has perished in the flames. All that is left of wooden houses is a pile of ashes and stove chimneys sticking up out of them. The many stone houses are burnt out, their windows and doors missing and their roofs caved in. Now and then a building collapses. People are rummaging about in the ruins, pulling out bundles, samovars and crockery, and carrying everything to the landing stage.[8]

On 25 August, the river Volga was used, belatedly, to evacuate women and children from the city, despite attacks on steamers by German aircraft. In all, between 200,000 and 280,000 civilians were evacuated across the river.[9] Others, almost unbelievably, remained in the town throughout the duration of the battle.

It looked as if Stalingrad would inevitably fall to the 6th Army. But the German troops met with fierce resistance from the Red Army, which immediately counterattacked from the north and maintained a foothold on the west bank of the Volga. German casualties were very high, but the army was still confident in early September that, as one soldier wrote home, 'Stalingrad will fall in the next few days.' But the bombing raids had reduced the city to bombed-out buildings and rubble, which would prove disastrous for a military operation that involved fighting at close quarters under the most appalling conditions for both sides. 'Not a house is left standing,' wrote a German lieutenant, 'there is only a burnt-out wasteland, a wilderness of rubble and ruins which is well-nigh impassable.'[10] In September, German troops moved into the city. One division attacked the slope of Mamaev Kurgan, the hill in the town which had formerly been a Tatar burial place. It was of vital strategic importance, as the German guns would have controlled the Volga if they had been able to establish themselves at this height. Troops fought each other to a standstill, but the Germans were unable to hold the hill. The other German division attacked the railway station. Resistance was much stronger than the Germans had expected. The station changed hands three times in two hours one morning; it changed hands fifteen times in five days, and eventually the Germans were left in charge of complete ruins. Fighting took place street by street, house by house, from cellars to the upper storeys, during the night as well as the day. The bloodiest battle of the Second World War had begun. It was to last for over five months.

After the first wave of attacks, the battle became one of deadly attrition. The fight for every building, bunker and sewer was soon termed the *Rattenkrieg* – Rat War – by German soldiers. Attacks were particularly deadly at night, when exhausted German soldiers were considered to be more vulnerable, and when they could be less protected by the air force, because the manoeuvrable Soviet U-2 biplanes had established night-time superiority. The Red Army included skilled snipers, who picked off German soldiers when they ventured outside buildings; snipers acquired almost a cult status, and some individuals killed up to 200 German soldiers. In late September, the Germans launched a desperate bombardment to try to take the city before the winter set in. Their troops reached the top of Mamaev Kurgan, but they were forced back and the summit became a no man's land of craters and abandoned military hardware separating the two sides. 'One more battle like that and we'll be in the Volga', stated General Vasilii Chuikov in some desperation.[11] On 9 November, the snow began to fall. Ice floes formed on the river, making it harder to transport fresh troops and the wounded across the water. German soldiers faced a new enemy: the intense cold.

The Volga was crucial for both sides. The Germans attacked shipping on the river, in certain knowledge that if they could control it, and force the Red Army back across it, they would control the city. The Red Army had to keep a foothold on the western bank to have any chance of saving Stalingrad. There could be no retreat across the river. Men had to be transported across it from the eastern bank, under fierce German bombing, in order to shore up the army on the narrow strip held by the Red Army on the western bank. The wounded were transported back across the river to hospitals on the eastern side. In all, it is estimated that over 500,000 people were transported across the Volga during the battle (soldiers and civilians). The river also brought vital supplies to the army from the north. The ship *Lastochka* (the *Swallow*), for example, transported 18,000 men and 20,000 tonnes of military equipment during the battle. The cost was high and many ships were sunk or damaged, and many lives lost. To give but a couple of examples: the transport ship *Kazanka* lost 14 crew on 31 July; the motor vessel *Tataria* lost

14 crew members on 4 August.[12] Airfields were situated to the east of the river. The 'Katiusha' rocket launchers (named after the popular Russian song of that name, in which Katiusha promises her fiancé that their love will endure while he defends the motherland) proved to be devastatingly effective. Operating from the eastern bank of the river, they could fire 16 rockets at a time from racks mounted on trucks, producing an ear-splitting screech that terrified German soldiers.

The river Volga took on a symbolic, as well as a logistical, importance and came to be seen by both sides as a dividing line between two peoples, two regimes – even two civilizations. To the Germans, the river was the extreme eastern edge of the Third Reich. A sergeant wrote home with no sense of irony:

> I am proud to number myself among the defenders of Stalingrad. Come what may, when it is time for me to die, I will have had the satisfaction of having taken part at the most eastern point of the great defensive battle on the Volga for my homeland, and given my life for our Führer and for the freedom of our nation.[13]

The German perception of the significance of the river was vividly captured in the novel *Stalingrad* by Vasily Grossman, who was a war correspondent with the Red Army:

> The last act of this drama – this epic drama being performed by grenadiers, tank crews and motor infantry on the huge stage of the steppe – would soon be concluded on the banks of the Volga. There was no precedent for this campaign in all annals of warfare and the thought of its imminent conclusion was profoundly exciting. The [German] general could sense the edge of the Russian lands; beyond the Volga lay Asia.[14]

To the Soviet soldier, the Volga was a source of protection, because the eastern bank was where the wounded were treated and from where the deadly rockets were launched. He had to cross the river in order to engage in the battle – and that could in itself be a terrifying experience; but he also knew that the river must not fall under German control, or else the battle, and the war, would be lost. As one Red Army soldier stated:

> It was pretty terrifying to cross over to Stalingrad, but once we got there we felt better. We knew beyond the Volga there was nothing, and that if we were to remain alive, we had to destroy the invaders.[15]

'For the defenders of Stalingrad there is no ground on the other side of the Volga' became the slogan of the Red Army. A wartime poster by Vladimir Serov shows Soviet soldiers advancing, with the words 'Let us defend the Mother Volga'.

Paulus realized that Stalingrad had become a trap for the German 6th Army. The town was simply too far away to be protected and supplied as the battle continued: as one German soldier wrote home in November, he was '2,053 miles from the German frontier'.[16] Nevertheless, the fate of Stalingrad was decided not within the town, but by a bold Soviet counter-attack in the rear. While the Soviet army continued to resist within the town, plans were made to encircle the Germans. The counter-attack was planned in the knowledge that the territory between the Don and the Volga was weakly defended, and much of it was guarded by Romanian, Hungarian and Italian divisions, which were less well equipped than German forces. The plan relied on Soviet forces holding out for 45 days in Stalingrad, while the troops, tanks and other military equipment were assembled to start the counter-offensive. On 19 November 1941, the attack began, with Soviet troops and then tanks overrunning the Romanian 4th Army Corps on the right bank of the Don,

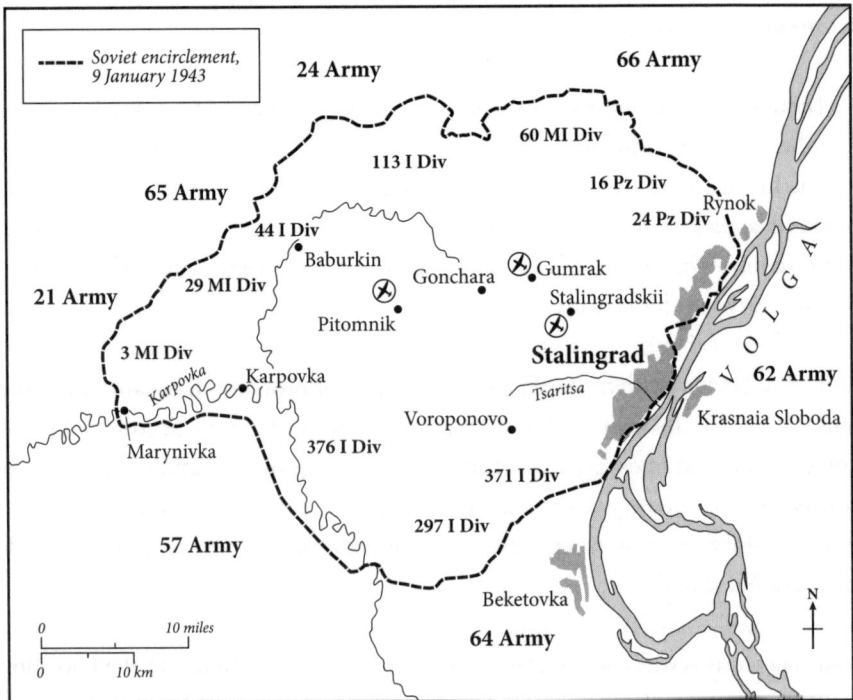

8. The Soviet Counter-Attack on Stalingrad

and then moving towards Stalingrad. By 22 November, the Red Army had taken the key town of Kalach on the river Don; four days later, the Axis troops had been forced back to the south of the Don. The army in Stalingrad was now sealed off between the Don and the Volga.

Even while the noose tightened around the town, Germans and Russians continued to fight each other over every block, and to launch raids and counter-raids on Mamaev Kurgan. A German lieutenant evocatively described the shocking conditions:

> We have fought for fifteen days for a single house, with mortars, machine guns, grenades and bayonets. The front is a corridor between burnt-out rooms: it is the thin ceiling between two floors ... imagine Stalingrad: eighty days and eighty nights of hand-to-hand struggles. The street is no longer measured in metres, but in corpses. Stalingrad is no longer a town. By day it is an enormous cloud of burning, blinding smoke; it is a vast furnace lit by the reflection of the flames. And when night arrives, one of those scorching, howling, bleeding nights, the dogs plunge into the Volga and swim desperately for the other bank. The nights of Stalingrad are a terror for them. Animals flee this hell; the hardest stones cannot bear it for long; only men endure.[17]

Men became weakened by the battle. The lack of fresh water resulted in typhus, and the insanitary conditions led to dysentery. German soldiers were not equipped for the intense cold; frostbite was rife and men lost toes and fingers. They wrapped themselves in any clothes and cloths they could find, and even skinned dead dogs to make improvised gloves. The stench of death and the dust from collapsed buildings made the air foul. Soldiers were plagued by lice, which inflamed their skin and made sleep difficult. Poor supplies of food lowered physical health, as well as morale. German soldiers began to die not from wounds or from obvious disease, but from 'exhaustion'. Starvation was never mentioned in the reports of German doctors, but rations were so poor that this cannot be discounted; nor could suicide, although there are no figures to support this.[18] 'Snow, wind, cold, all around us – sleet and rain ... since my leave I have never undressed. Lice. Mice at night', recorded Kurt Reuber, a 36-year-old German from Kassel, to his family.[19] By the time the Soviet troops had encircled Stalingrad, the German soldiers would not have had the physical capacity to break out of the town – even if Hitler had allowed them to do so. Of course, conditions were also poor for soldiers in the Red Army, and their experience of house-to-house fighting was the same as that of German soldiers. Hospitals were set up on the left bank of the Volga, but medical care was often inadequate and food supplies were frequently interrupted. Soviet soldiers, however, were better equipped for the winter; and importantly, while material conditions deteriorated for the Germans, for the Soviet soldiers they improved, making it easier to maintain morale.

By mid-December, the German army was so short of shells that it could no longer threaten the crossing of the Volga. Bit by bit, the town was encircled. On 16 January, Pitomnik airfield was taken, followed seven days later by the last German airfield west of Stalingrad. This meant that there was no longer any escape for the wounded, and no supplies could be brought in. Defeat was inevitable, but men fought to the death, fearing the worst if they were captured. By 30 January, the Soviet army was in control of most of the city centre; they attacked the Univermag department store, which was Paulus's headquarters, and took his formal surrender the next day. The remaining German soldiers were forced out of their buildings and cellars and rounded up. Aleksandr Sakulin, from Kalinin province, recalled 'hundreds of fascists' throwing down arms and discarding equipment and personal possessions – 'even a field kitchen where we later ate not badly'.[20] An unearthly quiet descended on the town.

Civilians who had somehow survived the battle emerged from cellars to gaze at their shattered town. In his novel *Life and Fate*, Vasily Grossman characterized the mood on the day the guns fell silent, through the eyes of a Red Army soldier gazing over the town and the river Volga:

It was dark. Both the east and the west were quiet. The silhouettes of factories, the ruined buildings, the trenches and dug-outs all merged into the calm, silent darkness of the earth, the sky and the Volga. This was the true expression of the people's victory. Not the ceremonial marches and orchestras, not the fireworks and artillery salutes, but this quietness – the quietness of a damp night in the country.[21]

In many ways, the battle of Stalingrad was a turning point in the Second World War. The invincibility of the German army had been shattered and its losses were staggering. For the Red Army, despite the horrendous loss of life, the victory gave soldiers the belief that they could win. It was not only the first time that the Germans had been defeated, but the defeat had been so comprehensive that it not only boosted the morale of the Red Army, but had a significant impact on the western allies as well. The heroism displayed at Stalingrad (any panic or cowardice was forgotten) became legendary. The German forces in the Soviet Union were still formidable, however, and it took more than three years for final victory to be achieved. As the Soviet forces moved across the steppe and through Ukraine, they were faced with resistance at almost every point.

The human cost of the battle of Stalingrad was horrific. The German 6th Army suffered at least 147,000 dead in the battle itself.[22] Overall, it is estimated that the Red Army may have lost almost 500,000 men, with over 650,000 wounded or sick. Soviet figures state that 64,224 civilians died in Stalingrad, including 1,744 who were shot and 108 who were hanged.[23] And the deaths continued after the battle. Some 91,000 German soldiers, most in a pitiable condition, were captured. By the following spring, almost half the prisoners of war had died from disease, malnutrition, poor medical attention – or had simply been killed: stragglers who could not keep up with the group on the march were often shot by the roadside. Prisoners of war were sent to camps in Central Asia and to towns on the Volga, including Volsk and Astrakhan; only some 5,000 eventually returned home. The German forces had included some 50,000 auxiliaries who were Soviet citizens, known as *Hiwis* (short for *Hilfswillige*, or voluntary helpers). These included Cossacks who were committed to a German victory, but also Russian prisoners of war who had been forced to undertake labouring duties. The *Hiwis* who survived the battle and were captured by the Red Army faced almost certain death – they were either summarily executed or sent to forced labour camps, where they were very likely worked to death.

At the time of the Soviet counter-offensive, there were 1,143,000 Soviet soldiers fighting in Stalingrad. Over 150,000 men were called up from Kuibyshev province (Samara);[24] we were told by our guide in Kazan a few years ago that 500,000 men were conscripted from the Tatar ASSR. It is hard to relate to such large numbers, but the impact on individual villages makes the extent of the sacrifice during the war clearer. Tatar villages and small towns in the Samara region, for example, lost an extraordinary number of young men (more, proportionally, than in the cities where the vital industries were located). The death rate of conscripted peasants was shocking: 166 of the 323 conscripted men from Teplyi Stan village; 136 of the 230 men from Balykla; 43 of the 77 men from Davletkulovo; 90 of the 176 men from Novoe Ermakovo and 316 of the 681 men from Staroe Ermakovo; 164 of the 367 men from Mochaleevka. Of the 6,000 or so men who were conscripted from Kamyshla region, only 2,353 returned.[25]

❖

Stalingrad, of course, was almost completely destroyed and had to be rebuilt. It has been estimated that at the end of the fighting over 41,000 buildings were in ruins and only 12 per cent of the housing was still standing. The battle had destroyed 110 schools, 15 hospitals and 68 healthcare centres. Of the town's 400,000 or so residents, only 7,655 remained in the town on its liberation. Mines, artillery shells and bodies had to be removed before reconstruction could take place. There were radical plans to rebuild the town as a completely new, model socialist city (built in part by volunteer Komsomol members); but planning was disrupted by internal disputes and by the sheer logistical problems of finding sufficient building material and labour to carry out the work, just when the returning population was putting enormous pressure on available housing and resources.[26] In the end, the decision was taken to rebuild many of the streets and buildings in the centre in the original style, as if the German presence should not be allowed to inflict any permanent change (much the same thinking applies in the post-Soviet period to the rebuilding of cathedrals that were destroyed in exactly the same, original style). In reality, much of the reconstruction of Stalingrad, and other cities, was haphazard and unplanned.

Reconstruction took place slowly. In January 1947, nearly four years after the battle, some 330,000 people were living in Stalingrad in very cramped conditions, and tens of thousands were still living in the basements and stairwells of ruined buildings or in makeshift dormitories.[27] In 1948, two British visitors found 'miles of stark ruins where people are still living in cellars', although they also noted that new 'showy public buildings' were being erected and construction materials were being brought in by cart and camel.[28] In 1951, there were still over 1,300 families living in dugouts and ruined buildings.[29] There was constant conflict between the need for housing, for public institutions, for factories and for better public space. The river embankment was constructed at great cost, in part so that parades and public displays could take place there; it was intended to be named after Stalin, but was only completed after his death and instead was called the 62nd Army Embankment, after the military unit that defended the river bank against the Germans.[30]

The battle, however, had to be commemorated – to honour the dead, but also to mark the turning point in the war. Plans for a memorial were developed in the 1950s and 1960s, and the complex finally opened in 1967. The design and implementation proved to be extraordinarily difficult and expensive.[31] The memorial complex occupies the whole of one side of the Mamaev Kurgan hill, some 2 square kilometres. At the top stands the colossal statue *The Motherland Calls*. At 52 metres, it was at the time of its unveiling the world's tallest free-standing statue (under the original plans, the monument would have been only about half that size, but Khrushchev apparently wanted it to be taller than the 46-metre figure of *Libertas* on the Statue of Liberty!). It portrays a woman in Grecian robes crying out and wielding an enormous (28 metre) sword. The visitor can then descend through a 'pantheon to the glory of the Soviet people', which has at its centre an enormous, guarded eternal flame; around the tiled walls are hung the names of 7,200 of the combatants who lost their lives. The memorial also comprises an artificial lake surrounded by towering sculptures of Soviet soldiers ('the square of heroes') and continues down through a rocky passageway (the 'wall ruins'), where loudspeakers play popular wartime songs and the sounds of battle. It is a moving experience for a visitor, even for a non-Soviet (and non-German) visitor such as myself, as a commemoration of the loss of so many young lives. *The Motherland Calls* is quite extraordinarily powerful, although the recent addition of an Orthodox church at the top of the complex appears incongruous. The river Volga plays a key part in the visual effect: the steps of the complex lead down to its banks. Furthermore, *The Motherland Calls* without doubt looks most impressive from the river below (the tourist boats sail past it and passengers throw flowers into the water just where the statue appears between the buildings to loom over the Volga). This does, however, mean that the figure with her enormous sword points east and not west – away from the German invader and calling to the Volga and the lands to the east of it: otherwise the view from the river would have been of the back of the statue, and the whole impact would have been lost!

The whole town is, however, a shrine to the battle. It contains several memorials, including the 'Pavlov house', which was defended by Soviet soldiers for two months before they were relieved. The eternal flame for fallen soldiers still draws its guard of honour from local school children, including girls.[32] And a solitary ruined building still stands outside the museum of the battle as a reminder of the destruction. The museum at Stalingrad describes the conflict, and at its apex is an enormous panorama of the battle (it was originally intended to be part of the Mamaev Kurgan complex, but this proved too problematic). Unlike the battle of Borodino (which has a panorama in Moscow) or the battle of Waterloo (which has a rotunda in Belgium), the battle of Stalingrad lasted months, rather than hours. The panorama in Stalingrad depicts the final days of the battle, when the 'ring' around the city was completed, when the Soviet army took Mamaev Kurgan and when the Germans were finally defeated (there is a depiction of German prisoners being led from the hill). It is also a collage of key events in the battle and acts of heroism

– including images of Mikhail Panikakha, who, though mortally wounded and engulfed in flames, used his last ounce of strength to throw himself at a German tank while holding a petrol-filled bottle, which exploded (on 2 October), and Gulia Koroleva, who carried 50 wounded soldiers to safety (on 23 November).[33] The original plan for the museum was for it to include the sounds of conflict, so that the visitor could experience the full horror of the battle; but this was clearly impractical. Nor, of course, could the stench, the dust, the shaking ground, the cold or simply the sheer terror of the battle be fully captured. The result is that the panorama does have a feeling of being an allegorical tableau (the sky is darkened above the German prisoners) – not so unlike the allegorical paintings that celebrated the victories of Catherine II over the Turks, where angels blow trumpets in the clouds above.

Stalingrad was, of course, the Volga town that suffered most. But Saratov, site of a major aircraft industry, was also bombed, and this is commemorated in the town's victory park. At the entrance is a line of tanks, Katiusha rocket launchers and other pieces of military hardware; and recordings of popular wartime songs are played. All the Volga towns, in fact, have parks which commemorate those who lost their lives in the war years. Both here and elsewhere in the country, the memory is kept very much alive.

The Volga towns became major industrial centres as a result of the need to supply the Red Army from factories away from the front; and that growth continued after the war ended. During the war, Saratov became a major manufacturing base for planes – something that town guides still mention with pride. The urban population was already expanding rapidly in the 1930s (driven in part by the expulsion of kulaks from villages), but this trend increased even further during and after the war, and especially in the 1950s and 1960s, when restrictions on leaving the collective farms were eased. This population movement occurred despite attempts by the Soviet government to improve life in the villages: to increase access to radio and cinema, to supply better (and more) material goods to village stores, to establish more medical services and more sports facilities and to provide greater opportunities for young people to acquire technical skills. Progress was, however, painfully slow. In 1956, only 21 of the 81 collective farms in Saratov province had electricity. The low level of education in the villages also remained a problem: again in Saratov province, only 1 per cent of collective farm members in the period 1953–64 had higher education, and only 18 per cent had finished middle school.[34] The collective farm controlled the movement of peasants – It was sometimes referred to as a 'second serfdom' by resentful peasants; but those – primarily young people – who could leave for the towns did so, and few returned. Military service also provided an escape route for young men, who disappeared into the towns after their service was complete. Immigration to the towns led to some empty villages, and certainly to villages with few young men and women. Meanwhile the population of the towns was young and disproportionately male.

The peasants flooding into the towns often had lower skills and had to settle for low-paid work; but within a generation they had become urbanized. The population of Saratov rose from 372,000 to 579,000 between 1939 and 1959; that of Kuibyshev grew from 390,000 to 806,000 in the same period.[35] Movement from the villages to the towns was particularly common among non-Russians: Tatars abandoned the villages to move to the towns – in particular to Kazan, where the proportion of Tatars vis-à-vis Russians increased. Chuvash also left the villages in large numbers – and at a greater rate than ethnic Russian peasants – to settle in Cheboksary and other towns, often taking the least skilled jobs in the first instance.[36]

The rapid growth of cities required new housing, communications and facilities. Soviet cities took shape in the 1950s and 1960s, with a vast housing programme of cheaply built, five-storey, mass-produced blocks made of pre-fabricated panels. (These buildings are still known as *khrushchevki*, after Khrushchev.) Millions of such blocks filled vast new suburbs. The flats were small and cramped, with tiny kitchens and bathrooms; and they were often badly constructed and badly maintained – but they proved a blessing for families who at the time were housed in barracks, or in communal flats in old buildings. The infrastructure for the vast new suburbs was often inadequate: blocks were often seemingly placed haphazardly, without any consideration of the main transport links; shops appeared to be located arbitrarily; parents might find there were no kindergartens or healthcare centres in the vicinity; the provision of clean water and sewers did not keep pace with the requirements; and it goes without saying that the new neighbourhoods had no churches or other places of worship. A new style of socialist city had been born.

The most dramatic change took place in Toliatti, west of Kuibyshev. The original settlement was called Stavropol ('city of the cross') and had been artificially created in 1737 to provide a base for Kalmyks who converted to Christianity. The town was renamed in 1964 in honour of Palmiro Togliatti, general secretary of the Italian Communist Party. The choice of an Italian communist was a result of the construction of an enormous Lada car factory in cooperation with Fiat. The construction of the car factory and then of a vast new hydroelectric power station and dam (see Chapter 17) flooded much of the original town, which had to be relocated. In the process, a small town grew into a great urban conurbation of

60,000 in 1957, and of over 700,000 people today. New housing and infrastructure were constructed for what was called at the time 'a new socialist city', including extensive sports facilities for the new, fit Soviet citizen.[37] Incidentally, the cars known in the West as Lada were called Zhiguli in the Soviet Union, after the Zhiguli hills on the bend of the river towards Kuibyshev (the very same hills that had inspired artists in the late nineteenth century).

Ulianovsk (former Simbirsk) had been a provincial backwater, but it was also the birthplace of Lenin (Vladimir Ilich Ulianov). In 1922–23, the cultural department in what was still called Simbirsk decided to open a museum to celebrate revolutionary events, which would have included a room devoted to Lenin and his family.[38] More grandiose plans evolved after the war, and in the 1960s the centre of what was now Ulianovsk, in Lenin's honour, was completely redesigned to celebrate the centenary of his birth in 1970. The centrepiece is the Lenin memorial building – a massive concrete, steel and glass structure with a permanent exhibition dedicated to the revolutionary leader; it includes the house where Lenin was born, which was moved and reconstructed in the open space in the centre of the building. The effort put into the planning was enormous – as were the costs of transporting material from the whole of the Soviet Union to build the complex, the construction costs and the labour required for the project.[39] In the process, the heart of old Simbirsk was ripped out. The new centre of the town was reconstructed around the memorial, with a broad avenue linking two massive squares and new large concrete buildings for administration, educational facilities, a library and a hotel. It was a utopian design, but the passage of time has not been kind to the concrete.[40] The Lenin complex remained a popular tourist attraction – 25,256 people visited it in 2001[41] – but it is currently (late 2019) being completely reconstructed. In April 2018, the Ulianovsk town council resolved to revert to the original name of the square – from Lenin Square back to Cathedral Square – despite the fact that the cathedral, built to commemorate those who died in the 1812 campaign in the Napoleonic Wars, was destroyed in 1936! In contrast, a street containing another house in which the Ulianov family lived has been preserved, and this part of the town now hosts a number of charming, small museums that capture the life of Simbirsk merchants and display the work of town architects in the late nineteenth and early twentieth centuries.

During the German invasion of the Soviet Union, the anti-religious policies were relaxed, because Stalin realized that he needed to mobilize the patriotic support of all the Soviet people to resist the invader. The Soviet Union was, however, an atheist state, and in the 1960s Khrushchev personally revived the campaign against religion.

This resulted in more churches, mosques and other places of worship across the Soviet Union being closed down or destroyed (and more priests and religious leaders being arrested and imprisoned). In the centre of towns, major churches (and mosques and synagogues) were converted into (or replaced with) cinemas, theatres or community centres. Despite the destruction of many old and beautiful churches, the centres of many Soviet towns remained untouched, graced by late-eighteenth- and nineteenth-century administrative buildings. The main streets of the Volga towns today – Saratov, Samara, Ulianovsk, Kazan, Nizhnii Novgorod, Tver – are still recognizable from photographs from the late nineteenth century.

❖

The Volga had already been sentimentalized in the Soviet Union before war broke out. Travel guides were published in the 1930s, which encouraged tourism on the Volga. One such book, published in 1930 and republished in 1933, opened with the outrageous statement that, after collectivization, the 'USSR has become the largest country in the world for agriculture' (the tourist's attention is also drawn to the delights of the cement factory in Stalingrad).[42] The satirical book *The Twelve Chairs* by Ilf and Petrov mocked the greed of the New Economic Policy in the 1920s, but also included a chapter entitled 'A Magic Night on the Volga'. In it, as the steamer sails gently down the river, the chairman of the very important committee involuntarily breaks into song, 'A ship sailed down the Volga/ Mother Volga, river Volga', and the other members join in with the chorus ('the resolution of the chairman's report was not recorded' is the dry comment by the authors).[43]

The film *Volga, Volga*, which came out in 1938, was an immensely popular musical comedy that is mainly set on a Volga steamship. It is about local people travelling to Moscow to compete in a music competition. The main song is 'Song of the Volga', which includes all the common perceptions of the river: deep, wide, powerful, place of suffering, but also freedom through the heroics of Razin and Pugachev. But the Volga in the song also glorifies the motherland, the Soviet Union, and includes a prescient warning to potential enemies of that state:

Long ages passed over our country
Like storm clouds full of grief,
And the Volga, our boundless river,
Flowed on in a torrent of tears.
She did not give in to tricks or chains,

THE VOLGA

The blue highway of our country.
Not in vain did Razin and Pugachev
Sail their longboats down the Volga.

O beautiful maid of our people.
Mighty with water like the sea,
And just as our motherland – free.

Our joy, like May, is youthful,
Our strength cannot be crushed,
Beneath the happy Soviet star
It is good to work and live.
Let our foes, like ravening wolves,
Leave their tracks at our country's borders;
They shall not see our beautiful Volga,
Nor drink from the Volga's waters.

O beautiful maid of our people,
Mighty with water like the sea,
And just as our motherland – free.
Wide,
Deep.
Strong!

'Life was very hard,' commented Galina Kosterina about the war years, 'but there were many happy events too.' These included films and social events: she loved watching *Volga, Volga*.[44] It was also said to be Stalin's favourite film. (It is claimed that during the Khrushchev period, the film was shown in edited form – minus the clip featuring the statues of Lenin and Stalin on the Moscow–Volga canal!) The film depicts an idyllic village on the Volga, crowded with livestock (cows, pigs and ducks), and one scene takes place in a very jolly and nicely furnished rural café. There is no sense of irony: the reality is that peasants were starving in the early 1930s, and in 1938 peasant villages were drab and demoralized, while the gulags were at their peak (see the previous chapter). The travel guides and the film were aimed at primarily an urban audience, and escapism is always attractive at a time of great economic and personal hardship.

The river Volga had been portrayed before the Revolution as the 'mother' of Russia, the 'protector' of her children, and was to an extent sacralized (see Chapter 11). The vision of the river as 'Mother Russia' was revived in the 1930s and 1940s, and the battle of Stalingrad served to enhance that depiction of the Volga as a defender of the motherland. In his novel *Stalingrad*, Vasily Grossman describes a scene before the battle, when the exhausted Soviet soldiers reach the Volga and wash themselves and their uniforms in its water, in symbolic and even religious terms, comparable to the conversion to Christianity of Kievan Rus in the late tenth century:

> We do not know whether he [Marshal Timoshenko], or any of the thousands of men throwing water over themselves, understood they were performing a symbolic ritual. This mass baptism, however, was a fateful moment for Russia. This mass baptism before the terrible battle for freedom on the high cliffs of the west bank of the Volga may have been as fateful a moment in the country's history as the mass baptism carried out in Kiev a thousand years earlier, on the banks of the Dnieper.[45]

In the event, the Germans were not able to control or cross the mighty river; their attempts to 'defile' the almost sacred river were repulsed. The hero of the book *The Tale of Egorka the Pilot* leads a convoy of ships up the Volga; he loves his 'native land and the free Mother Volga more than my own life'. 'She, the mother, hears only our entreaties, and brings death to our worst enemies.'[46]

In the 1950s and 1960s, the Volga continued to be sentimentalized through popular songs, such as 'The Volga Flows into my Heart' and 'I Just Can't Live Without the Volga'. The song 'The Volga River Flows' became immensely popular in the 1960s, when it was performed by the singer Liudmila Zykina (whose fame was so great and so enduring that she features on a *Russian* 15-rouble postage stamp).[47] 'We are Russians. We are children of the Volga', wrote the poet Evgenii Evtushenko.[48]

Where, however, did this leave the many non-ethnic Russians on the Volga? They had experienced the shocks of two world wars, revolutions, civil war, collectivization, and had been victims of the gulags to the same extent as ethnic Russians (if not more so). Many of the soldiers who died at Stalingrad were non-ethnic Russians, although the statues of the Soviet soldiers at the Stalingrad memorial complex are overwhelmingly European in their facial features. It is to the history and sense of identity of non-ethnic Russians in the Soviet and post-Soviet periods that we now turn.

16
NON-RUSSIANS ON THE VOLGA
Autonomy and identity

It is very easy for a Tatar to become a nationalist in the USSR. One just needs to attend history classes at school, when, at the description of the horrors of the Tatar-Mongol invasion, the whole class turns to look at you; watch films in which glorious Russians perish at the hands of savage Tatars; get used . . . to the idea that the Volga is a great Russian river, that it is Russian woods that surround you; take part every year in the festivals of seeing off the Russian winter . . . What can a Tatar be proud of?

Statement by a Tatar journalist in 1990[1]

The Bolshevik Revolution in 1917 seemed to offer the non-Russian peoples new opportunities. Lenin referred to the multi-ethnic Russian Empire as the 'prison of nations'. In place of an empire defined by its Russianness and Orthodox Christianity, the new Bolshevik state would be defined by 'socialism', in which all ethnic groups would be equal. One of the first acts of the new Bolshevik state, on 2 November 1917, was to issue the Declaration of the Rights of the Peoples of Russia, which proclaimed the equality of all the peoples of the former Russian Empire and promised them the right to self-determination, including the right of secession and the formation of separate states. The proclamation certainly helped the Bolsheviks gain the support of some non-ethnic Russians in the elections in 1917 and during the Civil War, in particular because Anton Denikin, general of the Whites, had proclaimed his cause to be for 'Russia, one and indivisible'.

The Soviet Union was, in principle, a voluntary federation of 15 equal socialist republics. The Volga region was part of the largest republic – the Russian Soviet Federative Socialist Republic (RSFSR) – but within that republic, autonomous regions were established for separate ethnic groups, including the non-Russians who lived on the Volga: Tatars, Chuvash, Mari, Udmurts, Mordvins, Germans.

286

Each autonomous republic within the RSFSR had its own flag, its own capital and its own administrative and higher education institutions, its own national theatre and national opera house. It proved more difficult in practice, however, either to exercise any meaningful degree of autonomy within these republics or to maintain a separate cultural identity in the Soviet Union during the massive economic and social changes described in the previous two chapters.

The dissolution of the Soviet Union in December 1991 resulted in the former socialist republics of the USSR becoming newly independent states, while the autonomous republics within the former RSFSR retained their separate status but remained part of the new Russian Federation. There were two major changes for the people of the Volga. First, the new country was less populous than the Soviet Union had been, but more predominantly ethnically Russian. In the last years of the Soviet Union, Russians had comprised only just over 50 per cent of the population of some 293 million. They had, however, been the dominant ethnic group in the RSFSR and retained that position in the Russian Federation – some 81 per cent of a population of some 144.5 million.

Second, the western and north-western borders of the new and independent post-Soviet country of Kazakhstan are close to Volga towns in the Russian Federation. The new border runs along the east side of the Volga from Astrakhan almost to Volgograd, and is less than 100km from the river in places; it then turns north-east, following the river almost as far as Samara, which is only 234km from the first major town across the Kazakh border – Oral (formerly Uralsk). Astrakhan has become a border town in the south of Russia, close to the Kazakh border (although not close to any major Kazakh towns: it is 373km from Atyrau, further along the north Caspian coast). The lower Volga and parts of the middle stretch have yet again become a frontier zone of the Russian state, a position that had not existed since the conquest of the Kazan and Astrakhan khanates in the mid-sixteenth century.

❖

The greatest potential threat to the unity of the Bolshevik state came from the Volga Tatars. By 1905, as we saw previously in Chapter 12, intellectual movements among the Kazan Tatars had led to a strong sense of national identity, at least among the urban Tatar elite; these became aspirations for political and cultural autonomy – and even independence – in 1917. In November of that year, the National Assembly of the Muslims of Inner Russia and Siberia met in Ufa and proclaimed a new Idel-Ural state (Idel being the Tatar word for the river Volga). If the new state had been established, it would have incorporated lands occupied by Chuvash, Mari, Udmurts and Bashkirs, and would have been fairly similar in extent to the lands that comprised the Kazan khanate before its conquest by Ivan IV. There was little opposition to the idea of being part of a Tatar state from the very small Chuvash or Mari intellectual elites, who in any event were more worried by the prospect of further Russification than by potential 'Tatarization'. Bashkirs and Volga Tatars shared many cultural and linguistic characteristics, as well as Islamic faith; but a small number of Bashkir intellectual nationalists feared being dominated by Tatars, and resented being regarded as almost a 'cultural subdivision' of the Volga Tatars,[2] although in reality historically most conflicts in Bashkir lands had been with Russians over encroachment on Bashkir grazing lands, rather than with Tatars. In 1917, one-third of the Bashkirs used Tatar as their native tongue.[3]

In January 1918, the Second Muslim Military Congress met in Kazan to try to set up the new state; but it encountered opposition from the Russian members of the Kazan soviet. In late February, the soviet ordered the arrest of the Tatar nationalists in Kazan, who fled to the Tatar suburb of the town and declared an Idel-Ural state. In March 1918, this stronghold was attacked and taken by the Red Army, and Kazan was put firmly under Bolshevik control. The Idel-Ural state had lasted only a month or so. It had had time to design (and fly) a new flag – a light blue background with a *tamga*, a Tatar heraldic symbol, in the top right corner – but little else.

In reality, decrees on independence were meaningless, as civil war broke out and the proposed territory of the new Idel-Ural republic became one of the battle grounds between Reds and Whites during the Civil War. At first, it looked as if a Tatar-Bashkir autonomous republic would be established – and a decree to that effect was passed by the People's Commissariat of Nationalities (known by the acronym Narkomnats) – with Kazan as its capital. After the Civil War was won, the new Soviet state was able to assert its authority more effectively, and used a policy of divide and rule to do so. The government had no desire to see strong independent or semi-independent Muslim states within the Soviet Union. The Kazan Tatars had led the movement for a large Tatar state, which would have incorporated Bashkirs in a subordinate position. To counter Tatar influence, the government in Moscow supported, at least in principle, the rights of Bashkirs and other non-Tatars, and in doing so divided the Muslim, and the Turkic, inhabitants of the Volga and Ural regions.

The middle and lower Volga was the first region in which Soviet policy towards non-ethnic Russians had to be defined, and it set the pattern for other ethnically mixed regions of the Soviet Union (as indeed had been the case when the region was first incorporated into Russia, in the mid-sixteenth century). When the autonomous republics were established after the end of the Civil War, Tatar hopes for a large, independent state were dashed. The Bashkir Autonomous Soviet Socialist Republic (ASSR) was set up in 1919, with its capital in Ufa, a year before the establishment of the Tatar ASSR. The new Tatar republic was far smaller than the proposed Idel-Ural state – just a third of the size, in fact. The split with the Bashkir ASSR, along with the establishment of other new autonomous republics on the Volga, meant that 75 per cent of ethnic Tatars were now outside the boundaries of the new Tatar ASSR and that Tatars constituted a majority in the Bashkir ASSR. In 1970, out of a total of nearly 6 million Tatars, some 4.7 million were in the RSFSR and over 900,000 in the Bashkir ASSR, with just over 1.5 million in the Tatar ASSR.[4]

Over the next few years, more autonomous republics were established in the Volga region. In 1920, the Votiak (the older name for Udmurts) Autonomous *Oblast* (a Russian word which means a region, but one smaller and of less significance than a republic) was established, with Izhevsk as its capital; it was renamed the Udmurt Autonomous Oblast in 1932, and became the Udmurt ASSR in 1934. In 1920, the Mari Autonomous Oblast was created; it was renamed the Mari ASSR in 1936. The capital was the former district town of Tsarevokokshaisk (on the Kokshaga river, so this means the 'tsar's town on the river Kokshaga'), renamed Krasnokokshaisk ('red Kokshaga') in 1919 and Ioshkar-Ola (which is 'red city' in Mari) in 1927. The Mari republic borders on the river Volga in the east. The Chuvash ASSR was established in 1925 (following the establishment of the Chuvash Autonomous Oblast), with its capital in Cheboksary on the Volga. Territory inhabited by Mordvins became first a region and then an oblast and finally a full autonomous republic in December 1934; the capital of the republic is the town of Saransk.

Perhaps the most remarkable autonomous republic was that of the Volga Germans. It was the first region of the Soviet Union to be declared an autonomous region, in 1918, and was given the formal status of the Volga German Autonomous Soviet Socialist Republic in February 1924. The republic was situated on the eastern bank of the Volga, and the capital was Pokrovsk or Kosakenstadt (the town was renamed Engels in 1931, and still retains that name). Pokrovsk lies on the opposite bank of the river Volga from Saratov. The Volga German ASSR did not cover all the territory where German colonists had settled, but was the centre of the 'meadow-side' colonists, on the eastern bank. The new republic had a majority of ethnic Volga Germans, along with large Russian and Ukrainian minorities. The 1939 census gave the number of Germans in the republic as 366,685.

It could be argued that the federal structure of the Soviet Union was only ever a temporary measure, because the Bolsheviks assumed that the Russian Revolution would be followed by world revolution, which would have eliminated all national borders. By 1924, when attempted revolutions elsewhere in Europe had been crushed and it had become clear that world revolution was not imminent, the Soviet Union moved instead to a policy of 'Socialism in One Country'. From this point, the existence of any autonomous *political* culture – either in the separate Soviet republics or in the autonomous republics within the RSFSR – was completely at odds with the central authority of the Communist Party of the Soviet Union. In 1921, the Communist Party banned factions within its ranks; this also had implications for regional communist parties within the separate Soviet republics and the autonomous republics within the RSFSR. Cadres were sent from the central Communist Party (many of whom were ethnic Russians) into the provinces to 'denationalize' local parties.[5]

The level of desire for independence, and the sense of a separate identity, had been higher among the Volga Tatars than among any other non-Russian people of the Volga at the time of the Revolution, and so the Soviet government (in effect the Communist Party of the Soviet Union) considered it all the more necessary to crush their pretensions. The way political control could be exercised from the centre to curb any regional independence can be seen in the experiences of Mirza Sultan-Galiev within the Tatar ASSR. Sultan-Galiev, the son of a Tatar schoolteacher, had risen rapidly through the ranks of the Communist Party. He argued that all classes of Muslims should be called 'proletarian', because they were all the victims of Russian colonization. He wanted to give Marxism a 'Muslim' face and to preserve the cohesion of Muslim society; in other words, in his view national liberation should take priority over class war. To this end, he campaigned to establish a Muslim Communist Party with its own elected central committee, and a Muslim Red Army with Muslim commanders and officers. He envisaged a new 'Republic of Turan' which would stretch from Kazan across Asia to the Pamirs. The downfall of Sultan-Galiev is symptomatic of the increase in centralization within the state in the 1920s and the intolerance of any deviance from the central Communist Party line. In May 1923, he was arrested on the personal initiative of Stalin and denounced for 'counter-revolutionary nationalist conspiracy against the power of the Soviets'. He was freed in 1924, but arrested again in 1928, tried as a 'traitor' and sentenced to 10 years' hard labour in the gulag. He was executed in December 1939.[6]

Between 1924 and 1939, the Communist Party of the Tatar ASSR was divided between those who supported the Galiev line (mainly ethnic Tatars) and the almost exclusively ethnic Russian communists in Kazan. The two sides disputed the extent of Russian 'chauvinism', the definition of imperialism and the possibility of world revolution against a background of brutal collectivization and purges. There was always only going to be one victor in this dispute. Sultan-Galiev's arrest was followed by a systematic purge of Tatar party officials, which was then extended to an all-out attack on Tatar intellectuals. In 1930 alone, 2,056 Tatar communists – over 13 per cent of the total membership – were expelled from the party; in all, 2,273 Tatar communists were executed for 'nationalistic deviation'.[7] The crushing of the independence of the Volga Tatar communists, who cherished hopes of independence, or at least a degree of genuine autonomy, was a *political* decision by the Communist Party of the Soviet Union, and coincided with the purges.

The Tatar ASSR, however, continued to exist until the final demise of the Soviet Union. By contrast, in 1941, after the German invasion of the Soviet Union, Joseph Stalin took the decision to deport the Volga Germans *en masse* to Siberia and Kazakhstan, having condemned them all as spies and saboteurs. The Volga German ASSR was abolished at the same time and removed from Soviet maps: the republic was not mentioned at all in Stalin's deportation order, which targeted the 'entire German population of the Volga district'. The expulsion was brutal and the colonists were treated as if they were criminals: some 600,000 people had their possessions and livestock confiscated, were packed into cattle trucks or freight trains with very little food or fresh water, and sent off east, where they were obliged to work in construction or felling timber. Those who died on the journey were simply thrown out of the railway carriages. No exceptions were made, even for Communist Party members. Indeed, many young male Volga Germans were serving at the time in the Red Army, and only learned of the fate of their families by chance, as this sad account testifies:

> While our company was halted for a rest in a Russian village one day . . . a resident of this community, upon overhearing several of us conversing in German, approached our little group out of curiosity to learn who we were. When told we were Volga Germans, he was quite surprised; then he informed us of the mass evacuation of our people, and that several families from this community in which we had halted already were *en route* to the Volga to occupy our homes – homes completely furnished, farmyards with domestic animals and machinery, potatoes to dig and cabbages to harvest – in fact, everything to start life anew there. We were shocked and could not believe this stranger. Yet, he roused a nauseating uneasiness in all of us, and so we wrote home at once.

After two months, our letters were returned marked 'Adresat Vybil' – Addressee Moved. Only one person in our group received a reply. It was from his sister who was married to a Russian, and therefore permitted to stay in her village.[8]

In 1955, after the death of Stalin, Konrad Adenauer, chancellor of West Germany, raised the issue of the Volga Germans with Khrushchev in Moscow, but they were not rehabilitated until 1964. Even then, they were not allowed to return home. In fact, since many of their villages had been destroyed in the war (on the 'mountain' or western side) or ruined by collectivization, few would have chosen to return to the Volga even had they been allowed to do so.[9] Many Volga Germans in Russia and Kazakhstan left for Germany after the collapse of the Soviet Union: there were some 17,000 ethnic Germans in Saratov province in 1989, but by 2010 that number had dropped to only 7,579.[10]

In reality, the Soviet Union was a highly centralized state, which allowed for only a limited autonomy of any institutions at any level, on the Volga or anywhere else. The decision, for example, to build hydroelectric plants and to change the depth and flow of the Volga (described in the next chapter) was taken in Moscow, and not locally. The physical results of the dominance of central economic planning were dramatic for a number of Volga towns – in particular Cheboksary, where the buildings by the shore of the river were flooded after the construction of the Cheboksary hydroelectric plant and reservoir. The level of the Volga was raised at Kazan (with the result that the memorial to those who died in the taking of Kazan in 1552 has been almost submerged). These were central (not local) Communist Party decisions, and were taken with no consideration of the local economic or environmental impact. The implementation of these grandiose plans led to vast construction sites and new towns on the Volga (such as Toliatti). Indeed, all *economic* planning was centralized: the government in Moscow determined which factories should be set up, what and how much they should produce, and where they should be located. Whole new regions in Volga towns were set aside for new industrial enterprises in the five-year plans.

The 1977 constitution retained the federal structure of the Soviet Union, but the federal powers of the separate Soviet republics and the autonomous republics within the RSFSR were poorly defined. In reality, any 'federalism' was illusory, because policy was determined centrally by the Communist Party of the Soviet Union. The fictional nature of power exercised by the autonomous republics within the RSFSR was obvious. As the Soviet Union faced collapse, Rafail Khakimov, vice-rector of the Kazan Institute of Culture and a leading nationalist, made this clear:

The notion of an autonomous republic as a government without sovereignty is a juridical and political nonsense . . . The RSFSR and autonomous republics are a political anachronism. . . . The USSR is built on the principle of the *matreshka* [the wooden nested dolls]: there is a federation within a federation, a republic within a republic, in a *krai* [region] we have an autonomous oblast, in an oblast an autonomous *okrug* [unit]. The submission of some nations to others, of some republics to others, sharply contradicts the principle of equality among nations and is a breeding ground for conflicts.[11]

❖

The dissolution of the Soviet Union exposed the sham of the official propaganda about the nature of the federal structure of the country. All the separate soviet republics broke away and formed separate countries (albeit some with less enthusiasm than others), using the federal constitution of the Soviet Union as the legal basis for their actions. That option was not available to the autonomous republics within the RSFSR (now the Russian Federation), but Boris Yeltsin encouraged the autonomous republics to assert their autonomy in a bid to gain their support in his power struggle with Mikhail Gorbachev. 'Take as much independence as you hold on to', he unwisely declared at a meeting in Kazan.[12] The Chechen ASSR declared independence in 1991, which led to bitter and bloody wars between the newly declared Chechen Republic and the Russian Federation in 1994–96 and 1999–2009. Chechnya is predominantly Muslim, and the Russian government feared another Muslim conflict with the Volga Tatars and other Muslims within the Russian Federation.

Tatar nationalist sentiment was expressed at the time through the foundation of the Tatar Public Centre, which became an organization that Tatar intellectuals rallied around. On 30 August 1990, the Tatars within the autonomous republic made a Declaration of the Sovereignty of the Tatar Republic. In 1991, after the dissolution of the Soviet Union, the Tatar Republic joined the Russian Federation, but asserted equal rights with Russians.[13] The 1992 constitution of the newly named Republic of Tatarstan (usually simply called Tatarstan) described the republic as a 'sovereign state', but that was declared unconstitutional by the Russian high court in Moscow. Tortuous negotiations then took place between Moscow and Kazan. The Tatar side raised the stakes by withholding federal taxes; the Russian side at one point threatened to bring the president to Moscow in an 'iron cage' – a clear reference to the fate of Pugachev in the late eighteenth century.[14] In 1994, a power-sharing

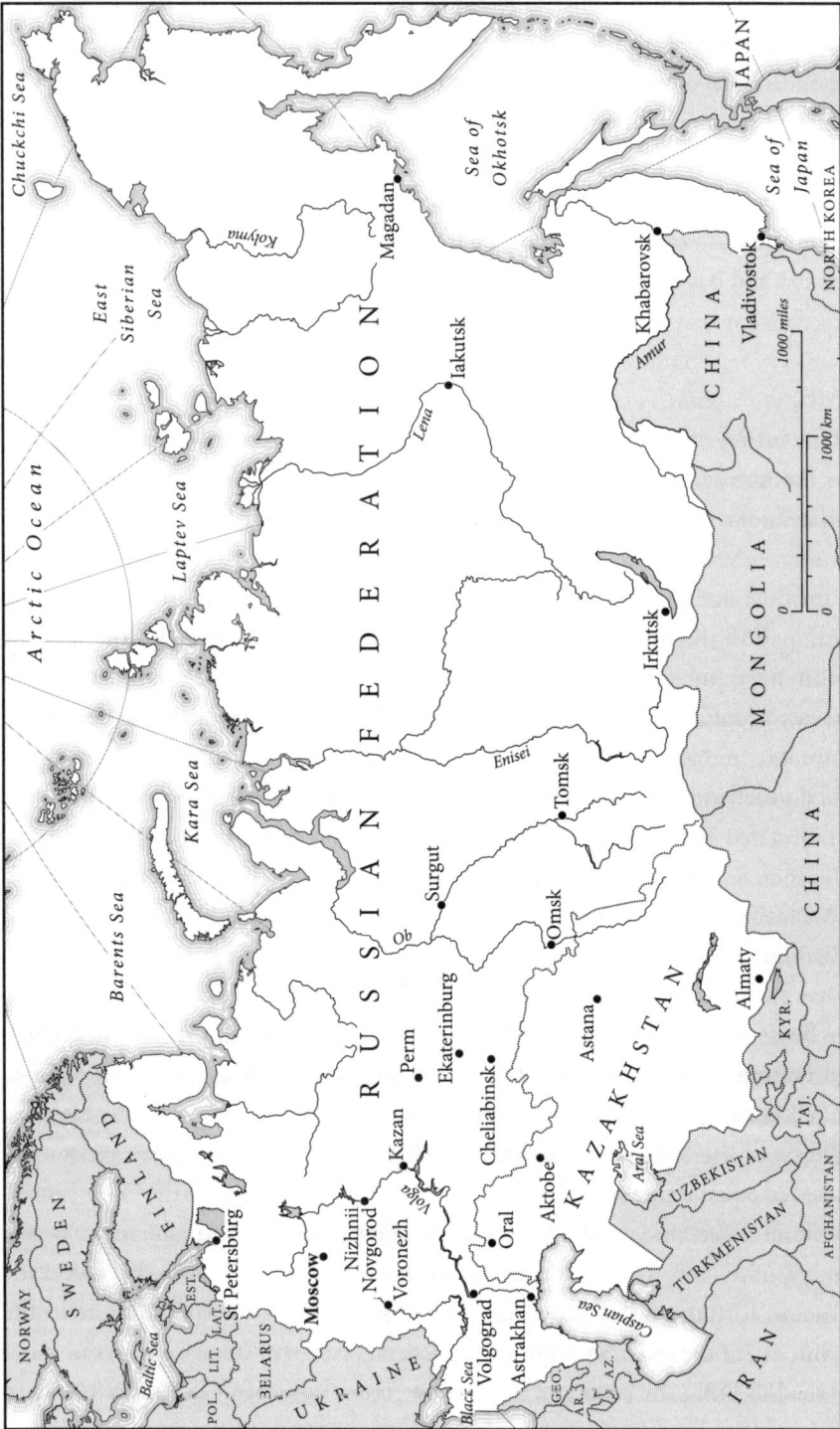

9. The Russian Federation and Kazakhstan

arrangement was agreed between the Russian Federation and Tatarstan (the first of 46 power-sharing treaties signed by Yeltsin), which gave the new republic control of resources and of 'citizenship', although what this meant in practice soon became a matter of dispute. Tatarstan had an advantage over other autonomous republics in terms of its wealth, which gave it some bargaining power; but the Russian government also had to handle the situation carefully, because it feared that the Russian Federation could disintegrate completely, as the Soviet Union had done.

Tatarstan today retains considerable autonomy and has its own national anthem and its own flag: two broad bands – one red (symbolizing the fight for happiness and the bravery of the Tatar people) and the other green (symbolizing hope, freedom and solidarity with Islam) – separated by a narrow strip of white (signifying the peaceful relations between the Tatars and the Russians). Between 1991 and 2010, Tatarstan under President Mintimer Shaimiev (the first president of Tatarstan) skilfully maintained a moderate stance, while asserting considerable autonomy from Moscow. Oil revenues have made the republic wealthy, which certainly gives it the ability to fund large projects and to assert itself economically vis-à-vis the Russian government; but it is still part of the Russian Federation, and the relationship between Tatarstan and the Russian government remains delicate.

After the collapse of the Soviet Union, between 1990 and 1992 all the former autonomous republics within the Russian Federation on the Volga dropped the words 'Soviet Socialist' and re-formed themselves as republics within the Russian Federation (the Mari ASSR renamed itself the Mari El Republic and the Mordovian ASSR renamed itself Mordovia). The only exception was the Volga German ASSR, which has never been re-created.

Vladimir Putin (who succeeded Yeltsin as president of Russia, 2000–08; was then prime minister, 2008–12; and since 2012 has again been president) has asserted more centralized control, and presidential power, over the whole Russian Federation, including Tatarstan. In 2000, by presidential decree, Putin divided the Russian Federation into seven federal districts (in 2010 the north Caucasus made it eight), in order to implement federal policies. These federal districts are larger than the autonomous republics (which they have subsumed within their boundaries). One is the Volga Federal District, headed by a presidential plenipotentiary envoy appointed by the Russian president; it is centred on Nizhnii Novgorod and incorporates Tatarstan, Mordovia and the Mari El, Chuvash and Udmurt republics. Also in 2000, the Russian government decreed that passports were to be issued for the whole Russian Federation, and that there could not be separate passports for Tatars (or any other ethnic group). A law of 2004 gave the Russian president the right to appoint the presidents of the republics and the right to dissolve their parliaments. In practice, Putin has put his own people into key posts in Tatarstan and elsewhere.[15] In 2007, the power-sharing agreement between Tatarstan and the Russian Federation was renewed, but gave fewer autonomous rights to Tatarstan. Shaimiev stepped down in 2010, and the Russian president (at that time Dmitrii Medvedev) appointed the successor suggested by Shaimiev: Rustam Minnikhanov. In July 2017, the power-sharing agreement between Tatarstan and the Russian Federation expired; it has not yet been renewed.

The republics have faced severe economic problems: heavy industry and uncompetitive factories have closed down in Cheboksary (including the Lenin textile factory, where the statue of Lenin now stands rather forlornly on uncut grass, in front of an abandoned building) and Ioshkar-Ola. Leaving aside Tatarstan, which has oil, the other non-Russian republics have struggled to adapt to the new economic conditions in Russia. Nothing, however, accounts for the bizarre construction in Ioshkar-Ola of a Disney-like theme park, but without the crowds. A whole complex has been built of, mainly, Dutch/Flemish-style buildings (one of which, bizarrely, houses a zoo), alongside a replica of the gates into the Moscow kremlin and a nineteenth-century style Orthodox cathedral. Across the river is an embankment of buildings (known as the Bruges embankment), also built in a pseudo-Dutch/Flemish style. It is strangely devoid of life – no bustle, no shops, no cafés – nothing, in fact, but an empty street. In another part of town, the National Art Gallery has been constructed in Dutch/Flemish style in red and white bricks, complete with an enormous clock tower (in which the 12 apostles appear on the hour). The whole thing is an obscene example of self-indulgent spending in a poor region, and bears no relation at all to Mari culture: outside the National Art Gallery is a statue of Ivan Obolensky-Nogotkov – the first governor of the town, appointed by the *Russians* in 1584. In September 2018, the mayor of the town, Pavel Plotnikov, was arrested for corruption and found guilty of bribe-taking on a massive scale. He was sentenced to 10 years' hard labour. (Plate 18).

The consequences of state economic planning dramatically changed the social composition of the towns and the lifestyles of urban citizens during the period of the Soviet Union, and this is a legacy that the current Volga autonomous republics have to deal with. The new industries resulted in mass migration from the countryside to the towns, and the vast new construction of identical housing blocks determined where and how people lived (as described in the previous chapter). At the same time, the new forms of media – cinema and television – and the opening of sports facilities, stadiums and parks meant that all Soviet citizens shared in popular, mass, Soviet culture. The Soviet state intervened in the leisure time of its citizens: from organizing mass demonstrations on special days to celebrating political and historical events to organizing the free time of children through the Komsomol. The shortage of consumer goods meant that Soviet citizens, apart from a few members of the elite, possessed the same material goods, wore similar-quality clothing and had access to roughly the same foodstuffs in shops and markets. The Volga region was no exception here, but non-Russians – whether within or outside the new autonomous republics – faced particular challenges in asserting a separate way of life (and still do).

A major challenge for all the non-Russians of the Volga was (and still is) to prevent themselves from being outnumbered in their towns by Russian migrants (or other national groups in the Soviet Union) as a result of mass industrialization. Russians were in a majority – or at least a significant minority – in all the new autonomous republics in the Soviet era. By 1986, ethnic Mari had fallen to 43 per cent of the Mari ASSR's population.[16] The majority of the population in the Chuvash ASSR was Chuvash – about two-thirds according to the 1989 census, with Russians accounting for only 27 per cent.[17] In 1959, 76 per cent of the population of the Udmurt ASSR were Udmurts; this had fallen to 66.6 per cent by 1989.

The Volga towns are today attractive to emigrants from the former socialist republics, which is also changing the ethnic composition of towns. Official figures from the last census, in 2010, show that ethnic Kazakhs comprised 5.4 per cent (32,783 people) of the population of Astrakhan, and many ethnic Russians have also moved from Kazakhstan to Russia. Another estimate is that there are over 76,000 Kazakhs in Saratov province – a small percentage of the overall population (which is 88 per cent Russian), but still enough to have an impact, given that most of these new migrants are young and urban. The conflict in the south Caucasus (in Nagorno-Karabakh, but also in Abkhazia) has led to a wave of recent Armenian emigration into Russia, estimated in 2010 to be particularly large in Saratov (23,831) and Volgograd (27,846).[18]

The Volga Tatars have had an advantage, in that they were already numerous in the original Tatar ASSR, and then their numbers rose further – from 4,765,000 in 1959 to 6,646,000 in 1989.[19] Tatars have a higher marriage rate, earlier marriages and a higher birth rate than Russians. Today the population of the town of Kazan is roughly 50:50 Russians and Tatars (the nineteenth-century town and the Soviet town had a significant Russian majority). According to the 2010 census, the population of the whole of Tatarstan is 3.7 million, of whom 53.2 per cent are Tatars and 39.7 per cent Russian.[20] Of course, in the Russian Federation as a whole (just as in the former Soviet Union) even the Tatars, the largest minority group, still make up a very small minority of the overwhelmingly Russian population.

One way in which a separate ethnic identity can be asserted is through language. Initially, the Soviet state encouraged the use of native languages within the autonomous republics. There were 129 newspapers in Tatar published in the Tatar ASSR in 1935, and 10 other newspapers in Tatar on the middle and lower Volga.[21] The Soviet government also encouraged the establishment of national theatres (which staged plays in the local language) and the publication of poetry and literature in local languages, along with translations into Russian. The script for non-Russian languages was always, however, controversial, and remains a national question today. The issue was most acute in the Tatar ASSR, where in 1917 Tatar had been written in Arabic script – something which drew the Tatars into the orbit of the Islamic world (although 'baptized' Tatars, who had converted to Christianity, used Cyrillic script and were treated as a separate ethnic group within the Soviet Union until 1928). The question of the alphabet of Tatar and other languages generated lively academic debate among Tatars and other Muslims in the Soviet Union,[22] but was ultimately decided by the Soviet government. In 1929, it decreed that Tatar (and the other Turkic languages in the Soviet Union) should be written in the Latin alphabet, which in effect modernized the language (just as, the year before, the Turkish language had been Latinized in Turkey). In 1934, a new push came from the centre to change the alphabet to Cyrillic, because Latin was now seen as the language of colonization and imperialism,[23] or as *Pravda* put it, using Cyrillic 'would lead to an even greater unification of the peoples of the USSR, to an even greater strengthening of the friendship of the peoples of the USSR'.[24] Put another way, this new policy asserted *Russian* control over non-Russians, by forcing them to adopt Cyrillic script (which, incidentally, is quite rigid in the sounds it represents and is poorly suited to Turkic languages). The main impact of the change of policies on the Volga was on the Volga Tatars, who had to change from Arabic to Latin and then to Cyrillic within a decade. Other Turkic languages in the region, such as Chuvash, and Finno-Ugric languages – like those of the Mari and Mordvins – were already written in Cyrillic.

In fact, the main concern of the non-Russian peoples of the Volga has not been the alphabet of the language, but the difficulty of maintaining the language at all in face of the dominance of Russian within the Soviet Union – and, post 1991, within the Russian Federation. Teaching in primary schools could be in native languages, but knowledge of Russian was (and still is) essential for higher education and for any professional occupation.[25] There was (and is) an overwhelming presence of Russian in the media: by the end of the Soviet period, local TV in the Tatar ASSR broadcast in Tatar for 4 hours a day; but this dropped to 1.8 hours of Chuvash-language broadcasting in the Chuvash ASSR, and 1.1 hours of Mari in the Mari ASSR.[26] Cheboksary grew massively after the Second World War and became the centre of almost all Chuvash manufacturing; but it was a Russian as well as a Chuvash town.[27] Teaching of Russian, in any event, was compulsory in all schools. When given a choice, parents in the Mari and Chuvash ASSRs chose to send their children to Russian schools, particularly in the towns, in order to help them in their future careers. At the same time, Russians, who lived mainly in the towns, were reluctant to learn the languages of the other ethnic groups. It was rare in the Soviet period to hear Tatar in Kazan or in the other major towns of the Tatar ASSR.

After the dissolution of the Soviet Union, the question of alphabets and the teaching of local languages resurfaced. In 1999, Tatarstan decreed that the alphabet for Tatar should be Latinized; but this was overridden by a Russian federal law in 2002, which ruled that all languages in the Russian Federation had to retain the Cyrillic script. The argument for Cyrillic is that the script can be read by all citizens of the Russian Federation; but it is clear that the alphabet is of much greater significance than that, and can be used to assert cultural dominance by the central, Russian, government. Schools in Tatarstan taught in both Russian and Tatar until very recently, but this led to tensions with the Russian community and with the Russian government, which put pressure on the Tatarstan government to end compulsory Tatar lessons in all schools.[28] In July 2018, the Russian duma (parliament) passed a law that protected the Russian language in regions with considerable numbers of non-Russians, which in practice meant the end of mandatory indigenous language teaching within the Russian Federation. The main impact of this has been felt on language teaching in schools, but it has also had an impact on the requirement for government officials to know indigenous languages.

In fact, Tatar is now heard regularly in Kazan and all signs are bilingual; it has also become popular to give children traditional Tatar names. Indeed, Kazan is becoming more and more a Tatar town, seen most visibly in the construction of many new mosques – something that has led to some unease among Russian professionals, who fear their careers may be restricted. Tatarstan simply has an advantage over the other former autonomous republics in that there are far more Tatars and it is more practical for them to use Tatar in a professional capacity. Even so, Russian remains the language of manufacturing, science, finance and commerce, and for professional jobs in Tatarstan and elsewhere. Instruction at the Federal University of Kazan today is in Russian for all but specifically Tatar subjects. There has been an attempt in other republics to support the teaching of indigenous languages, but such languages are far more likely to be spoken in the family than at work (and never used at work in towns, according to a survey undertaken in the mid-1990s).[29]

Cultural separateness can be expressed in other ways. It was not unusual in the Soviet Union for major family events – births, marriages, funerals – to be celebrated with local customs, rituals and foodstuffs, even when they had become secular, and for urban dwellers to go back to visit relatives in the countryside to celebrate these events. Some of these rituals were, however, weakened by the number of mixed marriages between Russians and non-Russians that took place after mass urbanization. In Kazan, for example, it was estimated in 1963 that 15 per cent of marriages were mixed (Russians and Tatars);[30] the percentage was similar for Chuvash and Russians in the Chuvash republic. Folk songs, dances and costumes can all be symbols of identity. Passengers on the cruise ships which call at Cheboksary are now met at the dock by Chuvash singers in national dress (and are given a drinks mat with an ancient Chuvash symbol of the sun on it).

It was also common to combine Soviet official holidays with former pagan or Muslim holidays. This was particularly important in the Tatar ASSR, because it meant that Tatars could assert a separate, Muslim identity at the same time. The Republic Day of the Tatar ASSR was 25 June, which was conveniently near the Tatar (and Bashkir) summer festival of *sabantuy*, so that the two could be celebrated at the same time.[31] It was possible for Tatars to take a day off work through sickness in order to celebrate a Tatar holiday ('we don't go to work that day; we celebrate Uraza-Bayran [Eid al-Fitr]', stated one report in 1955).[32] In fact, by the 1960s, *sabantuy* was being celebrated in major towns in the Tatar ASSR.[33] Today, Tatar identity and separateness are often projected through popular plays and concerts. I witnessed this myself in 2018, when a number of folk groups – Tatar and Russian – were performing by the Kazan kremlin one weekend. 'Glory to free Tatarstan', was the cry of the leader of a lively, and popular, Tatar folk group.

In the 1950s and 1960s, there was a concerted attack on religion in the Soviet Union, and more mosques were shut in the Tatar ASSR and the Volga region generally; meanwhile, a more vigorous propaganda campaign, in Russian and Tatar, attacked Islam and all forms of belief. There has been a revival of Islam in Tatarstan since the dissolution of the Soviet Union. Tatarstan now celebrates all the major Muslim festivals and Tatars have begun to observe Ramadan. Islamic identity has become important, even when Tatars do not actively attend the mosque. In 1983, shortly before the collapse of the Soviet Union, a mere 0.9 per cent of

Tatar students (that is, the educated elite) said they were believers; by 1990, the figure had risen to 20 per cent, a remarkable increase.[34] The physical presence of Islam is marked by the construction in the Kazan kremlin of the Kul Sharif mosque, named after a Muslim scholar who served in the mosque in the sixteenth century and who died, along with his students, defending the citadel and heart of Muslim Kazan against Russian forces in 1552. The mosque can hold 6,000 worshippers and is reputed to be the largest in the Russian Federation and one of the largest in Europe. It was opened in 2005, and now dominates the kremlin – quite extraordinary given that the kremlin was traditionally the visual manifestation of the power of the *Russian, Orthodox* regime, housing as it did the main cathedral and administrative buildings (all mosques in the citadel were destroyed in 1552). In 1994, a crescent was placed symbolically on the top of the Suyumbike tower in the kremlin. Legend has it that Ivan IV wanted to take the khan's niece, Suyumbike, as his bride, and she agreed if he could construct a tower higher than anything in Kazan. When he did so, she went to the top and jumped to her death, indicating that the tsar might take the town, but never her heart . . . 15 October has also been designated a day of mourning for the conquest of 1552.

Tatars have had an advantage over other non-Russian ethnic groups on the Volga, in that they can assert a Muslim identity which then links them with the wider Islamic world; although that in itself can be complex, of course. On the whole, Tatarstan today projects itself as a model of tolerant Islam and a supporter of multi-culturalism, and it presents itself as a possible bridge between Orthodox Russia and the Islamic world.[35] A physical manifestation of this is the rather bizarre 'temple of all religions', which has recently been constructed as a cultural centre on the initiative of a private individual (Ildar Khanov) on the outskirts of Kazan, and which features several types of religious architecture in one complex. The position taken by Tatarstan is delicate, given the more extreme Islamic movements not only in territories such as Dagestan, but in other parts of the Volga region as well – there is a more militant Muslim movement in Astrakhan, for example, that is influenced by Muslims in Dagestan.[36] Nor has Kazan been free from terrorist attacks: in 2012, two senior members of the Islamic clergy were assassinated.[37] The most deadly attack in the Volga region, however, took place in the predominantly Russian Volgograd, in December 2013, when two suicide bomb attacks – one on a bus and the other at the railway station – killed 34 people.

Other non-Russians – the Mari, Chuvash, Mordvins – cannot use religion to assert their separate, non-Russian identity, because the majority converted to Orthodoxy in the eighteenth century. There has, however, been a revival in the Mari

El Republic of interest in indigenous religion. A study in 2004 found that about 15 per cent of the Mari people in the republic regarded themselves as adherents of traditional beliefs. The intellectual, national movement in the Chuvash Republic supports the revival (but at the same time stresses the continuity) of Chuvash traditional beliefs, as well as the teaching and official use of the Chuvash language. In 1995, a sacred column for collective prayers and sacrifices was erected on the site of an unfinished communist monument and was dedicated to the 1,100th anniversary of the foundation of the Bolgar state. There were plans for an ethnographic museum with ceremonial buildings and shrines to the ancient Chuvash religion; however, these foundered in the wake of opposition from Orthodox Christian Chuvash, who opposed a return to what they regarded as 'paganism'. The site was vandalized and has been left incomplete.[38] The Day of the Chuvash Republic is celebrated along with public rituals, prayers and blessings, and ceremonies associated with traditional religion and Orthodox feast days. In recent years, there has also been a revival of interest in Islam in the Chuvash Republic, linked with a consciousness of ethnic and linguistic Turkic roots.[39]

A further example of the assertion of identity among the Tatars is the question of their origins. This debate dates back to the late nineteenth century, as we saw in Chapter 12, but it resurfaced in the 1950s, when Tatar academics presented the case that the Volga Tatars were descended from the ancient Bolgar people;[40] and that was the official line in the Tatar encyclopaedia of 1962. This interpretation is important for two political reasons: first, it separated the Volga Tatars from others, such as the Crimean Tatars; second, and most importantly, it projected the Volga Tatars as descended from a Turkic civilization that itself had some origins in Europe, rather than from what might be popularly regarded as the less civilized marauding Mongols from the east.

The Bolgar heritage now features strongly in Tatar identity. In 1989, thousands of Tatars celebrated the 1,100th anniversary of the adoption of Islam by the Bolgar Tatars.[41] The site of ancient Bolgar (on the river Volga south of Kazan) has become a tourist destination and holy site (although it was also used for an alternative pilgrimage to Mecca – sometimes known as the 'Little Hajj' – in the Soviet period, when foreign travel was not permitted). Bolgar became a UNESCO World Heritage site in 2014. The original archaeological remains have been reconstructed according to the plans of original mausoleums and minarets. There is an excellent museum that has an enormous collection of artefacts (both originals and copies of artefacts held in the Hermitage in St Petersburg). Extensive excavations of the Bolgar site are going on and scholarship on the Bolgar is very active today; there is an archaeological institute and seven museums on the site. The Bolgar site is also, however, important to project the identity of the Volga Tatars, and of Tatarstan, today. Next to the museum is a replica of a Bolgar mosque, which contains a museum of Islam, and also houses the largest Koran in the world – certified by a plaque on the wall from the *Guinness Book of Records*; this is a modern Koran, but one that deliberately asserts the spiritual importance of the site for the Muslims of the Volga and for Muslims world-wide. Furthermore, an enormous 'white mosque' has recently been built, just outside the walls of the original city of Bolgar, to assert the Muslim identity of present-day Tatarstan and to celebrate Bolgar as a spiritual site for Muslims.

Self-confidence in their history is also manifested in the very large number of publications on Tatar history in recent years. In particular, the seven-volume, English-language *The History of the Tatars since Ancient Times*, published in 2017 by the Academy of Sciences of Tatarstan, is an impressive piece of scholarship which highlights the work of contemporary Tatar scholars. The existence of a separate Tatarstan Academy of Sciences is in itself an assertion of intellectual self-confidence. Another manifestation of this confidence is the construction, north of Kazan, of a completely new technology park and university city – called Innopolis (written significantly in *Latin* script at the entrance to the complex). The new town was formally opened in 2012 and is intended to establish Tatarstan at the forefront of new scientific and technological developments. It is indicative of the ambition of the new Tatarstan, although it is too early to judge whether the project will be successful: when I visited in spring 2018, it still looked very empty.

❖

The Volga was portrayed in the nineteenth century as 'Mother Volga', but it was also assumed that it was primarily a *Russian* river (as expressed in the quotation at the beginning of the chapter by a Tatar journalist in 1990). In the Soviet period, and especially during the Second World War, it was redefined as the river that protected the 'motherland' – that is, the homeland of all Soviet citizens. The Tadzhik poet Mumin Kanoat (1932–2018) also makes the Volga a 'combatant':

I am the river. I was created by you, the earth.
Mother-earth! Today I meet the enemy.
My talk with the enemy is especially curt.
I have donned a steel breastplate.

I am the warrior river!
The Volga has arisen in defence of the Motherland.
And the people of the Volga have risen to defend the Motherland.[42]

The river Volga remains central to the identity of the non-Russian peoples who live on or near its banks. As poetry in native language remains a powerful expression of identity, this chapter will end with two poems from the Soviet period, originally in Chuvash and Tatar, both of which feature the river Volga. The first is by Sespel Mishi (Mikhail Sespel), born in 1899. He joined the Bolshevik party in 1918 and became president of the Chuvash Revolutionary Tribunal in Cheboksary in 1920. Much affected by the experience of famine in the Volga, and missing his native land, he became severely depressed and committed suicide in 1922. His poems were first published in 1928 and translated into Russian and Ukrainian, and into several other European languages. In 1967, the Chuvash government established a Sespel prize for literature and the arts. This extract from 'Hungry Psalm' is about the river Volga during the 1920–21 famine and echoes Nikolai Nekrasov's negative characterization of the Volga in the late nineteenth century (see Chapter 11):

The mouth of hunger, the dry teeth
Of famine gnaw, gnaw, gnaw my land.

Log huts, damp with sweat of martyrs,
Grow stiff in wretched poverty;
In agony lie the hungry fields,
Lie sick, dried up, hard as a bone.

The Volga stretches wide and groans,
Her yellow waves like tattered rags,
Moans with an endless, heart-sick moan,
Lamenting the land in dry-eyed grief.[43]

The second poem is by Iffat Tutash (the pen name of Zahida Burnasheva). She was born in the Russian town of Riazan in 1895 and educated at a Tatar girls' school, before working as a journalist and holding several government posts. Her poem 'The Volga' also contains several descriptive features that echo Russian-language poems about the river:

To the Volga this heart of mine I compare
For is it not true some resemblance they keep?
Like the Volga my heart is sweeping and wide;
Like the river's strong current it runs so deep.

. . .

At sunset the Volga is bathed in pure light;
When illuminated by love, then my heart is the same.
Like the waves of the Volga, caressed by the sun.
The flower that grows in my heart is the same.[44]

17
CONTROLLING AND PROTECTING THE RIVER VOLGA

Boris Pilniak's (Pilnyak) *The Volga Flows to the Caspian Sea* was written in 1929 and charts the 'battle' and ultimate triumph of socialism over nature, through the construction of new canals and dams which were intended to change the course of the rivers in Russia, and in the process alleviate any future risk of famine. The novel includes a conversation between father and son about the future of Russian rivers, including the Volga. The father, Nazar Syssoev, is an old man from the village of Akatievo; his sons are working on the construction site at Kolomna and are representatives of the 'new Soviet Russia'. The father has this conversation with his younger son, next to the underground furnace:

> The grey-haired old man was saying to his son:
> 'So this is how you live, in a cave?'
> 'Yes, this is how we live,' answered his son.
> 'Listen, my boy! Is it really true that the river is going to flow backwards?'
> 'Certainly it is.'
> 'Now listen! . . . Our grandfathers and our great-grandfathers lived here, and we people floated rafts from the Oka to the Volga; we have done so for a thousand years, and perhaps even longer. We were brought up to it from childhood, we know every shoal and every shallow that lie near Kolomna and near Kassimov. From the beginning of time we have lived by the river. Now, it seems, our life is coming to an end; there will no longer be an Oka at Riazan, Murom and Eletma. Just think of it! . . . What's going to become of us if, as they say, there will not only be no Oka, but even Akatievo itself is to vanish beneath the waters? That will be the end of the world! Exactly like Kiteshgrad [the mythical vanishing city of Kitezh]. Well, shall we all get drowned with Akatievo?'

'There will be no need to get drowned. The river is being created for a purpose. This revolution is taking place so that the river shall flow afresh, and Akatievo will simply be moved from the site of the new river to a fresh place. What has existed for a thousand years has gone, and must be made anew. That is revolution with a real purpose. The people of a revolution need not drown, father.'[1]

At the end of the novel, the river is successfully diverted, but the village of Akatievo is flooded. Pilniak (born Boris Vogau – he was descended from Volga Germans) glorified socialist achievements in this novel, but it did not save him from critics in the febrile atmosphere of the 1930s Soviet Union. He was arrested in October 1937 and accused of plotting to assassinate Stalin and of being a Japanese spy; he was tried on 21 April 1938, pleaded guilty and was shot the same day.

The Soviet Union was to be a modern state, one that would not be corrupted by the greed of capitalists and that would properly value scientists. The Bolsheviks embraced 'big science'. This was the dawn of a new world, which would be ruled by new scientific laws that would impact on every aspect of the state, the economy and the lives of Soviet citizens. The Soviet Union also believed that science could conquer nature: the depth and flow of the river Volga could be controlled by massive dams and hydroelectric plants. At the same time, they nationalized industries – including fishing – and centralized economic planning. The result of these policies was massive industrialization and the construction of vast new planned cities. The Soviet Union trained large numbers of engineers and scientists – relatively speaking, far more than western countries – in order to take state projects forward. The country had vast resources of power – water, timber and fossil fuels – and of precious metals, and the new state had the energy and confidence to harness these resources for the benefit of all the people, rather than for the rich few and foreign investors. So what were the results of this policy and why did this potentially wealthy country get it so wrong environmentally?

❖

The Soviet Union set out to control, to tame, the river Volga and in the process to tame nature itself. This was to be achieved by a series of hydroelectric station, dams and locks, which would control the water flow and depth, provide water for the new cities and stimulate industrial growth. The project was conceived in the 1930s, but was set back by the war;[2] construction continued into the 1980s. The major

10. Hydroelectric Plants and Reservoirs on the Volga

hydroelectric power stations (or GES, for short) built were: Uglich (constructed 1935–40); Rybinsk (1935–50); Gorkii (the Nizhnii Novgorod GES) (1948–59); Kuibyshev (the Zhiguli GES) (1950–57); Stalingrad (the Volga GES) (1950–61), which at the time was the largest such plant in Europe; Saratov (1956–71); Cheboksary (1968–86). At the same time, the Moscow–Volga canal (constructed 1932–37) and the Volga–Don canal (started before the war and opened in 1952) linked the rivers with the capital and with the Black Sea.

The new constructions were presented at the time as a great socialist achievement. Posters encouraged workers and members of the Komsomol to participate in this great enterprise. Contemporary paintings, such as those by Gleb Kun, portrayed workers, male and female, heroically constructing dams and in the process taming the river (not that this saved Kun – he was executed in 1938).[3] The network of power stations, dams and reservoirs was presented as the triumph of the socialist state, which had achieved what the tsarist regime had never managed to do: link the whole river network of the country and conquer nature in the process. Despite its location in the centre of European Russia, Moscow in the Soviet era was frequently called 'the city of the five seas', because the canal and river network linked it to seas in all directions. The Moscow–Volga canal was the key to this: the canal was said to have 'brought the Volga's waters to the Kremlin steps'.[4] The control that the new power stations could exercise over the depth of the river was greater than the control possible over the Mississippi – a fact that was highlighted at the time as evidence of the superiority of socialist planning over capitalist countries.[5] In 1953, the enormous monument *Mother Volga* was constructed on the shore of the Volga, near the Rybinsk reservoir and power station – the river, now controlled by the state, is yet again depicted in allegorical form as a woman. On the base of the sculpture is engraved the slogan: 'Communism is Soviet Power plus Electrification of the Whole Country.'

These constructions were great engineering achievements in their time. The flow and the depth of the Volga could be controlled by reservoirs and dams, and hydroelectricity powered new towns and new industries. It is only, however, in very recent publications that the human and environmental costs of taming the Volga have been discussed. The environmental problems faced on the Volga today are discussed below; but the root of the problem lies in the ambitious hydroelectric policies pursued by the Soviet Union. Little concern was shown for the villages and the historic buildings, including religious buildings, which were submerged by new reservoirs. The most famous example is the belfry tower of the church of Kaliazin – built at the end of the eighteenth century as part of the complex of the monastery of St Nicholas – which pokes out above the water level near Uglich and has become one of the sights for tourists on the cruise ships that ply between Moscow and St Petersburg. The hydroelectric plants were not as efficient as their European or North American equivalents – in part because of faults in their construction, and in part because central planning in the Soviet Union was inefficient. In practice, it proved difficult to ensure appropriate water levels on the middle Volga, and even now shipping can be disrupted by low water levels.[6]

Furthermore, only in recent years has it been acknowledged that forced labour was used in the construction of the hydroelectric plants, and indeed was essential because of the shortage of manpower. The numbers of prisoners used in the construction are staggering. As early as 1935, there were some 49,000 prisoners working at the Volgolag labour camp (on the Stalingrad plant), and this number had risen to over 80,000 by 1941. There were over 36,000 prisoners working on the Kuibyshev plant in 1939, and by the following year prisoners comprised over 65 per cent of the workforce. In 1939–41, almost 100,000 prisoners worked on the Rybinsk and Uglich sites.[7] In addition, some 200,000 gulag prisoners worked on the Moscow–Volga canal, and some 100,000 on the Volga–Don canal (with a similar number of German prisoners of war). The death rates on the hydroelectric stations were lower than the death rates experienced in the construction of the White Sea canal, where workers were given only basic tools and treated brutally, and where some 25,000 prisoners died. Conscious that the workforce had at least to be kept alive and fit to work, a minimal amount of care was given on the Volga projects, and prisoners had the incentive of a reduction in their sentence for good behaviour. The productivity of the prisoners remained low, however – partly because the physical capacity and skills of the prisoners were not necessarily what was required; partly because they were not given proper equipment; and partly because many prisoners were already in poor health. This was despite the use made of skilled prisoners in these enterprises: Vadim Livanov, for example, was a hydraulic engineer from Rybinsk who was arrested in 1929 and worked first on the White Sea canal as an engineer, and then on the hydroelectric plants at Uglich, Rybinsk and Kuibyshev as a prisoner.[8] The inefficient use of labour meant that it took longer in the Soviet Union than in North America to construct hydroelectric plants. After Stalin's death in 1953, tens of thousands of prisoners were released from the camps, which created a new labour shortage. There is no doubt that the use of forced labour tarnished the 'brave new world' on the Volga.

The mismanagement of the resources of the Soviet Union had many causes,[9] but the fundamental reason was the arrogance of the Soviet leadership in pushing forward great prestige projects, such as the hydroelectric power stations, without any environmental considerations and its refusal to tolerate any opposition to central planning. Scientists and engineers were, in principle, greatly valued in society; but in practice, their role was to implement the policies determined at the top, and they had no means to oppose them. Professional associations of engineers and scientists had little (if any) autonomy, and even the prestigious Academy of Sciences was unable to challenge state policies, at least until the very last years of the Soviet Union. Anyone who queried the plans could be accused of disloyalty to the regime, and of 'bourgeois wrecking'. A considerable number of scientists ended up in gulags, and their talents were used (or rather misused) to manage the gulag prisoners as they built canals connected to the river Volga. The target-driven nature of centralized state planning left little opportunity to reflect on the potential longer-term damage to the environment. The isolation of Soviet scientists from their counterparts in the West also inhibited discussion of the environmental consequences of the new hydroelectric plants and of resource exploitation. If the scientists were emasculated, then ordinary people had little opportunity to oppose Soviet economic policy. Local initiatives were simply impossible; in the 1930s, the few who expressed conservation concerns were arrested.

Of course, the Soviet Union was not the only country in the world to embark on great economic projects (the massive Hoover dam, for example, was constructed on the Colorado river in the 1930s)[10] or to want to harness its resources for the modernization of the country. Nor was it the only country – communist or capitalist – to take decisions that had harmful (albeit frequently unintentional) consequences for the environment.[11] However, the sheer scale of the plans of the Soviet leaders had particularly devastating consequences for the environment. The 'virgin lands' policy in the 1950s, for example, under Khrushchev, led to severe soil erosion in Kazakhstan, western Siberia and on the left bank of the Volga; and the exploitation of fossil fuels and nickel in the far north and north-east of Siberia has caused irreparable damage to the fragile tundra ecosystem.[12] The impact of Soviet policies was exceptionally severe on the Volga because of the construction of the vast new hydroelectric plants and dams and because rapid industrialization and urbanization took place along almost the entire length of the river. New towns like Toliatti sprang up, and others were physically transformed by the influx of tens of thousands of new inhabitants and by the new building and infrastructure that this required.

The dams and hydroelectric power stations changed the levels and flow of the Volga and affected the spawning grounds of sturgeon and other fish. The only stretches of the Volga which now flow freely are north of Tver and south of the Volga (formerly Stalingrad) GES and dam to the Volga delta. As a result, it is claimed that the flow from one end of the river to the other now takes 180 days, whereas previously it took 50.[13] The decrease in the flow of the river has also led to greater salinity of the water.[14] The level of the Caspian Sea has fallen because of the effect of the dams; it reached a low point in 1969, when it was estimated that the sea's level had fallen by almost 3 metres since 1929.[15] In the 1980s, there were ambitious plans to change the course of several Russian rivers, including diverting south the river Pechora in the Arctic, so that it flowed into the river Kama, which in turn flows into the Volga below Kazan; thus more water would flow into the Caspian Sea. These ideas were never implemented and were eventually abandoned by Mikhail Gorbachev. This has at least helped to preserve the delicate environment in the Arctic.

The river Volga gradually became a 'sewer', according to one historian.[16] The development of industry on the river led to raw industrial waste being discharged into it from chemical plants and paper mills. In the early 1950s, the mill at Balakhna (on the western bank of the Volga, 32km north of Nizhnii Novgorod) is said to have discharged 30 tonnes of fibres into the water every day.[17] Fertilizers used on the land, including dioxins, which did not break down, seeped into the river. Timber lay rotting on the bottom of the river, left behind as part of the wastage, which was a constant feature of the centralized five-year plans. The very slow development of water-treatment plants and sewerage systems in the rapidly expanding Volga towns also led to raw sewage being discharged into the river, which had a negative impact on the lives of urban citizens. Kazan suffered an outbreak of typhoid in 1943,[18] and in the 1970s fires flared up spontaneously on the Volga because of the heavy pollution near the surface of the river.

The dams, hydroelectric plants and new canals were intended to provide electricity for the cities, to improve agriculture for the countryside, and to enhance internal transportation. To an extent they were successful, although the projects never completely fulfilled their aims – the Volga–Don canal, for example, never produced the amount of increased shipping envisaged (in part because the narrowness of the canal required ships to queue for a long time at the locks). Advances were, however, achieved at a terrible environmental, economic and human cost. The combination of the dams and the pollution can be seen most acutely in the devastating impact on fishing in the Volga, and especially on the population of beluga sturgeon.

The Soviet Union used to produce most of the world's black caviar from beluga sturgeon, which spawned in the fresh water of the river Volga. The drop in sturgeon was particularly noticeable after the completion of the Volga (Stalingrad) dam in 1960–61. The spawning grounds of the sturgeon have been devastated by the dams: 98 per cent of the traditional spawning grounds were above the Volga (Stalingrad) dam, where the waters no longer run freely. The new vast reservoirs at Kuibyshev (Zhiguli) and elsewhere further disrupted spawning grounds, not least because of the enormous amount of waste dumped in them.[19] It is generally agreed that the fish catch dropped dramatically in the river Volga and the Caspian Sea as a result. One estimate is that in the 1960s the catch of beluga sturgeon dropped by about 80 per cent and stellate sturgeon by about 60 per cent.[20] Another estimate is that the fish catch in that decade dropped from about 500,000 tonnes to less than

100,000.[21] An assessment in 1977 of Soviet policy on the Volga noted the negative impact of dams and pollution and expressed doubts about whether proposed plans for creating hatcheries and artificial insemination would lead to a recovery in the stocks of sturgeon. The report at this time was pessimistic about the future, fearing the construction of more dams and increased pollution from industry; but it also noted that the important fishing industry of the Soviet Union was falling drastically in value.[22]

By the 1970s, environmental concerns were beginning to be aired in the Soviet Union both by prominent dissidents, such as the physicist Andrei Sakharov, and by student groups; but their activities had little impact on policy. There were attempts in 1972 to clean up the Volga and the river Ural, but with only modest success. In 1975, the Soviet Union signed the Helsinki Accords, which included cooperation on matters affecting the environment. Some factories were publicly criticized and fined for dumping waste in the 1980s. In 1988, for example, the metal plant at Cherepovets (on the river Sheksna, on the shores of the Rybinsk reservoir) was fined for dumping waste into the Rybinsk reservoir. But extensive criticism of policies, including in the Soviet press, occurred only during the reforms of Gorbachev, and gathered momentum only after the shock of the disaster at the Chernobyl nuclear power station (in northern Ukraine) in April 1986.

The dissolution of the Soviet Union in 1991 has allowed more scientific work, and more openness, about the impact on the environment of Soviet policies. Russian scientists and environmentalists can now publish freely, and in English. These studies have, however, mainly served to highlight the dire situation on the Volga, and have not always led to radical improvements. A study published in 2005 analysed the levels of sturgeon in the Caspian Sea between 1988 and 2000, and concluded not only that the stocks had declined, but that there had been a shift in the balance of species, and a disproportionate increase in sterlet at the expense of beluga sturgeon.[23] A study of fish stocks in the Kuibyshev (Zhiguli) reservoir between 1991 and 2009 came to much the same conclusion.[24] A study published in 2012 looked at the impact on sturgeon and sterlet between the years 1991 and 2009. It found that the river flow was crucial for the spawning season in the spring, and that spawning was higher in high-water years and poorer in low-water years, of which the worst was 2006, when water levels were very low. The study concluded, predictably, that the 'damming of the Volga near the city of Volgograd caused gradual decrease in the efficiency of natural reproduction of sturgeons, which has become extremely low during recent years'. The authors proposed that the river flow should be increased in the spring, and that the spawning grounds needed to be protected.[25]

Some improvements in water and air quality, and therefore in the quality of life in the cities, occurred because of the collapse of heavy industry following the dissolution of the Soviet Union. Despite this, in the 1990s it was estimated that half the surface water in the Russian Federation was polluted.[26] The level of pollution in the river Volga remains dangerously high: *The Times* reported on 21 April 2017, under the headline 'Volga is poisoned by pollution', that sewage, pesticides and heavy metals meant pollution had reached a critical level.

Unfortunately, environmental issues have all too often become political. One example of this is the discussion about a proposed pipeline, which is intended to carry oil from Kazakhstan across the Caspian Sea to Azerbaijan and Georgia, and from there to Romania and Austria. Vladimir Putin has invoked environmental concerns to stall the project, which he fears would diminish Russian power by reducing the dependence of these countries on Russian oil. The Russian delegation has expressed concerns about oil spillages in the Caspian, the dangers of seismic activity and landslides and terrorism. Putin stated at a 2007 summit of the heads of the states bordering on the Caspian Sea that 'ecological security must be the measure of security for all projects in the Caspian', and drew attention to a 'sharp fall in the population of sturgeon, in the rivers that empty into the Caspian as well as in the sea itself'.[27] He stated this with no sense of irony, considering that it had been the policies of the Soviet Union that had been responsible for the collapse in sturgeon stocks in the Volga and for the lower level of water in the Caspian Sea, which had affected bordering countries as much as the Russian Federation.

❖

In the immediate aftermath of the collapse of the Soviet Union, it was thought that more grassroots activities in the Russian Federation would thrive and that many of these would be concerned with the environment. In fact, membership of environmental groups peaked in 1988–90, at the time of the dissolution of the Soviet Union, but has declined since then. Environmental groups still maintain a presence on the river Volga, but it is a weak presence. Each year on 20 May, Friends of the Earth stage a number of activities concerning the environment along the river on 'Volga Day';[28] but the membership of Friends of the Earth and Greenpeace in Russia is very small compared with West European countries and North America. In 1992, a journalist in the United States wrote a piece entitled 'Poisoning Russia's River of Plenty', based on a visit he made to the village of Ikrainoe (*ikra* is the Russian word for 'caviar') near Astrakhan, where the fish were dying. The villagers were bemused, but seemed fatalistic: 'they say it's the ecology, they say the sturgeon is sick, that it has become kind of soft. We don't really know what this means.' The journalist rightly blamed the dead fish on pollution and the dams, and also noted that activists had tried to clean out chemicals from areas of still water in Sarepta, south of Volgograd.[29]

Why are environmental activist groups so weak in Russia, on the river Volga and elsewhere? One reason is that the immediate years after the dissolution of the Soviet Union were very difficult economically, as industry collapsed and the value of the rouble plummeted. In these uncertain circumstances, people were naturally more concerned with the day-to-day problems of survival – finding food, clothing for their children, jobs, etc. – than they were with environmental issues. They simply had other priorities, and that remains a factor today. However, the economy has improved and wealth has filtered down – albeit unevenly – to at least the middle classes in towns; and yet environmental groups have not become more active or popular. The fact is that Putin's government has become increasingly suspicious of aid from abroad for any groups with international membership in Russia, including environmental groups, and this has hindered the activities of organizations such as Greenpeace and Friends of the Earth in Russia. Environmental groups can all too easily look 'anti-Russian' if they receive money from abroad. Groups have been harassed by the police, and some members have been arrested. In general, Putin sees a very limited role for grassroots movements within Russia; indeed, it could be said that he sees a limited role for civil society and for pluralism generally in Russia.[30] The government still maintains control over the resources of the country and over all major economic initiatives.

There is, however, another fundamental problem in encouraging people to join grassroots organizations in Russia. The roots of this problem ran deep in the Soviet Union, and to an extent can be traced back to the imperial period: there is a strong feeling that nothing can be achieved at the local level and that ordinary people are helpless. This perception was very clear in a case study on civil engagement in environmental movements, conducted in Samara at the turn of the twenty-first century. The general finding was that people felt that the activities of these groups were ineffective and pointless, and would never achieve any change.

This feeling of impotence and powerlessness is not simply a reaction to the policies of Putin. Neither the imperial government nor the Soviet Union took account of local views when they implemented policies that had a major impact on the river Volga. Policies have always been top down – from the colonization of lands in the seventeenth century to railway construction in the late nineteenth century to collectivization and to the major dam and hydroelectric projects of the 1930s–60s. It was recognized, however, by a few of the respondents to the Samara survey that there are genuine environmental concerns about the river and pollution, and that this is a shared responsibility. One student commented:

I think all people are responsible for the situation because we don't think enough about what we do. We leave litter and when we go to the Volga, many people do not clean up afterwards . . . Everyone must be responsible for nature, we must understand it and know more about it, and then we can take care of it.[31]

The challenges surrounding the future of the river Volga do have to be part of serious discussion about the future environmental policies of Russia. The comment made by the student in Samara suggests that the younger generation might take more responsibility for the environment generally, and for the river Volga more specifically; but it is a matter that should concern all people in (and outside) the Russian Federation, as well as the Russian government.

A news report from June 2019 on the TV channel Russia 1, under a photograph labelled 'Volga-matushka [mother]', discussed the current low level of water in the river Volga after a dry spring. It included interviews with local residents in Gorodets and Astrakhan, who bemoaned the fact that the shallowness of the water had damaged their fishing. The programme discussed the environmental impact on the river of the construction of reservoirs and the discharge of industrial waste, and appealed for 'our national treasure' to be 'clean and cured'. The Volga, the commentator stated, is not just one of the greatest rivers – and the longest river in Europe – but is the 'most beautiful river' in the world. There is no more appropriate way to end this book than with the final words of the report:

Without the Volga, there would be no Russia.[32]

NOTES

INTRODUCTION

1. Robert Bremner, *Excursions in the Interior of Russia*, London, vol. 2, 1839, pp. 216–17.
2. ibid., p. 217.
3. *SIRIO*, vol. 10, p. 204.
4. I have followed A. Kappeler (*The Russian Empire: A Multiethnic History*, translated by Alfred Clayton, Harlow, 2001) and Nancy Shields Kollman (*The Russian Empire 1450–1801*, Oxford, 2017) in describing Russia as an empire before 1721. 'The history of the Russian multi-ethnic empire begins in 1552 with the conquest of Kazan by the Muscovite Tsar, Ivan IV, the Terrible' (Kappeler, *The Russian Empire*, p. 14).

CHAPTER 1

1. Quoted in F.S. Khakimzyanov and I.I. Izmailov, 'Language and Writing in Bolgar Town', in *Great Bolgar*, Kazan', 2015, p. 307.
2. Thomas Noonan, 'European Russia, *c.* 500 – *c.* 1050', in Timothy Reuter, ed., *The New Cambridge Medieval History*, vol. 3, *c. 900–1024*, Cambridge, 2000, pp. 491–92; Peter B. Golden, 'Aspects of the Nomadic Factor in the Economic Development of Kievan Rus'', in Peter B. Golden, *Nomads and their Neighbours in the Russian Steppe: Turks, Khazars and Qipchaqs*, Aldershot, 2003, p. 80.
3. See, for example, Barry Cunliffe, *By Steppe, Desert and Ocean: The Birth of Eurasia*, Oxford, 2015, especially chapters 7 and 12.
4. Douglas Dunlop, *The History of the Jewish Khazars*, Princeton, NJ, 1954, p. 96.
5. *Ibn Fadlān and the Land of Darkness: Arab Travellers in the Far North*, translated by Paul Lunde and Caroline Stone, London, 2012, p. 56. This is an anthology of travel accounts from the eighth to the thirteenth century.
6. Thomas Noonan, 'Why Dirhams First Reached Russia: The Role of Arab-Khazar Relations in the Development of the Earliest Islamic Trade with Eastern Europe', *Archivum Eurasiae Medii Aevi*, vol. 4, 1984, p. 278.
7. *Ibn Fadlān and the Land of Darkness*, p. 44.
8. For a thorough treatment of the subject see Peter B. Golden, 'The Conversion of the Khazars to Judaism', in Peter B. Golden, Haggai Ben-Shammai and András Róna-Tas, eds, *The World of the Khazars: New Perspectives*, Leiden and Boston, MA, 2007, pp. 123–62. For a denial of the conversion, see Shaul Stampfer, 'Did the Khazars Convert to Judaism?', *Jewish Social Studies*, vol. 19, no. 3, 2013, pp. 1–72. The subject remains controversial and has been used both by Zionists and by anti-Semites for their own purposes.
9. *Ibn Fadlān and the Land of Darkness*, pp. 116–17.
10. Noonan, 'European Russia, *c.* 500 – *c.* 1050', p. 502.
11. *Ibn Fadlān and the Land of Darkness*, p. 58.

12. Quoted in Thomas Noonan, 'Some Observations on the Economy of the Khazar Khaganate', in Peter B. Golden, Haggai Ben-Shammai and András Róna-Tas, eds, *The World of the Khazars: New Perspectives*, Leiden and Boston, MA, 2007, p. 207.
13. A.P. Novosel'tsev, *Khazarskoe gosudarstvo i ego rol' v istorii vostochnoi Evropy i Kavkaza*, Moscow, 1990, pp. 114–17.
14. Vadim Rossman, 'Lev Gumilev, Eurasianism and Khazaria', *East European Jewish Affairs*, vol. 32, no. 1, 2002, p. 37; Noonan, 'Some Observations on the Economy', pp. 232–33.
15. Janet Martin, *Treasures of the Land of Darkness: The Fur Trade and its Significance for Medieval Russia*, Cambridge, 1986, p. 36
16. Thomas Noonan, *The Islamic World, Russia and the Vikings, 750–900: The Numismatic Evidence*, Aldershot, 1998, pp. 322–42.
17. Noonan, 'Why Dirhams First Reached Russia', pp. 151–282.
18. *Ibn Fadlān and the Land of Darkness*, p. 57.
19. Noonan, 'Some Observations on the Economy', p. 211.
20. R.Kh. Bariev, *Volzhskie Bulgary: Istoriia i kul'tura*, St Petersburg, 2005, p. 48.
21. *Ibn Fadlān and the Land of Darkness*, p. 136.
22. István Zimonyi, *The Origins of the Volga Bulgars*, Szeged, 1990; Bariev, *Volzhskie Bulgary*, pp. 21–23.
23. *Ibn Fadlān and the Land of Darkness*, pp. 35–36.
24. Yahya G. Abdullin, 'Islam in the History of the Volga Kama Bulgars and Tatars', *Central Asian Survey*, vol. 9, no. 2, 1990, pp. 1–11.
25. *Ibn Fadlān and the Land of Darkness*, pp. 120–21.
26. Quoted in Florin Curta, 'Markets in Tenth-Century al-Andalus and Volga Bulghāria: Contrasting Views of Trade in Muslim Europe', *Al-Masaq*, vol. 25, no. 3, 2013, p. 311.
27. M.D. Poluboyarinova, 'Bolgar Trade', in *Great Bolgar*, Kazan', 2015, p. 110.
28. Quoted in Curta, 'Markets in Tenth-Century al-Andalus and Volga Bulghāria', p. 312.
29. Ingmar Jansson, '"Oriental Import" into Scandinavia in the 8th–12th Centuries and the Role of Volga Bulgaria', in *Mezhdunarodnye sviazi, torgovye puti i goroda Srednego Povolzh'ia IX – XII vekov: materialy mezhdunarodnogo simpoziuma Kazan', 8–19 sentiabria 1998 g.*, Kazan', 1999, pp. 116–22.
30. Quoted in Janet Martin, 'Trade on the Volga: The Commercial Relations of Bulgar with Central Asia and Iran in the 11th–12th Centuries', *International Journal of Turkish Studies*, vol. 1, no. 2, 1980, pp. 89–90.
31. Curta, 'Markets in Tenth-Century al-Andalus and Volga Bulghāria', p. 316.
32. Poluboyarinova, 'Bolgar Trade', pp. 105, 108.
33. R.G. Fakhrutdinov, *Ocherki po istorii Volzhskoi Bulgarii*, Moscow, 1984, p. 43.
34. Martin, 'Trade on the Volga', pp. 85–97.
35. M.D. Poluboiarinova, *Rus' i Volzhskaia Bolgaria v X – XV vv.*, Moscow, 1993, p. 31.
36. Thomas Noonan, 'Monetary Circulation in Early Medieval Rus': A Study of Volga Bulgar Dirham Finds', *Russian History*, vol. 7, no. 3, 1980, pp. 294–311; R.M. Valeev, 'K voprosu o tovarno-denezhnykh otnosheniiakh rannikh Bulgar (VIII–X vv.)', in *Iz istorii rannikh Bulgar*, Kazan', 1981, pp. 83–96; G.A. Fedorov-Davydov, 'Money and Currency', in *Great Bolgar*, Kazan', 2015, pp. 114–23.
37. I.V. Dubov, *Velikii Volzhskii put'*, Leningrad, 1989, pp. 151–60.
38. Noonan, 'Some Observations on the Economy', p. 235.
39. A.S. Bashkirov, *Pamiatniki Bulgaro-Tatarskoi kul'tury na Volge*, Kazan', 1928, pp. 66–70, quoting drawings and plans by, among others, A. Shmit.
40. A.M. Gubaidullin, *Fortifikatsiia gorodishch Volzhskoi Bulgarii*, Kazan', 2002.
41. Fakhrutdinov, *Ocherki po istorii Volzhskoi Bulgarii*, p. 56.
42. Curta, 'Markets in Tenth-Century al-Andalus and Volga Bulghāria', p. 319.
43. *Istoriia Tatarskoi ASSR*, Kazan', vol. 1, pp. 60–67; G.F. Polyakova, 'Non-Ferrous and Precious Metal Articles', in *Great Bolgar*, Kazan', 2015, pp. 132–37; M.D. Poluboyarinova, 'Glasswear', in *Great Bolgar*, Kazan', 2015, pp. 160–71; T.A. Khlebnikova, 'Tanning', in *Great Bolgar*, Kazan', 2015, pp. 168–71; A. Zakirova, 'Bone Carving', in *Great Bolgar*, Kazan', 2015, pp. 172–77.
44. Anna Kochkına, 'Prichernomorsko-sredizemnomorskie sviazi Volzhskoi Bulgarii v X – nachale XIII vv. (arkheologicheskie dannye o torgovykh putiakh)', in *Mezhdunarodnye sviazi, torgovye*

*puti i goroda Srednego Povolzh'ia IX – XII vekov: materialy mezhdunarodnogo simpoziuma Kazan',
8–19 sentiabria 1998 g.*, Kazan', 1999, pp. 132–38.

45. Curta, 'Markets in Tenth-Century al-Andalus and Volga Bulghāria', p. 317.
46. F. Donald Logan, *The Vikings in History*, 3rd edition, New York and London, 2005, p. 184.
    Logan outlines the debate on pp. 163, 184–85.
47. Discussed fully in Noonan, *The Islamic World, Russia and the Vikings.*
48. S.H. Cross, 'The Scandinavian Infiltration into Early Russia', *Speculum*, vol. 21, no. 4, 1946,
    pp. 505–14; Elena Mel'nikova, 'Baltiisko-Volzhskii put' v rannei istorii Vostochnoi Evropy', in
    *Mezhdunarodnye sviazi, torgovye puti i goroda Srednego Povolzh'ia IX – XII vekov: materialy mezh-
    dunarodnogo simpoziuma Kazan', 8–19 sentiabria 1998 g.*, Kazan', 1999, pp. 80–87.
49. I.V. Dubov, 'Velikii Volzhskii put' v istorii drevnei Rusi', in *Mezhdunarodnye sviazi, torgovye puti
    i goroda Srednego Povolzh'ia IX – XII vekov: materialy mezhdunarodnogo simpoziuma Kazan', 8–19
    sentiabria 1998 g.*, Kazan', 1999, pp. 88–93.
50. *Ibn Fadlān and the Land of Darkness*, pp. 45–47.
51. ibid., pp. 126–27.
52. Simon Franklin and Jonathan Shepard, *The Emergence of Rus 750–1200*, London and New York,
    1996, pp. 31–41.
53. Simon Franklin, 'Kievan Rus' (1015–1125)', in Maureen Perrie, ed., *The Cambridge History of
    Russia*, vol. 1, *From Early Rus' to 1698*, Cambridge, 2006, p. 74.
54. Simon Franklin, 'Rus', in David Abulafia, ed., *The New Cambridge Medieval History*, vol. 5,
    c. 1198–1300, Cambridge, 1999, p. 797, footnote 3.
55. Boris Zhivkov, *Khazaria in the Ninth and Tenth Centuries*, translated by Daria Manova, Leiden,
    2015, pp. 157–58.
56. M.I. Artamonov, *Istoriia khazar*, Leningrad, 1962, pp. 434–35.
57. Franklin and Shepard, *The Emergence of Rus*, p. 69.
58. Vladimir Petrukhin, 'Khazaria and Rus': An Examination of their Historical Relations', in Peter
    B. Golden, Haggai Ben-Shammai and András Róna-Tas, eds, *The World of the Khazars: New
    Perspectives*, Leiden and Boston, MA, 2007, p. 257; Golden, 'Aspects of the Nomadic Factor',
    p. 89.
59. *Ibn Fadlān and the Land of Darkness*, pp. 171–72.
60. Franklin and Shepard, *The Emergence of Rus*, p. 341.
61. John Fennell, *The Crisis of Medieval Russia 1200–1304*, London and New York, 1983, p. 2.
62. Martin, 'Trade on the Volga', p. 95.

CHAPTER 2

1. Quoted in G.A. Fyodorov-Davydov, *The Culture of the Golden Horde Cities*, translated by
   H. Bartlett Wells, Oxford, 1984, p. 16.
2. Quoted in Fennell, *The Crisis of Medieval Russia*, p. 76.
3. Quoted in ibid., p. 128.
4. The arguments for and against these claims are clearly outlined in Charles J. Halperin, *Russia and
   the Golden Horde: The Mongol Impact on Medieval Russian History*, Bloomington, IN, 1987, and
   Donald Ostrowski, *Muscovy and the Mongols: Cross-Cultural Influences on the Steppe Frontier,
   1304–1589*, Cambridge, 1998.
5. Quoting Friar Giovanni da Pian del Carpine in Janet Martin, *Medieval Russia 980–1584*,
   Cambridge, 1995, p. 145; Fennell, *The Crisis of Medieval Russia*, p. 87.
6. Quoted in J.J. Saunders, *The History of the Mongol Conquests*, London, 1971, p. 82.
7. Fennell, *The Crisis of Medieval Russia*, p. 87; more generally discussed in Charles J. Halperin,
   'Omissions of National Memory: Russian Historiography on the Golden Horde as Politics of
   Inclusion and Exclusion', *Ab Imperio*, vol. 3, 2004, pp. 131–44.
8. Fennell, *The Crisis of Medieval Russia*, p. 119.
9. Halperin, *Russia and the Golden Horde*, p. 26.
10. Michel Biran, 'The Mongol Empire and Inter-Civilization Exchange', in Benjamin Z. Kedar, ed.,
    The Cambridge World History, vol. 5, *Expanding Webs of Exchange and Conflict, 500 CE to 1500
    CE*, Cambridge, 2015, p. 540.

11. Leonid Nedashkovskii, 'Mezhdunarodnaia i vnutrenniaia torgovlia', in *Zolotaia Orda v mirovoi istorii*, Kazan', 2016, pp. 608–13.
12. Fyodorov-Davydov, *The Culture of the Golden Horde Cities*, pp. 22, 199–200.
13. B.D. Grekov and A.Iu. Iakubovskii, *Zolotaia Orda i ee padenie*, Moscow and Leningrad, 1950, pp. 149–51; Nedashkovskii, 'Mezhdunarodnaia i vnutrenniaia torgovlia', p. 610.
14. Michael Prawdin, *The Mongol Empire: Its Rise and Legacy*, translated by Eden and Cedar Paul, London, 1940, p. 278.
15. Quoted in Janet Martin, 'The Land of Darkness and the Golden Horde: The Fur Trade under the Mongols XIII – XIVth Centuries', *Cahiers du Monde Russe et Soviétique*, vol. 19, no. 4, 1978, p. 414.
16. Quoted in Fyodorov-Davydov, *The Culture of the Golden Horde Cities*, p. 16.
17. Martin, 'The Land of Darkness and the Golden Horde', pp. 409–12.
18. Ostrowski, *Muscovy and the Mongols*, p. 124.
19. Grekov and Iakubovskii, *Zolotaia Orda i ee padenie,* pp. 150–51.
20. Martin, *Medieval Russia*, p. 201.
21. István Vásáry, 'The Jochid Realm: The Western Steppe and Eastern Europe', in Nicola Di Cosmo, Allen J. Frank and Peter B. Golden, eds, *The Cambridge History of Inner Asia: The Chinggisid Age*, Cambridge, 2009, p. 81.
22. Halperin, *Russia and the Golden Horde*, p. 29.
23. Allen J. Frank, 'The Western Steppe in Volga-Ural Region, Siberia and the Crimea', in Nicola Di Cosmo, Allen J. Frank and Peter B. Golden, eds, *The Cambridge History of Inner Asia: The Chinggisid Age*, Cambridge, 2009, p. 246.
24. G.F. Valeeva-Suleimanova, 'Problemy izucheniia iskusstva Bulgar zolotoordskogo vremeni (vtoraia polovina XIII – nachalo XV vv.), in *Iz istorii Zolotoi Ordy*, edited by A.A. Arslanova and G.F. Valeeva-Suleimanova, Kazan', 1993, pp. 132–33.
25. Vásáry, 'The Jochid Realm', p. 74.
26. I. Zaitsev, *Astrakhanskoe khanstvo*, Moscow, 2004; I. Zaitsev, 'Astrakhanskii iurt', in *Zolotaia Orda v mirovoi istorii*, general editor V. Trepalov, Kazan', 2016, pp. 752–60.
27. Sigismund von Herberstein, *Notes upon Russia*, translated by R.H. Major, London, vol. 2, 1852, p. 73.
28. For understanding the complexities of the titles of princes and of their lands, see Christian A. Raffensperger, *Kingdom of Rus'*, Kalamazoo, MI, 2017.
29. See, in particular, Robert O. Crummey, *The Formation of Muscovy 1304–1613*, London and New York, 1987; and more specifically Henry R. Huttenbach, 'Muscovy's Conquest of Muslim Kazan and Astrakhan', 1552–56', in Michael Rywkin, ed., *Russian Colonial Expansion to 1917*, London, 1988, pp. 45–69; Christian Noack, 'The Western Steppe: The Volga-Ural Region, Siberia and the Crimea under Russian Rule', in Nicola Di Cosmo, Allen J. Frank and Peter B. Golden, eds, *The Cambridge History of Inner Asia: The Chinggisid Age*, Cambridge, 2009, pp. 303–08; and Michael Khodarkovsky, 'Taming the "Wild Steppe": Muscovy's Southern Frontier', *Russian History*, vol. 26, no. 3, 1999, pp. 241–97.
30. Janet Martin, 'North-Eastern Russia and the Golden Horde', in Maureen Perrie, ed., *The Cambridge History of Russia*, vol. 1, *From Early Rus' to 1689*, Cambridge, 2006, p. 143.
31. Janet Martin, 'The Emergence of Moscow (1359–1462)', in Maureen Perrie, ed., *The Cambridge History of Russia*, vol. 1, *From Early Rus' to 1698*, Cambridge, 2006, p. 163.
32. Crummey, *The Formation of Muscovy*, p. 114, from which the account of the struggle between Moscow and the principality of Tver has been drawn. The dispute is also outlined in Martin, *Medieval Russia*, pp. 169–77.
33. Isabel de Madariaga, *Ivan the Terrible*, New Haven, CT and London, 2005, p. 16.
34. Halperin, *Russia and the Golden Horde*, pp. 59, 70–71, 100; the coin is described on p. 100.
35. Jaroslaw Pelenski, *Russia and Kazan: Conquest and Imperial Ideology (1438–1560s)*, The Hague and Paris, 1974, p. 50.
36. Madariaga, *Ivan the Terrible*, pp. 49-51.
37. Quoted in Noack, 'The Western Steppe', p. 305.
38. Maureen Perrie, *The Image of Ivan the Terrible in Russian Folklore*, Cambridge, 1987, pp. 177–78.
39. Khodarkovsky, 'Taming the "Wild Steppe"', p. 282. See also D.A. Kotliarov, *Moskovskaia Rus' i narody Povolzh'ia v XV – XVI vekakh*, Izhevsk, 2005, pp. 275–78.

40. Richard Hakluyt, *The Principal Navigations, Voyages, Traffiques and Discoveries of the English Nation*, edited by Jack Beeching, London, 1972, p. 79.
41. Abdullin, 'Islam in the History of the Volga Kama Bulgars and Tatars', pp. 6–7.
42. *Historical Anthology of Kazan Tatar Verse: Voices of Eternity*, compiled and translated by David J. Matthews and Ravil Bukharev, Richmond, 2000, p. 112.
43. Quoted in Geoffrey Hosking, *Russia: People and Empire, 1552–1917*, London, 1998, p. 3.
44. Quoted in Pelenski, *Russia and Kazan*, p. 208.

CHAPTER 3

1. Discussed in Kollman, *The Russian Empire*, p. 2.
2. V.M. Kabuzan, *Narody Rossii v XVIII veke. Chislennost' i etnicheskii sostav*, Moscow, 1990, p. 226, from which census data of this period has been taken for non-Russians on the middle and lower Volga.
3. Johann Gottlieb Georgi, *Russia or a Compleat Historical Account of all the Nations which Compose that Empire*, translated by William Tooke, London, 4 vols, 1780–83, vol. 1, pp. 70, 113.
4. Adam Olearius, *The Travels of Olearius in Seventeenth-Century Russia*, translated and edited by Samuel H. Baron, Stanford, CA, 1967, p. 298; *The Travels of the Ambassadors from the Duke of Holstein*, London, 1669, p. 113.
5. Georgi, *Russia or a Compleat Historical Account*, vol. 1, p. 95.
6. Janet M. Hartley, *A Social History of the Russian Empire 1650–1825*, London and New York, 1999, p. 11.
7. Kabuzan, *Narody Rossii v XVIII veke*, p. 184.
8. Georgi, *Russia or a Compleat Historical Account*, vol. 2, pp. 22–23.
9. August von Haxthausen, *The Russian Empire: Its People, Institutions, and Resources*, translated by Robert Fairie, London, vol. 1, 1856, pp. 323–24.
10. Michael Khodarkovsky, *Russia's Steppe Frontier: The Making of a Colonial Empire, 1500–1800*, Bloomington, IN, 2002, pp. 9–11.
11. Mary Holderness, *New Russia: Journey from Riga to the Crimea, by the Way of Kiev*, London, 1823, p. 141.
12. Kappeler, *The Russian Empire*, p. 43.
13. Khodarkovsky, *Russia's Steppe Frontier*, p. 14.
14. Matthew P. Romaniello, 'Absolutism and Empire. Governance of Russia's Early Modern Frontier', PhD dissertation, Ohio State University, 2003, pp. 24–27.
15. D.J.B. Shaw, 'Southern Frontiers of Muscovy, 1550–1700', in J.H. Bater and R.A. French, eds, *Studies in Russian Historical Geography*, vol. 1, London, 1983, pp. 118–43.
16. *Istoriia Tatarskoi ASSR*, vol. 1, p. 149.
17. Olearius, *The Travels of Olearius*, pp. 314, 321.
18. Eduard Dubman, *Promyslovoe predprinimatel'stvo i osvoenie Ponizovogo Povolzh'ia v kontse XVI – XVII vv.*, Samara, 1999, pp. 58, 64, 69.
19. Janet Martin, 'Multiethnicity in Muscovy: A Consideration of Christian and Muslim Tatars in the 1550s–1560s', *Journal of Early Modern History*, vol. 5, no. 1, 2001, p. 20.
20. Romaniello, 'Absolutism and Empire', pp. 164–65.
21. S.B. Seniutkin, *Istoriia Tatar Nizhegorodskogo Povolzh'ia s poslednei treti XVI do nachala XX vv. (istoricheskaia sud'ba misharei Nizhegorodskogo kraia)*, Moscow and Nizhnii Novgorod, 2009, pp. 68, 86, 99.
22. Aider Normanov, *Tatary Srednego Povolzh'ia i Priural'ia v rossiiskom zakondatel'stve vtoroi poloviny XVI – XVIII vv.*, Kazan', 2002, p. 82.
23. Michael Rywkin, 'The Prikaz of the Kazan Court: First Russian Colonial Office', *Canadian Slavonic Papers*, vol. 18, no. 3, 1976, pp. 293–300.
24. Kollman, *The Russian Empire*, p. 174.
25. Ostrowski, *Muscovy and the Mongols*, pp. 60–61.
26. I.A. Kuznetsov, *Ocherki po istorii Chuvashskogo krest'ianstva*, Cheboksary, vol. 1, 1957, p. 79.
27. Romaniello, 'Absolutism and Empire', pp. 161–62.
28. *Ocherki istorii Mariiskoi ASSR*, Ioshkar-Ola, 1965, pp. 120–21.

29. Matthew P. Romaniello, *The Elusive Empire: Kazan and the Creation of Russia 1552–1671*, Madison, WI, and London, 2012, p. 125.
30. Martin, 'Multiethnicity in Muscovy', pp. 20–21.
31. Kuznetsov, *Ocherki po istorii Chuvashskogo krest'ianstva*, vol. 1, pp. 84, 86, 125.
32. S.Kh. Alishev, *Istoricheskie sud'by narodov Srednego Povolzh'ia XVI – nachalo XIX v.*, Moscow, 1990, p. 95.
33. Kabuzan, *Narody Rossii v XVIII veke*, pp. 127, 130.
34. G. Peretiatkovich, *Povolzh'e v XVII i nachale XVIII veka (ocherki iz istorii kolonizatsii kraia)*, Odessa, 1882, pp. 196, 281–83.
35. Kuznetsov, *Ocherki po istorii Chuvashskogo krest'ianstva*, vol. 1, p. 79.
36. Eduard Dubman, *Khoziaistvennoe osvoenie Srednego Povolzh'ia v XVI veka. Po materialam tserkov-no-monastyrskikh vladenii*, Kuibyshev, 1991, pp. 24–25.
37. N.B. Sokolova, 'Khoziaistvenno-torgovaia deiatel'nost' Makar'evskogo zheltovodskogo monastyria', in *Verkhnee i Srednee Povolzh'e v period feodalizma: mezhvuzovskii sbornik*, Gor'kii, 1985, pp. 43–49.
38. Matthew P. Romaniello, 'Controlling the Frontier: Monasteries and Infrastructure in the Volga Region, 1552–1682', *Central Asian Survey*, vol. 19, nos 3–4, 2000, pp. 426–40.
39. Dubman, *Promyslovoe predprinimatel'stvo*, pp. 108, 114.
40. Olearius, *Travels of Olearius*, p. 326; see also description of salt trade in early modern Russia in R.E.F. Smith and David Christian, *Bread and Salt: A Social and Economic History of Food and Drink in Russia*, Cambridge, 1984, pp. 57–60.
41. Romaniello, *The Elusive Empire*, p. 102.
42. E. Gur'ianov, *Drevnie vekhi Samary*, Kuibyshev, 1986, p. 48.
43. E.L. Dubman, P.S. Kabytov and O.B. Leont'ev, *Istoriia Samary (1586–1917 gg.)*, Samara, 2015, p. 81.
44. *Samara-Kuibyshev. Khronika sobytii 1886–1986*, Kuibyshev, 1985, pp. 11–12, 16; Gur'ianov, *Drevnie vekhi Samary*, p. 48; Dubman et al., *Istoriia Samary*, p. 81.
45. John Bell, *Travels from St Petersburg in Russia, to Diverse Parts of Asia*, Glasgow, 1763, p. 27.
46. *The History of the Tatars since Ancient Times*, Academy of Sciences of the Republic of Tatarstan, Kazan', 2017, vol. 4, pp. 247, 249.
47. M.A. Vodolagin, *Ocherki istorii Volgograda 1589–1967*, Moscow, 1968, pp. 21–22.

CHAPTER 4

1. Jonas Hanway, *An Historical Account of the British Trade over the Caspian Sea*, Dublin, 1754, vol. 1, p. 70.
2. Friedrich Christian Weber, *The Present State of Russia*, reprint, London, vol. 1, 1968, p. 128.
3. *PSZ*, no. 12187, vol. 16, p. 807, 27 June 1764 (abolishing this special regiment).
4. *PSZ*, no. 9707, vol. 13, p. 191, 6 February 1750.
5. Olearius, *The Travels of Olearius*, p. 139.
6. Quoted in M.D. Kurmacheva, *Goroda Urala i Povolzh'ia v krest'ianskoi voine 1773–1775 gg.*, Moscow, 1991, p. 47.
7. Afanasii Nikitin, *Khozhenie za tri moria Afanasiia Nikitina*, Tver', 2003, p. 112.
8. Olearius, *The Travels of Olearius*, pp. 292, 294.
9. V.D. Dimitriev and S.A. Selivanova, *Cheboksary: ocherki istorii goroda XVIII veka*, Cheboksary, 2011, p. 102.
10. James Spilman, *A Journey through Russia by Two Gentlemen who went in the Year 1739*, London, 1742, pp. 6–7.
11. M.A. Kirokos'ian, *Piraty Kaspiiskogo Moria*, Astrakhan', 2007, p. 105.
12. F.C. Koch, *The Volga Germans in Russia and the Americas from 1763 to the Present*, University Park, PA, and London, 1977, pp. 106–07.
13. Kirokos'ian, *Piraty Kaspiiskogo Moria*, p. 49.
14. Vodolagin, *Ocherki istorii Volgograda*, p. 19.
15. Kirokos'ian, *Piraty Kaspiiskogo Moria*, p. 56.
16. V.M. Tsybin and E.A. Ashanin, *Istoriia Volzhskogo kazachestva*, Saratov, 2002, p. 34.

17. Quoted in Terence Armstrong, ed., *Yermak's Campaign in Siberia*, London, Haklyut Society, 1975, p. 208.
18. F.N. Rodin, *Burlachestvo v Rossii*, Moscow, 1975, p. 53.
19. S. Konovalov, 'Ludvig Fabritius's Account of the Razin Rebellion', *Oxford Slavonic Papers*, vol. 6, 1955, p. 77.
20. Quoted in Paul Avrich, *Russian Rebels 1600–1800*, New York, 1972, p. 78.
21. Vodolagin, *Ocherki istorii Volgograda*, p. 28; Avrich, *Russian Rebels*, p. 79.
22. Quoted in Avrich, *Russian Rebels*, p. 87.
23. Konovalov, 'Ludvig Fabritius's Account of the Razin Rebellion', pp. 76, 85.
24. Vodolagin, *Ocherki istorii Volgograda*, p. 31.
25. *Ocherki istorii Mariiskoi ASSR*, p. 134.
26. Konovalov, 'Ludvig Fabritius's Account of the Razin Rebellion', p. 92.
27. N.B. Golikova, *Astrakhanskoe vosstanie 1705–1706*, Moscow, 1975, pp. 54, 76–78.
28. Vodolagin, *Ocherki istorii Volgograda*, p. 36.
29. Philip Longworth, 'The Pretender Phenomenon in Eighteenth-Century Russia', *Past and Present*, vol. 66, 1975, pp. 61–83.
30. *Krest'ianskaia voina 1773–1775 gg. v Rossii*, Moscow, 1973, pp. 109–11.
31. Quoted in Avrich, *Russian Rebels*, p. 227.
32. A.B. Bolotov, *Zhizn' i prikliucheniia Andreia Bolotova, opisannye samim im dlia svoikh potomkov*, Moscow–Leningrad, 1931, reprint Cambridge, MA, vol. 3, 1973, p. 145.
33. Quoted in John T. Alexander, *Autocratic Politics in a National Crisis: The Imperial Russian Government and Pugachev's Revolt, 1773–1775*, Bloomington, IN, 1969, p. 174.
34. ibid., p. 211.
35. *Krest'ianskaia voina*, pp. 165–69.
36. John T. Alexander, *Emperor of the Cossacks: Pugachev and the Frontier Jacquerie of 1773–1775*, Lawrence, KS, 1973, p. 177.
37. Kurmacheva, *Goroda Urala*, p. 161.
38. Guido Hausmann, *Mütterchen Wolga. Ein Fluss als Erinnerungsort vom 16. bis ins frühe 20. Jahrhundert*, Frankfurt and New York, 2009, p. 340.
39. M. Raeff, 'Pugachev's Rebellion', in R. Forster and J.P. Greene, eds, *Preconditions of Revolution in Early Modern Europe*, Baltimore, MD, and London, 1970, p. 195.
40. Avrich, *Russian Rebels, 1600-1800*, p. 251.
41. Philip Longworth, *The Cossacks*, London, 1971, p. 223.
42. Quoted in Sara Dickinson, *Breaking Ground: Travel and National Culture in Russia from Peter I to the Era of Revolution*, New York, 2003, p. 211.
43. Most thoroughly discussed with relation to Razin in Hausmann, *Mütterchen Wolga*, pp. 319–49.
44. Avrich, *Russian Rebels*, p. 121.
45. *An Anthology of Chuvash Poetry*, compiled by Gennady Aygi, translated by Peter France, London and Boston, 1991, pp. 40, 41.
46. Janet M. Hartley, 'Russia in 1812, Part 1: The French Presence in the *Gubernii* of Smolensk and Mogilev', *Jahrbücher für Geschichte Osteuropas*, vol. 38, 1990, p. 182.
47. Avrich, *Russian Rebels*, p. 263.
48. Simon Dixon, 'The "Mad Monk" Iliador in Tsaritsyn', *The Slavonic and East European Review*, vol. 88, nos 1–2, 2010, pp. 403–05.
49. Richard Stites, *Russian Popular Culture: Entertainment and Society since 1900*, Cambridge, 1992, pp. 18, 31.
50. Michael Khodarkovsky, 'The Stepan Razin Uprising: Was it a "Peasant War"?', *Jahrbücher für Geschichte Osteuropas*, vol. 42, no. 1, 1994, p. 3.
51. Boris Akunin, *Pelagia and the Red Rooster*, translated by Andrew Bromfield, London, 2008, pp. 3–4.

CHAPTER 5

1. Yusuke Toriyama, 'Images of the Volga River in Russian Poetry from the Reign of Catherine the Great to the End of the Napoleonic Wars', *Study Group on Eighteenth Century Russia*, 2013; www.sgecr.co.uk/newsletter2013/toriyama.html

2. *PSZ*, no. 14464, vol. 20, pp. 374–75, 5 April 1776.
3. *PSZ*, no. 16718, vol. 22, p. 1117, 7 October 1788.
4. 'Opis' 1000 del kazach'iago otdela', *Trudy Orenburgskoi uchenoi arkhivnoi komissii*, Orenburg, vol. 24, 1913, pp. 145–46, no. 717.
5. Quoted in Longworth, *The Cossacks*, p. 242.
6. Dubman, *Promyslovoe predprinimatel'stvo*, p. 69.
7. [T.C.], *The New Atlas: or, Travels and Voyages in Europe, Asia, Africa and America*, London, 1698, p. 172.
8. L.G. Beskrovnyi, *Russkaia armiia i flot v XVIII veke (ocherki)*, Moscow, 1958, pp. 311–22.
9. L.G. Beskrovnyi, *The Russian Army and Fleet in the Nineteenth Century*, Gulf Breeze, FL, 1996, p. 2.
10. Seniutkin, *Istoriia Tatar Nizhegorodskogo Povolzh'ia*, p. 226.
11. *PSZ*, no. 3884, vol. 6, p. 483, 19 January 1722.
12. S.V. Dzhundzhuzov, *Kalmyki v Srednem Povolzh'e i na iuzhnom Urale*, Orenburg, 2014, pp. 83, 119, 122, 155, 185.
13. R.F. Baumann, 'Subject Nationalities in the Military Service of Imperial Russia: The Case of the Bashkirs', *Slavic Review*, vol. 46, 1987, pp. 492–93.
14. M.G. Nersisian, ed., *Dekabristy ob Armenii i Zakavkaz'e*, Erevan, 1985, pp. 267–305.
15. John P. LeDonne, *The Grand Strategy of the Russian Empire 1650–1831*, Oxford, 2004, pp. 44–47, 124–29.
16. Beskrovnyi, *Russkaia armiia i flot*, pp. 326–27.
17. John Cook, *Voyages and Travels through the Russian Empire, Tartary, and Part of the Kingdom of Persia*, Edinburgh, vol. 1, 1770, p. 345.
18. Tsybin and Ashanin, *Istoriia Volzhskogo kazachestva*, pp. 29–30.
19. Alexander Pushkin, *The Complete Prose Tales of Alexander Sergeyevitch Pushkin*, translated by G.R. Aitken, London, 2008, pp. 355–56.
20. George Forster, *A Journey from Bengal to England . . . and into Russia by the Caspian Sea*, London, vol. 2, 1808, p. 303.
21. Janet M. Hartley, *Russia 1762–1825: Military Power, the State and the People*, Westport, CT, and London, 2008, p. 30; Janet M. Hartley, *Siberia: A History of the People*, London and New York, 2014, p. 93.
22. *PSZ*, no. 9771, vol. 12, pp. 317–19, 26 June 1750.
23. I.N. Pleshakov, 'Gardkouty v Saratovskom Povolzh'e: iz istorii rechnoi strazhi XVIII – pervoi poloviny XIX vv.', in *Voenno-istoricheskie issledovaniia v Povolzh'e*, Saratov, vol. 7, n.d., pp. 20–24.
24. William Glen, *Journal of a Tour from Astrachan to Karass*, Edinburgh, 1823, p. 7.
25. E.G. Istomina, *Vodnyi transport Rossii v doreformennyi period*, Moscow, 1991, p. 137.
26. Pleshakov, 'Gardkouty v Saratovskom Povolzh'e', pp. 25–27.
27. Quoted in Isabel de Madariaga, *Russia in the Age of Catherine the Great*, London, 1981, p. 279.
28. ibid., p. 144.
29. *SIRIO*, vol. 115, pp. 304, 307, 310, 353–54, 359, 513, 577.
30. Robert E. Jones, 'Catherine II and the Provincial Reform of 1775: A Question of Motivation', *Canadian Slavic Studies*, vol. 4, no. 3, 1970, p. 511.
31. E.P. Kuz'min, 'Povsednevnyi byt Tsarevokokshaiskikh voevod XVIII veka', in *Goroda Srednego Povolzh'ia: istoriia i sovremennost': sbornik statei*, Ioshkar-Ola, 2014, p. 44.
32. A.N. Biktasheva, *Kazanskie gubernatory v dialogakh vlastei (pervaia polovina XIX veka)*, Kazan', 2008, pp. 53–61, 65–71.
33. Information from an unpublished paper by Andrey Gornostaev, 'Eighteenth-Century Chichikovs and Purchasing Runaway Souls', paper given at the Study Group on Eighteenth-Century Russia 10th International Conference in Strasbourg, July 2018.
34. St Petersburg Institute of History, Russian Academy of Sciences, St Petersburg, fond 36, delo 477, ll. 612–13, Report from Iaroslavl', 1778.
35. P.L. Karabushchenko, *Astrakhanskaia guberniia i ee gubernatory v svete kul'turno-istoricheskikh traditsii XVIII – XIX stoletii*, Astrakhan', 2011, p. 161.
36. M. Wilmot and C. Wilmot, *The Russian Journals of Martha and Catherine Wilmot 1803–1808*, London, 1934, pp. 308–09.

37. See Janet M. Hartley, 'Bribery and Justice in the Provinces in the Reign of Catherine II', in Stephen Lovell, Alena Ledeneva and Andrei Rogachevskii, eds, *Bribery and* Blat *in Russia: Negotiating Reciprocity from the Middle Ages to the 1990s*, Basingstoke, 2000, pp. 48–64.
38. 'Zhaloba Saratovskikh krest'ian na zemskii sud', *Russkii arkhiv*, vol. 46, 1908, pp. 215–16.
39. F.Ia. Polianskii, *Gorodskoe remeslo i manufaktura v Rossii XVIII v.*, Moscow, 1960, p. 32.
40. *PSZ*, no. 7315, vol. 10, p. 207, 6 July 1737; no. 7571, vol. 10, p. 487, 30 April 1738; no. 7757, vol. 10, p. 729, 16 February 1739; no. 8623, vol. 11, pp. 664–65, 27 September 1742.
41. *PSZ*, no. 12174, vol. 16, pp. 786–90, 5 June 1764.
42. *PSZ*, no. 11902, vol. 16, p. 339, 22 August 1763.
43. Roger Bartlett, *Human Capital: The Settlement of Foreigners in Russia 1762–1804*, Cambridge, 1979, p. 75.
44. Koch, *The Volga Germans in Russia and the Americas*, p. 69.
45. Quoted in Michael Khodarkovsky, *Where Two Worlds Met: The Russian State and the Kalmyk Nomads, 1660–1771*, Ithaca, NY, and London, 1992, p. 230.
46. Natsional'nyi arkhiv Respubliki Tatarstan (hereafter NART), fond 1, opis' 2, delo 393, Chancellery of the Kazan' Governor, 1843–44.
47. Charles Scott, *The Baltic, the Black Sea and the Crimea*, London, 1854, pp. 93–94.
48. Gosudarstvennyi arkhiv Samarskoi oblasti (hereafter GASO), fond 1, opis' 8, tom 1, delo 108, Samara Provincial Chancellery, 1854.
49. GASO, fond 1, opis' 1, tom 1, delo 2060, Samara Provincial Chancellery, 1861.
50. Robert E. Jones, 'Urban Planning and the Development of Provincial Towns in Russia during the Reign of Catherine II', in J.G. Garrard, ed., *The Eighteenth Century in Russia*, Oxford, 1973, pp. 321–44.
51. Alexander, *Emperor of the Cossacks*, p. 196.
52. *PSZ*, no. 26633, vol. 34, pp. 32–33, 23 January 1817.
53. M. Rybushkin, *Zapiski ob Astrakhani*, 3rd edition, Astrakhan', 2008, pp. 133–35.
54. The best description of the tour is by G.B. Ibneeva, 'Puteshestvie Ekateriny II po Volge v 1767 godu: uznavanie imperiii', *Ab Imperio*, vol. 2, 2000, pp. 87–104; see also Hausmann, *Mütterchen Wolga*, pp. 173–88, and G.B. Ibneeva, *Imperskaia politika Ekateriny II v zerkale ventsenosnykh puteshestvii*, Moscow, 2009, pp. 182–95.
55. *SIRIO*, vol. 10, pp. 193, 207, 190.
56. *SIRIO*, vol. 10, p. 204.
57. Ibneeva, 'Puteshestvie Ekateriny II', p. 100.
58. This section is based on a synopsis of the paper by Toriyama, 'Images of the Volga River in Russian Poetry'.

CHAPTER 6

1. Kabuzan, *Narody Rossii v XVIII veka*, p. 230.
2. Paul W. Werth, *At the Margins of Orthodoxy: Mission, Governance, and Confessional Politics in Russia's Volga-Kama Region, 1827–1905*, Ithaca, NY, and London, 2002, p. 36.
3. L. Taimasov, 'Etnokonfessional'naia situatsiia v Kazanskoi gubernii nakanune burzhuaznykh reform', in K. Matsuzato, ed., *Novaia Volga i izuchenii etnopoliticheskoi istorii Volgo-Ural'skogo regiona: Sbornik statei*, Sapporo, 2003, p. 109.
4. ibid., p. 111.
5. NART, fond 1, opis' 1, delo 1107, Chancellery of the Kazan' Governor, 1754.
6. Michael Khodarkovsky, 'The Conversion of Non-Christians in Early Modern Russia', in R.P. Geraci and M. Khodarkovsky, eds, *Of Religion and Empire: Missions, Conversion, and Tolerance in Tsarist Russia*, Ithaca, NY, and London, 2001, p. 140.
7. ibid., pp. 122–23.
8. *PSZ*, no. 1, article 97, vol. 1, p. 133, 29 January 1649, the Law Code, or *Ulozhenie*, of 1649.
9. *PSZ*, no. 867, vol. 2, p. 313, 16 May 1681.
10. *PSZ*, no. 1117, vol. 2, p. 662, 5 April 1685.
11. Kuznetsov, *Ocherki po istorii Chuvashskogo krest'ianstva*, vol. 1, p. 86.
12. Khodarkovsky, *Russia's Steppe Frontier*, p. 193.

13. Quoted in ibid., p. 192.
14. ibid., p. 194.
15. *PSZ*, no. 9379, vol. 12, pp. 967–69, 11 March 1747.
16. Paul W. Werth, 'Coercion and Conversion: Violence and the Mass Baptism of the Volga Peoples, 1740–55', *Kritika*, vol. 4, no. 8, 2003, p. 551.
17. *PSZ*, no. 10597, vol. 14, pp. 607–12, 25 August 1756.
18. Seniutkin, *Istoriia Tatar Nizhegorskogo Povolzh'ia*, p. 321.
19. A.N. Kefeli, *Becoming a Muslim in Imperial Russia: Conversion, Apostasy, and Literacy*, Ithaca, NY, 2014, pp. 19–20.
20. A.G. Ivanov, *Mariitsy Povolzh'ia i Priural'ia*, Ioshkar-Ola, 1993, pp. 41, 45, 53.
21. *Treasures of Catherine the Great*, Catalogue of an Exhibition at Somerset House, London, 2000, p. 120.
22. Edward Daniel Clarke, *Travels in Various Countries of Europe, Asia and Africa*, vol. 1, *Russia Tartary and Turkey*, London, 1810, pp. 480–81.
23. Azade-Ayşe Rorlich, *The Volga Tatars: A Profile in National Resilience*, Stanford, CA, 2017, p. 44.
24. Taimasov, 'Etnokonfessional'naia situatsiia v Kazanskoi gubernii', p. 113.
25. Kefeli, *Becoming a Muslim in Imperial Russia*, p. 39.
26. Georgi, *Russia or a Compleat Historical Account*, vol. 1, pp. 120, 93, 107.
27. Quoted in Khodarkovsky, *Russia's Steppe Frontier*, p. 198.
28. NART, fond 1, opis' 1, delo 112, Chancellery of the Kazan' Governor, 1827.
29. Taimasov, 'Etnokonfessional'naia situatsiia v Kazanskoi gubernii', pp. 121, 125.
30. NART, fond 1, opis' 2, delo 294, Chancellery of the Kazan' Governor, 10 May 1840.
31. Rorlich, *The Volga Tatars*, p. 45.
32. NART, fond 1, opis' 3, delo 218, Chancellery of the Kazan' Governor, 17 June 1866.
33. GASO, fond 1, opis' 1, tom 1, delo 528, Samara Provincial Chancellery, 1837–54.
34. NART, fond 1, opis' 2, delo 399, Chancellery of the Kazan' Governor, 1843.
35. NART, fond 1, opis' 2, delo 1632, Chancellery of the Kazan' Governor, 1860–61.
36. Quoted in Oxana Zemtsova, 'Russification and Educational Policies in the Middle Volga Region (1860–1914)', PhD dissertation, European University of Florence, 2014, p. 67.
37. NART, fond 2, opis' 1, delo 1920, Kazan' Provincial Office, 1845–47.
38. Werth, *At the Margins of Orthodoxy*, p. 160.
39. Janet M. Hartley, 'Education and the East: The Omsk Asiatic School', in Maria Di Salvo, Daniel H. Kaiser and Valerie A. Kivelson, eds, *Word and Image in Russian History: Essays in Honor of Gary Marker*, Boston, MA, 2015, pp. 253–68.
40. Samuel Collins, *The Present State of Russia*, London, 1671, pp. 3–4.
41. Wilmot and Wilmot, *The Russian Journals of Martha and Catherine Wilmot*, p. 176.
42. K.A. Papmehl, *Metropolitan Platon of Moscow (Peter Levshin), 1737–1812: The Enlightened Prelate, Scholar and Educator*, Newtonville, MA, 1983, p. 25.
43. John Parkinson, *A Tour of Russia, Siberia and the Crimea 1792–1794*, London, 1971, p. 178.
44. NART, fond 1, opis' 3, delo 5196, Chancellery of the Kazan' Governor, 1881–82.
45. L.Iu. Braslavskii, 'Raskol Russkoi pravoslavnoi tserkvi i ego posledstviia v istorii narodov Srednego Povolzh'ia', in *Istoriia khristianizatsii narodov Srednego Povolzh'ia: kriticheskie suzhdenie i otsenka: mezhvuzovskii sbornik nauchnykh trudov*, Cheboksary, 1988, p. 40.
46. A.A. Vinogradov, *Staroobriadtsy Simbirsko-Ul'ianovskogo Povolzh'ia serediny XIX – pervoi treti XX veka*, Ul'ianovsk, 2010, pp. 18–19.
47. Braslavskii, 'Raskol russkoi pravoslavnoi tserkvi', p. 40.
48. Gosudarstvennyi arkhiv Iaroslavskoi oblasti (hereafter GAIO), fond 72, opis' 1, delo 122, Chancellery of the Iaroslavl' Governor-General, 1778.
49. William Spottiswoode, *Tarantasse Journey through Eastern Russia in the Autumn of 1856*, London, 1857, p. 18.
50. L.V. Burdina, 'Staroobriadchestvo v Kostromskom krae', in *Staroobriadchestvo: istoriia, kul'tura, sovremennost'. Materialy*, Moscow, 2000, p. 220.
51. D.D. Prlmako, 'Staroobriadcheskaia obshchina goroda Rzheva v XIX v.', in *Staroobriadchestvo v Tverskom krae: proshloe i nastoiashchee*, Tver' and Rzhev, 2007, p. 56.

52. E.V. Potapova, 'Vlast' i staroobriadtsy: iz istorii staroobriadcheskoi obshchiny goroda Rzheva v pervoi polovine XIX v.', in *Staroobriadchestvo v Tverskom krae: proshloe i nastoiashchee*, Tver' and Rzhev, 2007, p. 62.
53. The policy is described fully in Thomas Marsden, *The Crisis of Religious Toleration in Imperial Russia: Bibikov's System for the Old Believers, 1841–1855*, Oxford, 2015.
54. A.V. Morokhin, 'Prikhodskoe dukhovenstvo i staroobriadchestvo v Nizhegorodskom Povolzh'e v pervoi polovine XVIII v.', in *Staroobriadchestvo: istoriia, kul'tura, sovremennost'. Materialy*, Moscow, 2000, p. 70.
55. Marsden, *The Crisis of Religious Toleration*, pp. 95–98.
56. NART, fond 1, opis' 2, delo 1231, Chancellery of the Kazan' Governor, 1856.
57. *PSZ*, series 2, no. 35020, vol. 34, part 2, pp. 178–79, 23 October 1859.
58. GASO, fond 1, opis' 1, tom 1, delo 2201, Samara Provincial Chancellery, 1862.
59. *PSZ*, series 2, no. 39547, vol. 38, pp. 384–85, 26 April 1863.
60. NART, fond 1, opis' 3, delo 856, Chancellery of the Kazan' Governor, 1868.
61. GAIO, fond 73, opis' 1, tom 2, delo 4399, Chancellery of the Iaroslavl' Governor, 1852.
62. GASO, fond 1, tom 1, opis' 1, delo 198, Samara Provincial Chancellery, 1856.
63. NART, fond 2, opis' 2, delo 1966, Kazan' Provincial Office, 1881.
64. NART, fond 1, opis' 3, delo 9400, Chancellery of the Kazan' Governor, 1894.
65. NART, fond 2, opis' 1, delo 2264, Kazan' Provincial Office, 1867.

CHAPTER 7

1. Quoted in Charlotte Henze, *Disease, Health Care and Government in Late Imperial Russia: Life and Death on the Volga, 1823–1914*, London and New York, 2011, p. 126 (report from 1908).
2. Rodin, *Burlachestvo v Rossii*, p. 151.
3. John T. Alexander, *Bubonic Plague in Early Modern Russia: Public Health and Urban Disorder*, Oxford, 2003, p. 105.
4. ibid., pp. 13–15, 19, 23–24.
5. Hanway, *An Historical Account of the British Trade over the Caspian Sea*, vol. 1, p. 82.
6. Quoted in Alexander, *Bubonic Plague*, p. 232.
7. David L. Ransel, *A Russian Merchant's Tale: The Life and Adventures of Ivan Alekseevich Tolchenëv, based on his Diary*, Bloomington, IN, 2009, p. 18.
8. Alexander, *Bubonic Plague*, p. 244.
9. *PSZ*, nos. 22737–8, vol. 30, p. 27, 17 January 1808.
10. *PSZ*, no. 29623, vol. 38, pp. 1233–34, 6 October 1823.
11. E.A. Vishlenkova, S.Iu. Malysheva and A.A. Sal'nikova, *Kul'tura povsednevnosti provintsial'nogo goroda: Kazan' i Kazantsy v XIX–XX vekakh*, Kazan', 2008, p. 95.
12. Nathan M. Gerth, 'A Model Town: Tver', the Classical Imperial Order, and the Rise of Civic Society in the Russian Provinces, 1763–1861', PhD dissertation, University of Notre Dame, 2014. The account of the cholera in Tver province is taken from this thesis.
13. Pastor Guber, 'Dnevnik Pastora Gubera', *Russkaia starina*, vol. 22, no. 8, 1878, pp. 581–90.
14. *PSZ*, series 2, no. 3889, vol. 5, part 2, pp. 6–7, 3 September 1830.
15. Vodolagin, *Ocherki istorii Volgograda*, p.74.
16. Bremner, *Excursions in the Interior of Russia*, vol. 2, p. 222.
17. Roderick E. McGrew, *Russia and the Cholera 1823–1832*, Madison, WI, 1965, pp. 21, 52, 54, 64.
18. Vishlenkova et al., *Kul'tura povsednevnosti provintsial'nogo goroda*, p. 92.
19. N.A. Tolmachev, *Putevye zametki N.A. Tolmacheva o zhizni i byte krest'ian Kazanskoi gubernii v seredine XIX v. Sbornik dokumentov i materialov*, compiled by Kh.Z. Bagautdinova, Kazan', 2019, pp. 24, 26.
20. Gerth, 'A Model Town: Tver', pp. 204–05, 214–15.
21. Henze, *Disease, Health Care and Government*, pp. 27–28.
22. Hans Heilbronner, 'The Russian Plague of 1878–79', *Slavic Review*, vol. 21, no. 1, 1962, pp. 89–112.
23. London, National Archives, PC1/2673, Privy Council papers.

24. John P. Davis, 'The Struggle with Cholera in Tsarist Russia and the Soviet Union, 1892–1927', PhD dissertation, University of Kentucky, 2012, pp. 112, 116.

25. Henze, *Disease, Health Care and Government*, pp. 31–43.

26. Quoted in ibid., p. 82.

27. Quoted in ibid., p. 86.

28. ibid., pp. 64–65.

29. Vodolagin, *Ocherki istorii Volgograda*, p. 112.

30. Davis, 'The Struggle with Cholera in Tsarist Russia and the Soviet Union', p. 199.

31. Alexander, *Bubonic Plague*, p. 37.

32. Vishlenkova et al., *Kul'tura povsednevnosti provintsial'nogo goroda*, p. 97.

33. E.V. Cherniak and A.B. Madiiarov, *Gorodskoe samoupravlenie v Kazani 1870–1892 gg.*, Kazan', 2003, p. 87.

34. H.A. Munro-Butler-Johnstone, *A Trip up the Volga to the Fair of Nijni-Novgorod*, Oxford and London, 1875, p. 80.

35. P.V. Alabin, *Samara: 1586–1886 gody*, compiled by P.S. Kabytov, Samara, 1992, p. 138.

36. The best account of these developments can be found in Davis, 'The Struggle with Cholera in Tsarist Russia and the Soviet Union'.

37. Rachel Koroloff, 'Seeds of Exchange: Collecting for Russia's Apothecary and Botanical Gardens in the Seventeenth and Eighteenth Centuries', PhD dissertation, University of Illinois at Urbana-Champaign, 2014, pp. 71–72.

38. John Perry, *The State of Russia under the Present Czar*, reprint of the 1716 edition, London, 1967, p. 94.

39. *PSZ*, no. 3609, vol. 6, p. 214, 6 July 1720.

40. Koroloff, 'Seeds of Exchange', pp. 101–03.

41. *PSZ*, no. 8304, vol. 11, p. 331, 22 December 1740.

42. See David Moon, 'The Russian Academy of Sciences Expeditions to the Steppes in the Late Eighteenth Century', in *Personality and Place in Russian Culture: Essays in Memory of Lindsey Hughes*, edited by Simon Dixon, London, 2010, pp. 204–25.

43. Perry, *The State of Russia under the Present* Czar, p. 82.

44. Quoted in Moon, 'The Russian Academy of Sciences Expeditions to the Steppes', p. 215.

45. P.S. Pallas, *Travels through the Southern Provinces of the Russian Empire in the Years 1793 and 1794*, London, vol. 1, 1802, pp. 187, 215–19, 242–43.

46. David Moon, *The Plough that Broke the Steppes: Agriculture and Environment on Russia's Grasslands, 1700–1914*, Oxford, 2013, pp. 53–54.

47. On my last visit (in 2018) this section of the museum was one of the few that had not been modernized, and the accompanying text to the costumes was frozen in time in the Soviet period.

CHAPTER 8

1. Guzel Yakhina, *Deti moi: roman*, printed by Amazon, 2018, p. 5.

2. Hartley, *A Social History of the Russian Empire*, p. 20.

3. Ramil R. Khayrutdinov, 'The System of the State Village Government of the Kazan Governorate in the Early 18th – the First Third of the 19th Centuries', *Journal of Sustainable Development*, vol. 8, no. 5, 2015, p. 2.

4. *Istoriia Chuvashskoi ASSR*, Cheboksary, vol. 1, 1966, p. 92.

5. A.G. Cross, ed., *An English Lady at the Court of Catherine the Great*, Cambridge, 1989, p. 57.

6. Dominic Lieven, *The Aristocracy in Europe 1815–1914*, Basingstoke and London, 1992, pp. 44–45.

7. G. Gorevoi, T. Kirillova and A. Shitokov, 'Byvshie dvorianskie usad'by po beregam rek Tudovki i Volgi', in *Po Volge pod flagom 'Tverskoi zhizni'. Sbornik statei*, Staritsa, 2008, pp. 84–86.

8. GAIO, fond 72, opis' 1, delo 122, ll. 3–3v, Chancellery of the Iaroslavl' Governor-General.

9. I.M. Kataev, *Na beregakh Volgi i istorii usol'skoi votchiny grafov Orlovykh*, Cheliabinsk, 1948, p. 12.

10. S.A. Aleksandrova and T.I. Vedernikova, *Sel'skie dvorianskie usad'by Samarskogo zavolzh'ia v XIX – XX vv.*, Samara, 2015, pp. 14–16.

11. Wilmot and Wilmot, *The Russian Journals of Martha and Catherine Wilmot*, p. 147.

12. E.S. Kogan, *Ocherki istorii krepostnogo khoziaistva po materialam votchin Kurakinykh 2 – i poloviny XVIII veka*, Moscow, 1960, pp. 121–22.
13. E. Melton, 'Enlightened Seignioralism and its Dilemmas in Serf Russia, 1750–1830', *Journal of Modern History*, vol. 62, no. 4, 1990, p. 700.
14. Rodney D. Bohac, 'The Mir and the Military Draft', *Slavic Review*, vol. 47, no. 4, 1988, p. 653.
15. L.S. Prokof'eva, *Krest'ianskaia obshchina v Rossii vo vtoroi polovine XVIII – pervoi polovine XIX veka (na materialakh votchin Sheremetevykh)*, Leningrad, 1981, p. 152.
16. London, British Library, Add. MS 47431, f. 67v, Baki estate papers, 1819–25.
17. Alison K. Smith, 'Peasant Agriculture in Pre-Reform Kostroma and Kazan' Provinces', *Russian History*, vol. 26, no. 4, 1999, p. 372.
18. Prokofe'va, *Krest'ianskaia obshchina v Rossii*, p. 48.
19. Tracy K. Dennison, 'Serfdom and Household Structure in Central Russia: Voshchazhnikovo, 1816–1858', *Continuity and Change*, vol. 18, no. 3, 2003, p. 416.
20. Gennady Nikolaev, 'The World of a Multiconfessional Village', in *The History of the Tatars since Ancient Times*, vol. 6, Kazan', 2017, p. 612.
21. Alishev, *Istoricheskie sud'by narodov*, p. 110.
22. GASO, fond 1, opis' 1, tom 1, delo 528, Samara Provincial Chancellery, 1854.
23. Tolmachev, *Putevye zametki*, p. 72.
24. E.P. Busygin, N.V. Zorin and E.V. Mikhailichenko, *Obshchestvennyi i semeinyi byt russkogo sel'skogo naseleniia Srednego Povolzh'ia*, Kazan', 1973, p. 61.
25. Seniutkin, *Istoriia Tatar Nizhegorodskogo Povolzh'ia*, p. 278.
26. *PSZ*, no. 10597, vol. 14, pp. 607–12, 25 August 1756.
27. *PSZ*, no. 13490, vol. 19, pp. 101–04, 2 August 1770.
28. NART, fond 1, opis' 3, delo 3068, ll. 17–26ob, Chancellery of the Kazan' Governor.
29. NART, fond 1, opis' 3, delo 3272, l. 25, Chancellery of the Kazan' Governor.
30. Nikolaev, 'The World of a Multi-confessional Village', p. 614.
31. Spottiswoode, *Tarantasse Journey*, pp. 148–49.
32. Pallas, *Travels through the Southern Provinces*, vol. 1, p. 99.
33. Weber, *The Present State of Russia*, vol. 1, pp. 118–19.
34. E.P. Busygin, *Russkoe naselenie Srednego Povolzh'ia*, Kazan', 1966, p. 271.
35. Stephen D. Watrous, ed., *John Ledyard's Journey through Russia and Siberia 1787–1788*, Madison, WI, and London, 1966, p. 145.
36. Haxthausen, *The Russian Empire*, vol. 1, p. 331.
37. Spottiswoode, *Tarantasse Journey*, p. 161.
38. Haxthausen, *The Russian Empire*, vol. 1, p. 349.
39. Moon, *The Plough that Broke the Steppes*, p. 65.
40. Perry, *The State of Russia under the Present Czar*, p. 76.
41. Quoted in Moon, *The Plough that Broke the Steppes*, p. 66.
42. Smith, 'Peasant Agriculture in Pre-Reform Kostroma and Kazan' Provinces', pp. 366–67.
43. Haxthausen, *The Russian Empire*, vol. 1, p. 325.
44. Wilmot and Wilmot, *The Russian Journals of Martha and Catherine Wilmot*, pp. 81, 146.
45. Weber, *The Present State of Russia*, vol. 1, p. 121.
46. Bremner, *Excursions in the Interior of Russia*, vol. 2, p. 219.
47. Boris Mironov and Brian A'Hearn, 'Russian Living Standards under the Tsars: Anthropometric Evidence from the Volga', *Journal of Economic History*, vol. 68, no. 3, 2008, pp. 900–09.
48. A.-L.-L. de Custine, *The Empire of the Czar*, 2nd edition, London, vol. 3, 1844, p. 187.
49. I.I. Vorob'ev, *Tatary Srednego Povolzh'ia i Priural'ia*, Moscow, 1967, p. 294.
50. I.A. Zetkina, *Natsional'noe prosvetitel'stvo Povolzh'ia: formirovanie i razvitie*, Saransk, 2003, p. 55.
51. Quoted in Nikolaev, 'The World of a Multiconfessional Village', p. 612.
52. Raufa Urazmanova, 'Ceremonies and Festivals', in *The History of the Tatars since Ancient Times*, vol. 6, Kazan', 2017, p. 693.
53. Gennady Nikolaev, 'Ethnocultural Interaction of the Chuvash and Tatars', in *The History of the Tatars since Ancient Times*, vol. 6, Kazan', 2017, pp. 626–27.

CHAPTER 9

1. Robert E. Jones, *Provincial Development in Russia: Catherine II and Jacob Sievers*, New Brunswick, NJ, 1984.
2. Pallas, *Travels through the Southern Provinces*, vol. 1, p. 7.
3. Haxthausen, *The Russian Empire*, vol. 1, p. 93.
4. Polianskii, *Gorodskoe remeslo i manfaktura*, p. 48.
5. de Custine, *The Empire of the Czar*, vol. 3, p. 174.
6. Laurence Oliphant, *The Russian Shores of the Black Sea in the Autumn of 1852, with a Voyage down the Volga and a Tour through the Country of the Don Cossacks*, Edinburgh and London, 1854, pp. 82, 98.
7. V. Butyrkin, 'Razskazy iz sluzhby na Volge', *Morskoi sbornik*, vol. 69, 1863, pp. 405–18, vol. 71, 1864, p. 410.
8. Reginald G. Burton, *Tropics and Snow: A Record of Travel and Adventure*, London, 1898, p. 175.
9. Vodolagin, *Ocherki istorii Volgograda*, p. 104.
10. Gur'ianov, *Drevnie vekhi Samary*, p. 108.
11. Vodolagin, *Ocherki istorii Volgograda*, p. 71.
12. *Istoriia Tatarskoi ASSR*, vol. 1, p. 236.
13. Haxthausen, *The Russian Empire*, vol. 1, p. 329.
14. *Istoriia Tatarskoi ASSR*, vol. 1, p. 237.
15. Seniutkin, *Istoriia Tatar Nizhegorodskogo Povolzh'ia*, p. 289.
16. Haxthausen, *The Russian Empire*, vol. 1, p. 356.
17. NART, fond 22, opis' 2, delo 28, 38, opis' 3, delo 12, 36, 45, Kazan' Tatar ratusha.
18. GAIO, fond 73, opis' 1, delo 4399, Chancellery of the Iaroslavl' Governor, 1852.
19. L.M. Sverdlova, *Kazanskoe kupechestvo: sotsial'no-ekonomicheskii portret (kon. XVIII – nach. XX v.)*, Kazan', 2011, p. 173.
20. Catherine Evtuhov, *Portrait of a Province: Economy, Society and Civilization in Nineteenth-Century Nizhnii Novgorod*, Pittsburgh, PA, 2011, p. 113.
21. Boris Rotermel', *Tverskie nemtsy. Die Russlanddeutschen von Twer*, Tver', 2011, p. 14.
22. Evtuhov, *Portrait of a Province*, p. 113.
23. NART, fond 1, opis' 2, delo 1107, Chancellery of the Kazan' Governor, 1855.
24. S. Lisovskaia, 'Istoriia obrazovaniia, formirovaniia i razvitiia evreiskoi obshchiny Iaroslavlia', in *Proshloe i nastoiashchee evreiskikh obshchin Povolzh'ia i Tsentral'noi Rossii*, Nizhnii Novgorod, 2011, p. 37; E.L. Derechinskaia, 'Istoriia Nizhegorodskoi evreiskoi obshchiny v kontekste istorii evreev Rossii', in *Proshloe i nastoiashchee evreiskikh obshchin Povolzh'ia i Tsentral'noi Rossii*, Nizhnii Novgorod, 2011, p. 62.
25. Evtuhov, *Portrait of a Province*, pp. 112–14.
26. E.E. Kalinina, 'Evrei Udmurtii', in *Material'naia i dukhovnaia kul'tura narodov Urala i Povolzh'ia: istoriia i sovremennost'*, Glazov, 2016, pp. 288–90.
27. NART, fond 2, opis' 2, delo 755, Kazan' Provincial Office, 1875.
28. GAIO, fond 73, opis' 1, tom 3, delo 6372, Chancellery of the Iaroslavl' Governor, 1882.
29. GASO, fond 1, opis' 1, tom 2, delo 5502, Samara Provincial Chancellery, 1914.
30. Polianskii, *Gorodskoe remeslo i manufaktura*, p. 34.
31. Alabin, *Samara*, p. 56.
32. Dimitriev and Selivanova, *Cheboksary*, p. 228.
33. GAIO, fond 72, opis' 2, delo 1970, Chancellery of the Iaroslavl' Governor-General, 1810.
34. NART, fond 22, opis' 2, delo 578, 838, Kazan' Tatar Ratusha records, 1790–1795.
35. NART, fond 1, opis' 2, delo 63, Chancellery of the Kazan' Governor, 1834.
36. NART, fond 1, opis' 2, delo 1003, Chancellery of the Kazan' Governor, 1855–56.
37. GAIO, fond 73, opis' 1, tom 3, delo 6372, Chancellery of the Iaroslavl Governor, 1882.
38. Dubman et al., *Istoriia Samary*, pp. 137, 200.
39. *Istoriia gubernskogo goroda Iaroslavlia*, compiled by A.M. Rutman, Iaroslavl', 2006, p. 285.
40. Spottiswoode, *Tarantasse Journey*, p. 22.
41. N.V. Dutov, *Iaroslavl': istoriia i toponimika ulits i ploshchadei goroda*, 2nd edition, Iaroslavl', 2015, p. 16.
42. Cherniak and Madiiarov, *Gorodskoe samoupravlenie v Kazani*, p. 87.

43. Vishlenkova et al., *Kul'tura povsednevnosti provintsial'nogo goroda*, p. 98.
44. Quoted in A.N. Zorin, *Goroda i posady dorevoliutsionnogo Povolzh'ia*, Kazan', 2001, p. 557.
45. A.A. Demchenko, *Literaturnaia i obshchestvennaia zhizn' Saratova (iz arkhivnykh razyskanii)*, Saratov, 2008, p. 39.
46. Vishlenkova et al., *Kul'tura povsednevnosti provintsial'nogo goroda*, p. 182.
47. ibid., p. 55.
48. Zorin, *Goroda i posady dorevoliutsionnogo Povolzh'ia*, p. 319.
49. *PSZ*, no. 21716, vol. 28, pp. 981–82, 15 April 1805.
50. Vishlenkova et al., *Kul'tura povsednevnosti provintsial'nogo goroda*, p. 136.
51. Munro-Butler-Johnstone, *A Trip up the Volga*, p. 60.
52. Haxthausen, *The Russian Empire*, vol. 1, p. 322.
53. Edward Tracy Turnerelli, *Russia on the Borders of Asia: Kazan, Ancient Capital of the Tatar Khans*, 2 vols, London, 1854, vol. 2, p. 17.
54. Haxthausen, *The Russian Empire*, vol. 1, p. 143.
55. Turnerelli, *Russia on the Borders of Asia*, vol. 2, p. 30.
56. Max J. Okenfuss, 'The Jesuit Origins of Petrine Education', in J.G. Garrard, *The Eighteenth Century in Russia*, Oxford, 1973, p. 122.
57. *PSZ*, no. 6695, vol. 9, p. 483, 26 February 1735.
58. L.P. Burmistrova, 'Publichnye lektsii v Kazanskom Universitete (30 – 60-e gody XIX v.)', in *Stranitsy istorii goroda Kazani*, Kazan', 1981, p. 36.
59. E.L. Dubman and P.S. Kabytov, eds, *Povolzh'e – 'vnutrenniaia okraina' Rossii: gosudarstvo i obshchestvo v osvoenii novykh territorii (konets XVI – nachalo XX vv.)*, Samara, 2007, p. 216.
60. Dubman et al., *Istoriia Samary*, p. 180.
61. M.V. Zaitsev, *Saratovskaia gorodskaia duma (1871–1917)*, Saratov, 2017, p. 245.
62. Demchenko, *Literaturnaia i obshchestvennaia zhizn' Saratova*, p. 97.
63. S.M. Mikhailova, *Kazanskii Universitet i prosveshchenie narodov Povolzh'ia i Priural'ia (XIX vek)*, Kazan', 1979, p. 45–46, 51.
64. Zorin, *Goroda i posady dorevoliutsionnogo Povolzh'ia*, p. 136.
65. M.A. Kirokos'ian, *Astrakhanskii kupets G.V. Tetiushinov*, Astrakhan', 2014, pp. 61–62.
66. Patrick O'Meara, *The Russian Nobility in the Age of Alexander I*, London, 2019, pp. 82–83.
67. ibid., p. 108.
68. Seymour Becker, *Nobility and Privilege in Late Imperial Russia*, DeKalb, IL, 1985, pp. 140–41.
69. E.G. Bushkanets, *Iunost' L'va Tolstogo: kazanskie gody*, Kazan', 2008, pp. 79, 89.
70. Richard Stites, *Serfdom, Society, and the Arts in Imperial Russia: The Pleasure and the Power*, New Haven, CT, and London, 2005, p. 170.
71. Haxthausen, *The Russian Empire*, vol. 1, p. 227.
72. Turnerelli, *Russia on the Borders of Asia*, vol. 1, p. 230.
73. E. Anthony Swift, *Popular Theater and Society in Tsarist Russia*, Berkeley, CA, and London, 2002, pp. 22–23.
74. Zaitsev, *Saratovskaia gorodskaia duma*, p. 297.
75. Stites, *Serfdom, Society, and the Arts in Imperial Russia*, p. 108.
76. ibid., p. 268.
77. Murray Frame, *School for Citizens: Theatre and Civil Society in Imperial Russia*, New Haven, CT, and London, 2006, pp. 123–25.
78. Stites, *Serfdom, Society, and the Arts in Imperial Russia*, p. 266.
79. Stites, *Russian Popular Culture*, p. 10.
80. Vishlenkova et al., *Kul'tura povsednevnosti provintsial'nogo goroda*, p. 153.
81. [T.C.], *The New Atlas*, p. 172.
82. Spottiswoode, *Tarantasse Journey*, p. 174.
83. Forster, *A Journey from Bengal to England*, vol. 2, p. 303.
84. A.I. Iukht, *Torgovlia s vostochnymi stranami i vnutrennii rynok Rossii (20–60-e gody XVIII veka)*, Moscow, 1994, p. 57.
85. L.K. Ermolaeva, 'Krupnoe kupechestvo Rossii v XVII – pervoi chetverti XVIII v. (po materialam astrakhanskoi torgovli)', *Istoricheskie zapiski*, vol. 114, 1986, pp. 302–25.
86. V.A. Cholakhian, *Armiane na Saratovskoi zemle*, Saratov, 2018, p. 89.

87. Rybushkin, *Zapiski ob Astrakhani*, p. 92.
88. Cholakhian, *Armiane na Saratovskoi zemle*, p. 91.
89. Zh.A. Ananian, 'Armianskoe kupechestvo v Rossii', in *Kupechestvo v Rossii XV – pervaia polovina XIX veka*, Moscow, 1997, p. 238.
90. Jarmo T. Kotilaine, *Russia's Foreign Trade and Economic Expansion in the Seventeenth Century: Windows on the World*, Leiden and Boston, MA, 2005, pp. 480–81.
91. A.L. Riabtsev, *Gosudarstvennoe regulirovanie vostochnoi torgovli Rossii v XVII – XVIII vekov*, Astrakhan', 2012, pp. 22–23.
92. Spilman, *A Journey through Russia*, p. 11.
93. A.V. Syzranov, *Islam v Astrakhanskom krae: istoriia i sovremennost'*, Astrakhan', 2007, p. 32.
94. Rybushkin, *Zapiski ob Astrakhani*, pp. 91–96.
95. Jacques Margeret, *The Russian Empire and Grand Duchy of Muscovy: A 17th-Century French Account*, translated and edited by Chester S.L. Dunning, Pittsburgh, PA, 1983, p. 10.
96. *PSZ*, no. 2/8481, vol. 10, part 2, pp. 1030–32, 15 October 1835.
97. *PSZ*, no. 4304, vol. 7, p. 115, 20 September 1723; no. 9311, vol. 12, pp. 581–83, 5 August 1746; no. 19169, vol. 25, pp. 840–42, 28 October 1799; no. 19186, vol. 25, p. 863, 11 November 1799.
98. *PSZ*, no. 8919, vol. 12, pp. 77–78, 14 April 1744; no. 8991, vol. 12, pp. 169–70, 13 July 1744.
99. *PSZ*, no. 14853, vol. 20, p. 805, 11 November 1779.
100. Evelina Kugrysheva, *Istoriia Armian v Astrakhani*, Astrakhan', 2007, pp. 161–64.
101. *PSZ*, no. 19028, vol. 25, pp. 695–99, 3 July 1799.
102. *PSZ*, no. 19656, vol. 26, pp. 392–95, 19 November 1800.
103. Spottiswoode, *Tarantasse Journey*, pp. 187–88.

CHAPTER 10

1. Katherine Blanche Guthrie, *Through Russia: From St Petersburg to Astrakhan and the Crimea*, London, 1874, vol. 1, p. 321.
2. N.F. Tagirova, 'Khlebnyi rynok Povolzh'ia vo vtoroi polovine XIX – nachale XX vv.', dissertation, Samara State University, 1999, p. 74.
3. James Abbott, *Narrative of a Journey from Heraut to Khiva, Moscow and St Petersburgh, during the late Russian Invasion of Khiva*, London, 1884, vol. 2, p. 115.
4. Oliphant, *The Russian Shores of the Black Sea*, p. 37.
5. Pallas, *Travels through the Southern Provinces*, vol. 1, p. 97.
6. Perry, *The State of Russia under the Present Czar*, pp. 102–03.
7. Olearius, *The Travels of Olearius*, pp. 301–02.
8. Hausmann, *Mütterchen Wolga*, pp. 302–03.
9. *PSZ*, no. 8598, vol. 11, p. 638, 13 August 1742.
10. Richard M. Haywood, 'The Development of Steamboats on the Volga and its Tributaries, 1817–1865', *Research in Economic History*, vol. 6, 1981, p. 139 and *passim*.
11. Rodin, *Burlachestvo v Rossii*, pp. 75, 141.
12. F.Kh. Kissel', *Istoriia goroda Uglicha*, reprint of 1844 edition, Uglich, 1994, p. 370.
13. Bell, *Travels from St Petersburg*, p. 10.
14. Munro-Butler-Johnstone, *A Trip up the Volga*, p. 10.
15. *PSZ*, no. 8922, vol. 12, p. 85, 14 April 1744.
16. Kotilaine, *Russia's Foreign Trade and Economic Expansion*, p. 458.
17. Iukht, *Torgovlia s vostochnymi stranami*, pp. 58, 102–26; Riabtsev, *Gosudarstvennoe regulirovanie vostochnoi torgovli Rossii*, p. 64.
18. E.G. Istomina, *Vodnye puti Rossii: vo vtoroi polovine XVIII – nachale XIX veka*, Moscow, 1982, p. 117.
19. Tsybin and Ashanin, *Istoriia Volzhskogo kazachestva*, p. 52.
20. Smith and Christian, *Bread and Salt*, pp. 185, 57.
21. Dubman et al., *Istoriia Samary*, p. 125.
22. Oliphant, *The Russian Shores of the Black Sea*, p. 85.
23. Spottiswoode, *Tarantasse Journey*, p. 104.

24. Istomina, *Vodnye puti Rossi*, p. 116.
25. Rodin, *Burlachestvo v Rossii*, p. 63.
26. Istomina, *Vodnyi transport Rossii,* p. 145.
27. M.Iu. Volkov, *Goroda Verkhnego Povolzh'ia i severo-zapada Rossii. Pervaia chetvert' XVIII v.*, Moscow, 1994, p. 127.
28. Istomina, *Vodnyi transport Rossii*, p. 139.
29. Haxthausen, *The Russian Empire*, vol. 1, p. 144.
30. Discussed fully in Robert E. Jones, *Bread upon the Waters: the St Petersburg Grain Trade and the Russian Economy, 1703–1811*, Pittsburgh, PA, 2013.
31. Istomina, *Vodnye puti Rossii*, p. 35.
32. GAIO, fond 72, opis' 2, delo 837, Chancellery of the Iaroslavl' Governor-General.
33. R.A. French, 'Canals in Pre-Revolutionary Russia', in J.H. Bater and R.A. French, eds, *Studies in Russian Historical Geography*, vol. 2, London, 1983, pp. 451–81.
34. Spottiswoode, *Tarantasse Journey*, p. 7.
35. Rodin, *Burlachestvo v Rossii*, p. 110.
36. *Narodnoe sudostroenie v Rossii: entsiklopedicheskii slovar' sudov narodnoi postroiki*, edited by P.A. Filin and S.P. Kurnoskin, St Petersburg, 2016, pp. 49, 281–88.
37. Istomin, *Vodnyi puti Rossii*, pp. 54–58, Haywood, 'The Development of Steamboats', pp. 138–39.
38. Dorothy Zeisler-Vralsted, *Rivers, Memory and Nation-Building: A History of the Volga and the Mississippi Rivers*, New York, 2014, p. 69.
39. Haywood, 'The Development of Steamboats', pp. 143–52, 159.
40. Oliphant, *The Russian Shores of the Black Sea*, p. 38.
41. Spottiswoode, *Tarantasse Journey*, p. 13.
42. Haywood, 'The Development of Steamboats', pp. 155, 177.
43. ibid., p. 158.
44. A.A. Khalin, *Istoriia volzhskogo rechnogo parokhodstva (seredina XIX – nachalo XX v.)*, Nizhnii Novgorod, 2017, p. 129.
45. Guthrie, *Through Russia*, vol. 2, pp. 15, 18.
46. I.N. Slepnev, 'Vliianie sozdaniia zheleznodorozhnoi seti na tovarizatsiiu zernovogo proizvodstva Rossii (vtoraia polovina XIX v.)', in *Povolzh'e v sisteme vserossiiskogo rynka: istoriia i sovremennost'*, Cheboksary, 2000, pp. 53–68.
47. N. Andreev, *Illiustrirovannyi putevoditel' po Volge i eia pritokam Oke i Kame*, 2nd edition, Moscow, 1914, p. 226.
48. N.F. Tagirova, *Rynok Povolzh'ia (vtoraia polovina XIX – nachalo XX vv.)*, Moscow, 1999, pp. 87–106.
49. French, 'Canals in Pre-Revolutionary Russia', p. 476.
50. Rodin, *Burlachestvo v Rossii*, pp. 56, 82, 125, 174.
51. Istomina, *Vodnye puti Rossii*, p. 92; Istomina, *Vodnyi transport Rossii*, p. 74.
52. Rodin, *Burlachestvo v Rossii*, p. 145.
53. ibid., p. 97.
54. Maksim Gorky, *My Childhood*, translated by Ronald Wilks, London, 1966, p. 36.
55. Rodin, *Burlachestvo v Rossii*, pp. 34, 77.
56. N.B. Golikova, 'Iz istorii formirovaniia kadrov naemnykh rabotnikov v pervoi chetverti XVIII v.', *Istoriia SSSR*, 1965, no. 1, pp. 78–79.
57. Istomina, *Vodnyi transport Rossii*, p. 68.
58. M.F. Prokhorov, 'Otkhodnichestvo krest'ian v gorodakh Verkhnego Povolzh'ia v seredine XVIII veka', *Russkii gorod*, Moscow, vol. 9, 1990, p. 151–53.
59. N.B. Golikova, *Naemnyi trud v gorodakh Povolzh'ia v pervoi chetverti XVIII veka*, Moscow, 1965, p. 69.
60. Istomina, *Vodnye puti Rossii*, p. 76.
61. Rodin, *Burlachestvo v Rossii*, p. 129.
62. Hausmann, *Mütterchen Wolga*, pp. 295–308.
63. Butyrkin, 'Razskazy iz sluzhby na Volge', p. 102.
64. Rodin, *Burlachestvo v Rossii*, p. 159.
65. ibid., p. 187.
66. Tagirova, *Rynok Povolzh'ia,* pp. 142–44.

67. Oliphant, *The Russian Shores of the Black Sea,* p. 27.
68. Dubman et al., *Istoriia Samary,* pp. 124–30.
69. N.F. Filatov, *Goroda i posady Nizhegorodskogo Povolzh'ia v XVIII veke,* Gor'kii, 1989, p. 48.
70. Ann Lincoln Fitzpatrick, *The Great Russian Fair: Nizhnii Novgorod 1846–90,* London and New York, 1990, p. 18, but much of this section is based on this monograph.
71. *PSZ,* no. 27302, vol. 35, pp. 143–44, 6 March 1818.
72. William Forsyth, *The Great Fair of Nijni Novgorod and How We Got There,* London, 1865, p. 114.
73. Oliphant, *The Russian Shores of the Black Sea,* p. 23.
74. Munro-Butler-Johnstone, *A Trip up the Volga,* p. 76.
75. Fitzpatrick, *The Great Russian Fair,* p. 48.
76. Forsyth, *The Great Fair of Nijni Novgorod,* pp. 110–11.
77. Bremner, *Excursions in the Interior of Russia,* vol. 2, pp. 226–27.
78. Fitzpatrick, *The Great Russian Fair,* p. 139.
79. Bremner, *Excursions in the Interior of Russia,* vol. 2, pp. 228–31.
80. Fitzpatrick, *The Great Russian Fair,* p. 37.
81. Bremner, *Excursions in the Interior of Russia,* vol. 2, p. 235.
82. Oliphant, *The Russian Shores of the Black Sea,* p. 9.

CHAPTER 11

1. Quoted in Christopher Ely, *This Meager Nature: Landscape and National Identity in Imperial Russia,* DeKalb, IL, 2002, p. 63.
2. Quoted in ibid., p. 68.
3. Ivan Goncharov, *Oblomov,* translated by David Magarshack, Harmondsworth, 1954, p. 104. Also quoted in Ely, *This Meager Nature,* p. 126.
4. G.P. Dem'ianov, *Putevoditel' po Volge ot Tveri do Astrakhani,* 6th edition, Nizhnii Novgorod, 1900, no page numbers.
5. 'The Storm' in Aleksandr Ostrovsky, *Four Plays,* translated by Stephen Mulrine, London, 1997, p. 11.
6. Cited in the *Slovar' russkogo iazyka XI–XII vv.,* vol. 9, Moscow, 1982, under *matka.*
7. *Poety 1790–1810-kh godov,* Leningrad, 1971, pp. 673–74. See also Toriyama, 'Images of the Volga River in Russian Poetry'.
8. Dem'ianov, *Putevoditel' po Volge ot Tveri do Astrakhani.* This poem is undated and attributed to 'Vl.Iu', no page numbers.
9. ibid., poem by A. Lugovoi, no page numbers.
10. Oleg Riabov, ' "Mother Volga" and "Mother Russia": On the Role of the River in Gendering Russianness', in Jane Costlow and Arja Rosenholm, eds, *Meanings and Values of Water in Russian Culture,* Abingdon, 2017, p. 88.
11. Butyrkin, 'Razskazy iz sluzhby na Volge', p. 105.
12. Aleksandr Nikolaevich Naumov, *Iz utselevshikh vospominanii, 1868–1917,* 2 vols, New York, 1954, vol. 1, p. 294.
13. Riabov, ' "Mother Volga" and "Mother Russia" ', pp. 81–98.
14. E.A. Bazhanov, *Sviashchennye reki Rossii,* Samara, 2008, p. 96.
15. Author's own translation.
16. 'The Ice is Moving' in Maksim Gorky, *Selected Short Stories,* New Delhi, 2000, p. 306.
17. 'One Autumn', in Gorky, *Selected Short Stories,* p. 155.
18. Evgenii Chirikov, *Marka of the Pits,* translated by L. Zarine, London, n.d. [1930?], pp. 1–2.
19. Evgenii Chirikov, *Otchii dom: semeinaia khronika,* Moscow, 2010, p. 229.
20. Christopher Ely, 'The Origins of Russian Scenery: Volga River Tourism and Russian Landscape Aesthetics', *Slavic Review,* vol. 62, no. 4, 2003, p. 668. Much of this section is drawn from this article.
21. ibid., p. 669.
22. *Khudozhniki brat'ia Grigorii i Nikanor Chernetsovy: grecheskii mir v russkom iskusstve,* St Petersburg, 2000, fig. 51. Sketches of other Volga towns are at 2, 18, 23, 47, 48, 49, 50. I thank Andrew Curtin for this reference.

23. John E. Bowlt, 'Russian Painting in the Nineteenth Century', in Theofanis G. Stavrou, ed., *Art and Culture in Nineteenth-Century Russia*, Bloomington, IN, 1983, p. 130.
24. Ely, *This Meager Nature*, p. 71.
25. Rosalind P. Gray, *Russian Genre Painting in the Nineteenth Century*, Oxford, 2000, p. 86.
26. Stites, *Serfdom, Society, and the Arts in Imperial Russia*, p. 359.
27. Il'ia Repin, *Dalekoe blizkoe i vospominaniia*, Moscow, 2002, pp. 238–40.
28. Molly Brunson, 'Wandering Greeks: How Repin Discovers the People', *Ab Imperio*, vol. 2, 2012, p. 98.
29. ibid., p. 92.
30. This is well documented in Andrey Shabanov, *Art and Commerce in Late Imperial Russia: The Peredvizhniki, a Partnership of Artists*, London, 2019.
31. Elizabeth Valkenier, *Ilya Repin and the World of Russian Art*, New York, 1990, p. 83.
32. Ivan Evdokimov, *Levitan: povest'*, Moscow, 1958, pp. 282–87.
33. Ely, 'The Origins of Russian Scenery', p. 681. The painting is reproduced on p. 667, fig. 4.
34. Riabov, '"Mother Volga" and "Mother Russia"', p. 85.
35. Serge Gregory, *Antosha and Levitasha: The Shared Lives and Art of Anton Chekhov and Isaac Levitan*, DeKalb, IL, 2015, p. 89.
36. Anton Chekhov, 'The Grasshopper', from *The Wife and Other Stories*, translated by Constance Garnett, New York, 1918, p. 42.
37. Ivan Bunin, *Sunstroke: Selected Stories of Ivan Bunin*, translated by Graham Hettlinger, Chicago, IL, 2002, p. 3.
38. K.S. Petrov-Vodkin, *Moia povest' (Khlynovsk)*, St Petersburg, 2013, pp. 234, 281.
39. Quoted in Riabov, '"Mother Volga" and "Mother Russia"', p. 84.
40. Quoted in Ely, *This Meager Nature*, p. 68.
41. Quoted in ibid., p. 65.
42. Guido Hausmann, 'The Volga Source: Sacralization of a Place of Memory', in *Velikii volzhskii put'*, Kazan', 2001, pp. 340–46; also Hausmann, *Mütterchen Wolga*, pp. 100–10.
43. Haywood, 'The Development of Steamboats', p. 176.
44. This section draws on Ely, 'The Origins of Russian Scenery', here p. 671.
45. ibid., p. 674.
46. ibid., pp. 675–77.
47. Dem'ianov, *Putevoditel' po Volge ot Tveri do Astrakhani*, pp. 279, 325.
48. Andreev, *Illiustrirovannyi putevoditel' po Volge*, pp. 66, 158, 226, 234.
49. S.V. Starikov, *Velikaia reka Rossii na rubezhe XIX–XX vekov: Volga ot Nizhnego Novgoroda do Kazani na starinnykh otkrytkakh*, Ioshkar-Ola, 2009.
50. Anton Chekhov, *Letters of Anton Chekhov*, selected and edited by Avrahm Yarmolinsky, New York, 1973, p. 136.
51. Karl Baedeker, *Karl Baedeker's Russia 1914*, London, 1971.
52. Quoted in Ely, 'The Origins of Russian Scenery', p. 676.

CHAPTER 12

1. P.M Poteten'kin, *Krest'ianskie volneniia v Saratovskoi gubernii v 1861–1863 gg.*, Saratov, 1940, pp. 55–57.
2. L.V. Tushkanov, *Chastnovladel'cheskoe khoziaistvo Saratovskoi gubernii v poreformennyi period (1861–1904 gg.)*, Volgograd, 2010, pp. 28, 35, 45, 59.
3. Timothy Mixter, 'Of Grandfather-Beaters and Fat-Heeled Pacifists: Perceptions of Agricultural Labor and Hiring Market Disturbances in Saratov, 1872–1905', *Russian History*, vol. 7, nos 1–2, 1980, p. 141.
4. Moon, *The Plough that Broke the Steppes*, p. 216.
5. 'Autobiographical Sketch' in Alexey Tolstoy, *Collected Works*, vol. 6, *Ordeal*, Moscow, 1982, p. 15.
6. James Simms, 'The Economic Impact of the Russian Famine of 1891–92', *The Slavonic and East European Review*, vol. 60, no. 1, 1982, p. 69.
7. James Long, 'Agricultural Conditions in the German Colonies of Novouzensk District, Samara Province, 1861–1914', *The Slavonic and East European Review*, vol. 57, no. 4, 1979, p. 540.

8. Richard G. Robbins, *Famine in Russia 1891–1892: An Imperial Government Responds to a Crisis*, New York and London, 1975, p. 171.

9. *Krest'ianskoe dvizhenie v Simbirskoi gubernii v period revoliutsii 1905–1907 godov: dokumenty i materialy*, Ul'ianovsk, 1955, pp. 46, 53.

10. Timothy Mixter, 'Peasant Collective Action in Saratov Province, 1902–1906', in Rex A. Wade and Scott J. Seregny, *Politics and Society in Provincial Russia: Saratov, 1590–1917*, Columbus, OH, 1989, p. 214.

11. Naumov, *Iz utselevshikh vospominanii*, vol. 2, pp. 72–74.

12. I.V. Volkov, 'Stolypinskaia agrarnaia reforma v Iaroslavskoi gubernii', in *Ocherki istorii Iaroslavskogo kraia*, Iaroslavl', 1974, p. 166.

13. Dubman and Kabytov, *Povolzh'e – 'vnutrenniaia okraina' Rossii*, p. 274.

14. Quoted in Judith Pallot, *Land Reform in Russia, 1906–17: Peasant Responses to Stolypin's Project of Rural Transformation*, Oxford, 1999, p. 76.

15. Dubman and Kabytov, *Povolzh'e – 'vnutrenniaia okraina' Rossii*, p. 273.

16. Quoted in Judith Pallot, *Land Reform in Russia, 1906–17*, p. 170.

17. Judith Pallot, 'Agrarian Modernisation on Peasant Farms in the Era of Capitalism', in J.H. Bater and R.A. French, eds, *Studies in Russian Historical Geography*, vol. 2, London, 1983, pp. 423–29.

18. Quoted in Pallot, *Land Reform in Russia, 1906–17*, p. 231.

19. Kuznetsov, *Ocherki po istorii Chuvashskogo krest'ianstva*, vol. 1, pp. 16, 322, 326.

20. *Volneniia urzhumskikh Mariitsev v 1889 godu*, compiled by L. Shemier et al., Ioshkar-Ola, 2017, pp. 33, 54.

21. Long, 'Agricultural Conditions in the German Colonies', pp. 531–51.

22. Hartley, *Siberia*, p. 170.

23. Munro-Butler-Johnstone, *A Trip up the Volga*, pp. 37–38.

24. David Macey, 'Reflections on Peasant Adaptation in Rural Russia of the Beginning of the Twentieth Century: The Stolypin Agrarian Reforms', *Journal of Peasant Studies*, vol. 31, nos 3–4, 2004, pp. 400–26.

25. Peter Waldron, *Between Two Revolutions: Stolypin and the Politics of Renewal in Russia*, London, 1998, p. 176.

26. Dubman et al., *Istoriia Samary*, p. 234; V.A. Cholakhian, *Sotsial'no-demograficheskie posledstviia industrial'nogo razvitiia Nizhnego Povolzh'ia (konets XIX v. – 1930-e gg.)*, Saratov, 2008, pp. 19–24.

27. Donald J. Raleigh, 'Revolutionary Politics in Provincial Russia: The Tsaritsyn "Republic" in 1917', *Slavic Review*, vol. 40, no. 2, 1981, p. 195.

28. Abraham Ascher, *The Revolution of 1905*, vol. 1, *Russia in Disarray*, Stanford, CA, 2004, p. 138.

29. *1905 god v Tsaritsyne (vospominaniia i dokumenty)*, compiled by V.I. Tomarev and E.N. Shkodina, Stalingrad, 1960, pp. 46–49.

30. *Revoliutsionnoe dvizhenie v Astrakhani i Astrakhanskoi gubernii v 1905–1907 godakh: sbornik dokumentov i materialov*, Astrakhan', 1957, p. 124.

31. *Revoliutsionnoe dvizhenie v Chuvashii v pervoi russkoi revoliutsii 1905–1907; dokumenty i materialy*, Cheboksary, 1965, pp. 32–34.

32. This section is based on Hugh Phillips, 'Riots, Strikes and Soviets: The City of Tver in 1905', *Revolutionary Russia*, vol. 17, no. 2, 2004, pp. 49–66.

33. M.L. Razmolodin, *Chernosotennoe dvizhenie v Iaroslavle i guberniiakh Verkhnego Povolzh'ia v 1905–1915 gg.*, Iaroslavl', 2001, pp. 23–24.

34. I.A. Bushuev, 'Pogrom v provintsii: o sobytiiakh 19 oktiabria 1905 g. v Kostrome', in *Gosudarstvo, obshchestvo, tserkov' v istorii Rossii XX veka*, part 2, Ivanovo, 2014, pp. 398–402.

35. Mustafa Tuna, *Imperial Russia's Muslims: Islam, Empire, and European Modernity 1788–1914*, Cambridge, 2015, p. 135.

36. Kh.Kh. Khasanov, 'Iz istorii formirovaniia Tatarskoi natsii', in *Tatariia v proshlom i nastoiashchem: sbornik statei*, Kazan', 1975, pp. 191–93.

37. NART, fond 1, opis' 6, delo 91, Chancellery of the Kazan' Governor, 1901.

38. NART, fond 1, opis' 3, delo 1834, Chancellery of the Kazan' Governor, 1868.

39. Aleksandr Kaplunovskii, 'Tatary, musul'mane i russkie v meshchanskikh obshchinakh Srednego Povolzh'ia v kontse XIX – nachale XX veka', *Ab Imperio*, vol. 1, 2000, pp. 101–15.

40. NART, fond 2, opis' 2, delo 4432, Kazan' Provincial Office, 1892.
41. Kefeli, *Becoming Muslim in Imperial Russia*, p. 139.
42. Quoted in Zemtsova, 'Russification and Educational Policies in the Middle Volga Region', p. 155.
43. D.M. Iskhakov, 'Kriasheny (istoriko-etnograficheskii ocherk)', in *Tatarskaia natsiia: istoriia i sovremennost'*, Kazan', 2002, pp. 121–23.
44. NART, fond 1, opis' 6, delo 612, Chancellery of the Kazan' Governor.
45. Norihiro Naganawa, 'Holidays in Kazan: The Public Sphere and the Politics of Religious Authority among Tatars in 1914', *Slavic Review*, vol. 71, no. 1, 2012, pp. 25–48.
46. Quoted in Zemtsova, 'Russification and Educational Policies in the Middle Volga Region', p. 115.
47. Ch.Kh. Samatova, *Imperskaia vlast' i Tatarskaia shkola vo vtoroi polovine XIX – nachale XX veka*, Kazan', 2013, p. 232.
48. Quoted in Zemtsova, 'Russification and Educational Policies in the Middle Volga Region', p. 116.
49. Samatova, *Imperskaia vlast' i Tatarskaia shkola*, p. 231.
50. Syzranov, *Islam v Astrakhanskom krae*, p. 60.
51. Quoted in Samatova, *Imperskaia vlast' i Tatarskaia shkola*, p. 124.
52. Quoted in Zemtsova, 'Russification and Educational Policies in the Middle Volga Region', p. 97.
53. Koch, *The Volga Germans in Russia and the Americas*, p. 155.
54. James Long, *From Privileged to Dispossessed: The Volga Germans 1860–1917*, Lincoln, NE, and London, 1988, p. 190.
55. Rorlich, *The Volga Tatars*, pp. 49–64.
56. ibid., p. 71.
57. ibid., pp. 110–12.
58. Christian Noack, 'State Policy and the Impact on the Formation of a Muslim Identity in the Volga-Urals', in *Islam in Politics in Russia and Central Asia (Early Eighteenth to Late Twentieth Centuries)*, edited by S.A. Dudoignon and Komatsu Hisau, London, New York and Bahrain, 2001, pp. 3–24.
59. *Historical Anthology of Kazan Tatar Verse*, p. 112.

CHAPTER 13

1. Quoted in W.E. Mosse, 'Revolution in Saratov (October – November 1910)', *The Slavonic and East European Review*, vol. 49, 1971, p. 594.
2. L.R. Gabdrafikova and Kh.M. Abdulin, *Tatary v gody Pervoi Mirovoi Voiny (1914–1918 gg.)*, Kazan', 2015, p. 260.
3. Quoted in Hugh Phillips, ' "A Bad Business" – The February Revolution in Tver' ', *The Soviet and Post-Soviet Review*, vol. 23, no. 2, 1996, p. 127.
4. E.Iu. Semenova, *Mirovozzrenie gorodskogo naseleniia Povolzh'ia v gody Pervoi Mirovoi Voiny (1914 – nachalo 1918 gg.): sotsial'nyi, ekonomicheskii, politicheskii aspekty*, Samara, 2012, pp. 156–64, 174, 176, 220–21.
5. ibid., pp. 103, 91.
6. Cholakhian, *Sotsial'no-demograficheskie posledstviia*, pp. 35–38.
7. E.Iu. Semenova, *Rossiiskii gorod v gody Pervoi Mirovoi Voiny (na materialakh Povolzh'ia)*, Samara, 2016, p. 42; Zaitsev, *Saratovskaia gorodskaia duma*, p. 171.
8. Gabdrafikova and Abdulin, *Tatary v gody Pervoi Mirovoi Voiny*, pp. 132, 177.
9. V. Denningkhaus, 'Russkie nemtsy i obshchestvennye nastroeniia v Povolzh'e v period Pervoi Mirovoi Voiny', in *Voenno-istoricheskie issledovaniia v Povolzh'e*, Saratov University, vol. 7, 2006, p. 176.
10. Semenova, *Mirovozzrenie gorodskogo naseleniia Povolzh'ia*, pp. 95, 126, 434.
11. ibid., p. 415.
12. S.S. Pliutsinskii, 'Pereselennye nemtsy na territorii Astrakhanskoi gubernii v gody Pervoi Mirovoi Voiny (1914–1918 gg.)', in *Istoricheskaia i sovremennaia regionalistika Verkhovnogo Dona i Nizhnego Povolzh'ia: sbornik nauchnykh statei*, Volgograd, 2005, pp. 218–31.

13. Long, *From Privileged to Dispossessed*, p. 232.
14. Semenova, *Rossiiskii gorod v gody Pervoi Mirovoi Voiny*, pp. 162–63.
15. *Provintsial v Velikoi Rossiiskoi Revoliutsii. Sbornik dokumentov. Simbirskaia guberniia v ianvare 1917 – marte 1918 gg.*, edited by N.V. Lipatova, Ul'ianovsk, 2017, p. 294.
16. Quoted in Sarah Badcock, *Politics and the People in Revolutionary Russia: A Provincial History*, Cambridge, 2007, pp. 32–33.
17. ibid., p. 34.
18. Donald J. Raleigh, *Revolution on the Volga: 1917 in Saratov*, Ithaca, NY, and London, 1986, p. 85.
19. *Provintsial v Velikoi Rossiiskoi Revoliutsii*, p. 284.
20. Quoted in Badcock, *Politics and the People in Revolutionary Russia*, p. 37.
21. This section is drawn from Phillips, ' "A Bad Business" ', pp. 123–41.
22. Badcock, *Politics and the People in Revolutionary Russia*, p. 20.
23. Quoted in ibid., pp. 103, 118.
24. Quoted in Mosse, 'Revolution in Saratov', p. 598.
25. Sarah Badcock, 'From Saviour to Pariah: A Study of the Role of Karl Ianovich Grasis in Cheboksary during 1917', *Revolutionary Russia*, vol. 15, no. 1, 2002, pp. 69–96.
26. Raleigh, 'Revolutionary Politics in Provincial Russia', pp. 198–207.
27. Mosse, 'Revolution in Saratov', pp. 586–602.
28. ibid., p. 599.
29. Quoted in Laura Engelstein, *Russia in Flux: War, Revolution, Civil War 1914–21*, Oxford, 2018, p. 416.
30. This section is drawn primarily from Stephen M. Berk, 'The Democratic Counterrevolution: Komuch and the Civil War on the Volga', *Russian History*, vol. 7, 1980, pp. 176–90.
31. Orlando Figes, *Peasant Russia, Civil War: The Volga Countryside in Revolution 1917–1921*, London, 1989, p. 178.
32. A.A. Khalin et al., *Ocherki istorii volzhskogo rechnogo parokhodstva v XX – nachale XXI v.*, Nizhnii Novgorod, 2018, pp. 19–20, 31–33.
33. Peter Kenez, *Civil War in South Russia, 1918*, Berkeley, CA, and London, vol. 2, 1977, p. 38.
34. S.V. Mamaeva, *Promyshlennost' Nizhnego Povolzh'ia v period voennogo kommunizma (1918 – vesna 1921 g.)*, Astrakhan', 2007, pp. 101, 110.
35. Cholakhian, *Sotsial'no-demograficheskie posledstviia*, pp. 62–63.
36. K.I. Sokolov, *Proletarii protiv 'Proletarskoi' vlasti: protestnoe dvizhenie rabochikh v Tverskoi gubernii v kontse 1917–1922 gg.*, Tver', 2017, p. 50.
37. Donald J. Raleigh, *Experiencing Russia's Civil War: Politics, Society, and Revolutionary Culture in Saratov, 1917–1922*, Princeton, NJ, and Oxford, 2002, pp. 250, 261–64.
38. ibid., pp. 187–89, 194, 196.
39. Alexis Babine, *A Russian Civil War Diary: Alexis Babine in Saratov, 1917–1922*, edited by Donald J. Raleigh, Durham, NC, and London, 1989, pp. 30, 56, 92, 182, 188.
40. Raleigh, *Revolution on the Volga: 1917 in Saratov*, Ithaca, NY, and London, 1986, p. 186.
41. Irina Koznova, *Stalinskaia epokha v pamiati krest'ianstva Rossii*, Moscow, 2016, p. 92.
42. Aleksandrova and Vedernikova, *Sel'skie dvorianskie usad'by Samarskogo Zavolzh'ia*, pp. 41–46.
43. Badcock, *Politics and the People in Revolutionary Russia*, p. 186.
44. ibid., pp. 230–31.
45. E.I. Medvedev, *Grazhdanskaia voina v Srednem Povolzh'e (1918–1919)*, Saratov, 1974, pp. 199, 239.
46. Raleigh, *Experiencing Russia's Civil War*, p. 319.
47. *Oktiabr' v Povolzh'e*, Saratov, 1967, p. 478.
48. Quoted in Figes, *Peasant Russia, Civil War*, pp. 266–67.
49. Raleigh, *Experiencing Russia's Civil War*, p. 338.
50. C.E. Bechhofer, *Through Starving Russia. Being the Record of a Journey to Moscow and the Volga Provinces in August and September 1921*, Westport, CT, 1921, pp. 39, 43, 47, 56.
51. Wendy Z. Goldman, *Women, the State and the Revolution: Soviet Family Policy and Social Life 1917–36*, Cambridge, 1993, pp. 67–69.
52. Rafael Shaydullin, 'Peasantry and the State and the Tatar Autonomous Soviet Socialist Republic', in *The History of the Tatars since Ancient Times*, vol. 7, Kazan', 2017, pp. 283–84.

53. V.V. Alekseev, *Zemlia Borskaia: vekhi istorii*, Samara, 2016, pp. 334, 336.
54. This section is drawn from James Long, 'The Volga Germans and the Famine of 1921', *Russian Review*, vol. 51, no. 4, 1992, pp. 510–25, here pp. 512, 518.
55. Quoted in ibid., p. 521.
56. Guzel Yakhina, *Zuleikha: A Novel*, translated by Lisa C. Hayden, London, 2015, p. 43.

CHAPTER 14

1. Quoted in Lynne Viola, *Peasant Rebels under Stalin: Collectivization and the Culture of Peasant Resistance*, Oxford, 1996, p. 56.
2. Anne Appelbaum, *Gulag: A History*, London, 2003, p. 103.
3. I.F. Ialtaev, *Derevnia Mariiskoi avtonomnoi oblasti v gody kollektivizatsiia (1929–1936)*, Ioshkar-Ola, 2015, p. 86.
4. *Kollektivizatsiia sel'skogo khoziaistva v srednem Povolzh'e (1927–1937)*, edited by N.N. Panov and F.A. Karevskii, Kuibyshev, 1970, p. 208.
5. *Istoriia Tatarskoi ASSR*, vol. 2, p. 267.
6. Anatolii Golovkin, *Zhernova. Kniga pamiati Tverskikh Karel*, Tver', 2017, pp. 22–23.
7. Alekseev, *Zemlia Borskaia*, pp. 357–58.
8. A.A. German, *Bol'shevistskaia vlast' i nemetskaia avtonomiia na Volge (1918–1941)*, Saratov, 2004, pp. 215, 224.
9. Ialtaev, *Derevnia Mariiskoi avtonomnoi oblasti*, pp. 58, 176, 88.
10. *The History of the Tatars since Ancient Times*, vol. 7, Kazan', 2017, appendix, documents, pp. 797–98.
11. Quoted in Viola, *Peasant Rebels under Stalin*, p. 151.
12. Koznova, *Stalinskaia epokha v pamiati krest'ianstva*, pp. 266, 268.
13. Viola, *Peasant Rebels under Stalin*, pp. 140, 162.
14. Sheila Fitzpatrick, *Stalin's Peasants: Resistance and Survival in the Russian Village after Collectivization*, Oxford and New York, 1994, p. 17.
15. Koznova, *Stalinskaia epokha v pamiati krest'ianstva*, p. 179.
16. Quoted in Viola, *Peasant Rebels under Stalin*, p. 62.
17. Quoted in ibid., pp. 144, 157–58.
18. ibid., p. 82.
19. Ialtaev, *Derevnia Mariiskoi avtonomnoi oblasti*, p. 92.
20. Viola, *Peasant Rebels under Stalin*, p. 88.
21. ibid., p. 161.
22. *The History of the Tatars since Ancient Times*, vol. 7, appendix, documents, pp. 803–05.
23. Quoted in German, *Bol'shevistskaia vlast' i nemetskaia avtonomiia na Volge*, p. 219.
24. Fitzpatrick, *Stalin's Peasants*, p. 75.
25. Koznova, *Stalinskaia epokha v pamiati krest'ianstva*, pp. 312, 317, 319.
26. Fitzpatrick, *Stalin's Peasants*, p. 76.
27. Quoted in Robert Conquest, *The Harvest of Sorrow: Soviet Collectivisation and the Terror-Famine*, London, 1986, pp. 281–82.
28. Andrew Cairns, *The Soviet Famine 1932–33: An Eye Witness Account of Conditions in the Spring and Summer of 1932*, Edmonton, Canadian Institute of Ukrainian Studies, 1989, pp. 114–15.
29. *Golod v SSSR 1930–1934 gg.*, Moscow, 2009, pp. 342, 344–46.
30. Ialtaev, *Derevnia Mariiskoi avtonomoi oblasti*, pp. 178–79, 204.
31. Koznova, *Stalinskaia epokha v pamiati krest'ianstva*, p. 183.
32. Yakhina, *Zuleikha*, pp. 134–35.
33. Aleksandr Beliakov and Stanislav Smirnov, *Lishentsy Nizhegorodskogo kraia (1918–1936 gg.)*, Nizhnii Novgorod, 2018, pp. 64, 44, 97.
34. *Kak eto bylo: dokumental'nyi sbornik*, compiled by V.A. Ugriumov, Nizhnii Novgorod, 2011, pp. 171–72.
35. Eugenia Semyonovna Ginzburg, *Journey into the Whirlwind*, translated by Paul Stevenson and Max Hayward, New York and London, 1967, pp. 52–53.
36. *Provintsial v Velikoi Rossiiskoi Revoliutsii*, pp. 190–91.

37. Ayslu Kabirova, 'Political Repression in the TASSR in the 1930s', in *The History of the Tatars since Ancient Times*, vol. 7, Kazan', 2017, pp. 366–70.
38. Z.G. Garipova, *Kazan': obshchestvo, politika, kul'tura*, Kazan', 2004, pp. 33–36, 48–49.
39. Ildikó Lehtinen, *From the Volga to Siberia: The Finno-Ugric Peoples in Today's Russia*, Helsinki, 2002, p. 32.
40. *An Anthology of Chuvash Poetry*, p. 129.
41. Jacob Weber [Iakov Veber], *Iakov Iakovlevich Veber; Katalog k 125-letiu so dnia rozhdeniia. Zhivopis'*, Saratov, 1995. I am grateful to Professor Roger Bartlett for this source.
42. Rotermel', *Tverskie nemtsy*, p. 97.
43. Golovkin, *Zhernova. Kniga pamiati Tverskikh Karel*, pp. 384, 392–96.
44. *Iaroslavskii krai: istoriia, traditsii, liudi*, Iaroslavl', 2017, p. 135.
45. Garipova, *Kazan'*, p. 60.
46. *Russkie monastyri: Sredniaia i Nizhniaia Volga*, Moscow, 2004, pp. 213–14.
47. L.E. Koroleva, A.I. Lomovtsev and A.A. Korolev, *Vlast' i musul'mane Srednego Povolzh'ia (vtoraia polovina 1940-kh – pervaia polovina 1980-kh gg.*, Penza, 2001, pp. 65–66.
48. Gosudarstvennyi arkhiv Astrakhanskoi oblasti, fond 1, delo 743, ll. 31–38, Astrakhan' Provincial Soviet, acts on the handing over of religious communities, 1924.
49. Golovkin, *Zhernova. Kniga pamiati Tverskikh Karel*, pp. 389–90.
50. Beliakov and Smirnov, *Lishentsy Nizhegorodskogo kraia*, pp. 129–30.
51. Ruslan R. Ibragimov, *Vlast' i religiia v Tatarstane v 1940–1980-e gg.*, Kazan', 2005, pp. 61–62.
52. Nikolai Semenov, *O chem ne uspel*, Saratov, 2016, p. 111.
53. Jehanne M. Gheith and Katherine R. Jolluck, eds, *Gulag Voices: Oral Histories of Soviet Incarceration and Exile*, New York, 2011, pp. 134–40, 143.
54. Eugenia Semyonovna Ginzburg, *Within the Whirlwind*, translated by Ian Boland, San Diego, London and New York, 1981, p. 228.

CHAPTER 15

1. Riabov, ' "Mother Volga" and "Mother Russia" ', p. 90.
2. A.M. Samsonov, *Stalingradskaia bitva*, 2nd edition, Moscow, 1968, pp. 281, 307.
3. Richard Overy, *Russia's War*, London, 1997, pp. 72, 76, 94.
4. Quoted in Janet M. Hartley, *Alexander I*, London and New York, 1994, p. 118.
5. Quoted in Melanie Ilic, *Life Stories of Soviet Women: The Interwar Generation*, London and New York, 2013, p. 39.
6. Overy, *Russia's War*, p. 117.
7. Antony Beevor, *Stalingrad*, London, 1999, p. 107. Much of this section is based on this brilliant account.
8. Quoted in Catherine Merridale, *Ivan's War: The Red Army 1939–45*, London, 2005, p. 150.
9. *Rechnoi transport SSSR 1917–1957: sbornik statei o razvitii rechnogo transporta SSSR za 40 let*, Moscow, 1957, p. 33; Khalin et al., *Ocherki istorii volzhskogo rechnogo parokhodstva*, p. 75.
10. Quoted in Beevor, *Stalingrad*, p. 149.
11. Quoted in ibid., p. 162.
12. Khalin et al., *Ocherki istorii volzhskogo rechnogo parokhodstva*, pp. 70, 74–75; V. Kazarin, *Volzhskie stranitsy Samarskoi istorii*, Samara, 2011, pp. 197–98.
13. Quoted in Beevor, *Stalingrad*, p. 364.
14. Vasily Grossman, *Stalingrad*, translated by Robert and Elizabeth Chandler, London, 2019, p. 478.
15. Quoted in Merridale, *Ivan's War*, p. 153.
16. Quoted in Beevor, *Stalingrad*, p. 235.
17. Quoted in Chris Bellamy, *Absolute War: Soviet Russia in the Second World War*, London, 2009, pp. 525–26.
18. Beevor, *Stalingrad*, pp. 304, 338.
19. Quoted in Merridale, *Ivan's War*, p. 154.
20. A.F. Raikov, *Voinu glazami ochevidtsa: vospominaniia o Velikoi Otechestvennoi Voine*, Tver', 2010, p. 92.

21. Vasily Grossman, *Life and Fate*, translated by Robert Chandler, London, 2006, p. 645.

22. Overy, *Russia's War,* p. 185.

23. Samsonov, *Stalingradskaia bitva*, p. 316.

24. A.V. Khramkov, *Trudiashchiesia Kuibyshevskoi oblasti v gody Velikoi Otechestvennoi Voiny 1941–1945*, Kuibyshev, 1985, pp. 11–12.

25. *Tatary Samarskogo kraia: istoriko-etnograficheskie i sotsial'no-ekonomicheskie ocherki*, edited by Sh.Kh. Galimov, Samara, 2017, pp. 499, 504, 507, 511, 514, 523, 565.

26. Robert Dale, 'Divided We Stand: Cities, Social Unity and Post-War Reconstruction in Soviet Russia, 1945–1953', *Contemporary European History*, vol. 24, no. 4, 2015, pp. 498–99. This was not only the case in Stalingrad, but also in other Soviet towns, where ambitious plans to build new socialist cities were watered down by the realities of the lack of resources and the desperate need for new housing; for an account of Nizhnii Novgorod/Gorkii, see Heather D. Dehaan, *Stalinist City Planning: Professionals, Performance, and Power*, Toronto, Buffalo, NY, and London, 2013.

27. Dale, 'Divided we Stand', pp. 499–500.

28. National Archives, Kew, FO 371/71659, Foreign Office Northern/Soviet Union, 'The trip by G.M. Warr and P. J. Kelly to Stalingrad and Astrakhan in 1948', pp. 10, 11, 14.

29. Dale, 'Divided we Stand', p. 506.

30. Elena Trubina, 'The Reconstructed City as Rhetorical Space: The Case of Volgograd', in Tovi Fenster and Haim Yacobi, *Remembering, Forgetting and City Builders*, Farnham, 2010, p. 113.

31. Scott W. Palmer, 'How Memory was Made: The Construction of the Memorial to the Heroes of the Battle of Stalingrad', *Russian Review*, vol. 68, no. 3, 2009, pp. 373–407.

32. Trubina, 'The Reconstructed City as Rhetorical Space', pp. 114–15.

33. I should like to thank Aleksandr Kiselev, of Volgograd University, for helping me to understand better the content of the panorama.

34. These attempts are described, for example, in O.U. El'chaninova, *Sel'skoe naselenie Srednego Povolzh'ia v period reform 1953–1964 gg.*, Samara, 2006, and V.S. Gorokhov, *Po zakonam kolkhoznoi zhizni*, Saratov, 1979.

35. Gregory D. Andrusz, *Housing and Urban Development in the USSR*, London, 1984, pp. 232–33.

36. V.P. Ivanov, 'Vliianie migratsii iz sela na demograficheskie i etnicheskie kharakteristiki gorodskikh semei', in *Sel'skoe khoziaistvo i krest'ianstvo Srednego Povolzh'ia v usloviiakh razvitogo sotsializma*, Cheboksary, 1982, pp. 100–24.

37. *Gorod Tol'iatti*, Kuibyshev, 1967, pp. 68, 79.

38. *Kul'tura Simbirskogo-Ul'ianovskogo kraia: sbornik dokumentov i materialov*, Ul'ianovsk, 2014, p. 54.

39. L.F. Khlopina, *Memorial nad Volgoi*, Ul'ianovsk, 2010. This book, published in 2010, presents the project in heroic terms, without considering the cost or the destruction; but there are good plans and photographs of the complex.

40. V. Samogorov, V. Pastushenko, A. Kapitonov and M. Kapitonov, *Iubileinyi Ul'ianovsk*, Ekaterinburg, 2013.

41. *Kul'tura Simbirskogo-Ul'ianovskogo kraia*, p. 315.

42. *Povolzh'e: spravochnik-putevoditel' po Volge, Kame, Oke . . . na 1933*, compiled by A.S. Insarov, G.G. Sitnikov and I.I. Fedenko, 2nd edition, Moscow, 1933, pp. 7, 73.

43. Ilf and Petrov (Il'ia Fainzil'berg and Evgenii Kataev), *The Twelve Chairs*, translated by John Richardson, London, 1965, pp. 220–21.

44. Ilic, *Life Stories of Soviet Women*, p. 40.

45. Grossman, *Stalingrad*, p. 144.

46. Riabov, ' "Mother Volga" and "Mother Russia" ', pp. 90–91. See also Oleg Riabov, ' "Let us Defend Mother Volga": The Material Symbol of the River in the Discourse of the Stalinist Battle', *Women in Russian Society*, no. 2, 2015, pp. 11–27.

47. Riabov, ' "Mother Volga and "Mother Russia" ', p. 87.

48. ibid., p. 89.

CHAPTER 16

1. Quoted in Sergei Kondrashov, *Nationalism and the Drive for Sovereignty in Tatarstan, 1988–92: Origins and Development*, London and Basingstoke, 2000, p. 64.

2. Allen Frank and Ronald Wixman, 'The Middle Volga: Exploring the Limits of Sovereignty', in Ian Bremmer and Ray Taras, eds, *New States, New Politics: Building the Post-Soviet Nations*, Cambridge, 1996, p. 147.

3. Tomila Lankina, *Governing the Locals: Local Self-Government and Ethnic Mobilization in Russia*, Lanham, MD, and Oxford, 2004, p. 53.

4. Alexandre Bennigsen and S. Enders Wimbush, *Muslims of the Soviet Empire: A Guide*, C. Hurst & Company, Bloomington, IN, 1986, pp. 225–26.

5. Hélène Carrière d'Encausse, 'Party and Federation in the USSR: The Problem of the Nationalities and Power in the USSR', *Government and Opposition*, vol. 13, no. 2, 1978, pp. 138–40, 145.

6. Marie Benningsen Broxup, 'Volga Tatars', in Graham Smith, ed., *The Nationalities Question in the Soviet Union*, London and New York, 1990, pp. 280–82.

7. Rorlich, *The Volga Tatars*, p. 155.

8. Koch, *The Volga Germans in Russia and the Americas*, p. 288.

9. Ann Sheehy, *The Crimean Tatars, Volga Germans and Meskhetians: Soviet Treatment of Some National Minorities*, Minority Rights Group, report no. 6, 1971, pp. 25–28.

10. Cholakhian, *Armiane na Saratovskoi zemle*, p. 98.

11. Quoted in Broxup, 'Volga Tatars', p. 284.

12. Jeffrey Kahn, *Federalism, Democratization, and the Rule of Law in Russia*, Oxford, 2002, p. 95.

13. Azade-Ayşe Rorlich, 'History, Collective Memory and Identity: The Tatars of Sovereign Tatarstan', *Communist and Post-Communist Studies*, vol. 32, no. 4, 1999, pp. 379–96.

14. Kahn, *Federalism, Democratization and the Rule of Law in Russia*, p. 153.

15. See, for example, Izmail Sharifzhanov, 'The Parliament of Tatarstan, 1990–2005: Vain Hopes, or the Russian Way towards Parliamentary Democracy in a Regional Dimension', *Parliaments, Estates and Representation*, vol. 27, no. 1, 2007, pp. 239–50; Christopher Williams, 'Tatar Nation Building since 1991: Ethnic Mobilisation in Historical Perspective', *Journal of Ethnopolitics and Minority Issues in Europe*, vol. 10, no. 1, 2011, pp. 94–123.

16. *Narody Povolzh'ia i Priural'ia: Komi-zyriane, Komi permiaki, Mariitsy, Mordva, Udmurty*, edited by N.F. Mokshin, T.F. Fedianovich and L.S. Khristoliubova, Moscow, 2000, p. 196.

17. Ivan Boiko, Iuri Markov and Valentina Kharitonova, 'The Chuvash Republic', *Anthropology & Archaeology of Eurasia*, vol. 44, no. 2, 2005, p. 41.

18. Cholakhian, *Armiane na Saratovskoi zemle*, pp. 97–98.

19. Frank and Wixman, 'The Middle Volga', p. 158.

20. Teresa Wigglesworth-Baker, 'Language Policy and Post-Soviet Identities in Tatarstan', *Nationalities Papers*, vol. 44, no. 1, 2016, p. 21.

21. Bennigsen and Wimbush, *Muslims of the Soviet Union*, p. 235.

22. Rorlich, *The Volga Tatars*, pp. 150–52.

23. Ben Fowkes, *The Disintegration of the Soviet Union: A Study in the Rise and Triumph of Nationalism*, London and Basingstoke, 1997, p. 67.

24. Quoted in Mark Sebba, *Ideology and Alphabets in the Former USSR*, Lancaster University Working papers, 2003, p. 4.

25. Wigglesworth-Baker, 'Language Policy and Post-Soviet Identities in Tatarstan', p. 23.

26. Anatoly M. Khazanov, *After the USSR: Ethnicity, Nationalism, and Politics in the Commonwealth of Independent States*, Madison, WI, 1995, p. 12.

27. A.S. Nikitin, 'Deiatel'nost' Cheboksarskogo gorodskogo soveta narodnykh deputatov po upravlenii sotsial'no-kul'turnym razvitiem goroda (1917–1980 gg.)', in *Voprosy istorii politicheskogo, ekonomicheskogo i sotsial'no-kul'turnogo razvitiia Chuvashskoi ASSR*, Cheboksary, 1983, pp. 87–108.

28. Max Seddon, 'Tatar Culture Feels Putin's Squeeze as Russian Election Nears', *Financial Times*, 25 January 2018.

29. Boiko et al., 'The Chuvash Republic', p. 52.

30. Koroleva et al., *Vlast' i musul'mane Srednego Povolzh'ia*, p. 57.

31. Raufa Urazmanova, 'Festive Culture of Tatars in Soviet Times', in *The History of the Tatars since Ancient Times*, vol. 7, Kazan', 2017, pp. 464–72; Raufa Urazmanova, 'The Transformation of the Tatar Holiday Culture in the Post Soviet Period', *The History of the Tatars since Ancient Times*, vol. 7, Kazan', 2017, pp. 688–96.

32. Koroleva et al., *Vlast' i musul'mane Srednego Povolzh'ia*, p. 51.

33. *Narody Povolzh'ia i Priural'ia*, p. 61.
34. Ruslan R. Ibragimov, 'Islam among the Tatars in the 1940s–1980s', in *The History of the Tatars since Ancient Times*, vol. 7, Kazan', 2017, p. 463.
35. Guzal Yusupova, 'The Islamic Representation of Tatarstan as an Answer to the Equalization of the Russian Regions', *Nationalities Papers*, vol. 44, no. 1, 2016, pp. 38–54.
36. Kimitaka Matsuzato, 'The Regional Context of Islam in Russia: Diversities along the Volga', *Eurasian Geography and Economics*, vol. 47, no. 4, 2006, pp. 452–54.
37. Renat Shaykhudinov, 'The Terrorist Attacks on the Volga Region, 2012–13: Hegemonic Narratives and Everyday Understandings of (In)Security', *Central Asian Survey*, vol. 37, no. 1, 2018, pp. 50–67.
38. Olessia B. Vovina, 'Building the Road to the Temple: Religion and National Revival in the Chuvash Republic', *Nationalities Papers*, vol. 28, no. 4, 2000, p. 695.
39. Durmuş Arik, 'Islam among the Chuvashes and its Role in the Change of Chuvash Ethnicity', *Journal of Muslim Minority Affairs*, vol. 27, no. 1, 2010, pp. 37–54.
40. D.M. Iskhakov, *Problemy stanovleniia i transformatsii Tatarskoi natsii*, Kazan', 1997, p. 93.
41. Urazmanova, 'The Transformation of the Tatar Holiday Culture', p. 693.
42. Quoted in Khalin et al., *Ocherki istorii volzhskogo rechnogo parokhodstva*, p. 77.
43. *An Anthology of Chuvash Poetry*, p. 156.
44. *Historical Anthology of Kazan Tatar Verse*, p. 156.

CHAPTER 17

1. Boris Pilnyak, *The Volga Flows to the Caspian Sea*, London, 1932, pp. 227–28.
2. *Rechnoi transport SSSR 1917-1957*, p. 37.
3. Dorothy Zeisler-Vralsted, 'The Aesthetics of the Volga and the National Narrative in Russia', *Environment and History*, vol. 20, 2014, p. 114.
4. Cynthia Ruder, 'Water and Power: The Moscow Canal and the "Port of Five Seas"', in Jane Costlow and Arja Rosenholm, eds, *Meanings and Values of Water in Russian Culture*, Abingdon, 2017, p. 174.
5. A. Voliani, *Elektrogigant na Volge*, Leningrad, 1934, p. 63.
6. This is discussed more fully in Evgenii Burdin, *Volzhskii kaskad GES: Triumf i tragediia Rossii*, Moscow, 2011.
7. See, for example, ibid., pp. 22, 58, 65, 66, 242–43. But this is also acknowledged in recent studies: *Bol'shaia Volga: iz istorii stroitel'stva Verkhnevolzhskikh GES*, Rybinsk, 2015, p. 64 and *Iaroslavskii krai*, p. 132.
8. Introduction to Konstantin Livanov, *Zapiski doktora (1926–1929)*, Rybinsk, 2017, p. 21.
9. The best overall study of environmental policies in the Soviet Union is P. Josephson et al., *An Environmental History of Russia*, Cambridge, 2013.
10. For a good comparative study of great river projects, see Zeisler-Vralsted, *Rivers, Memory and Nation-Building*, pp. 89–94.
11. See, for example, the account of the impact of pollution on the Rhine in Marc Cioc, *The Rhine: An Eco-Biography, 1815–2000*, Seattle, WA, and London, 2002, pp. 109, 148, 158.
12. Hartley, *Siberia*, pp. 227–28, 233–34.
13. Jeffrey Hays, 'Water Pollution in Russia', http://factsanddetails.com/russia/Nature_Science_Animals/sub9_8c/
14. Josephson et al., *An Environmental History of Russia*, p. 120.
15. Philip P. Micklin, 'International Environmental Implications of Soviet Development of the Volga River', *Human Ecology*, vol. 5, no. 2, 1977, p. 119.
16. Josephson et al., *An Environmental History of Russia*, p. 222.
17. Donald Filtzer, *The Hazards of Urban Life in Late Stalinist Russia: Health, Hygiene and Living Conditions 1943–53*, Cambridge, 2010, pp. 80–81.
18. Filtzer, *The Hazards of Urban Life in Late Stalinist Russia*, p. 88.
19. Josephson et al., *An Environmental History of Russia*, p. 199.
20. Burdin, *Volzhskii kaskad GES*, p. 223.
21. Josephson et al., *An Environmental History of Russia*, p. 171.

22. Micklin, 'International Environmental Implications of Soviet Development of the Volga River', pp. 113–35.
23. T.V. Usova, 'Composition of Sturgeon Fry Migrating from Spawning Areas in the Lower Volga', *Russian Journal of Ecology*, vol. 36, no. 4, 2005, pp. 288–90.
24. I.F. Galanin, A.N. Ananin, V.A. Kuznetsov and A.S. Sergeev, 'Changes in the Species Composition and Abundance of Young-of-the-Year Fishes in the Upper Volga Stretch of the Kuibyshev Reservoir during the Period of 1991 to 2009', *Russian Journal of Ecology*, vol. 45, no. 5, 2014, pp. 407–13.
25. P.V. Veshchev, G.I. Guteneva and R.S. Mukhanova, 'Efficiency of Natural Reproduction of Sturgeons in the Lower Volga under Current Conditions', *Russian Journal of Ecology*, vol. 43, no. 2, 2012, pp. 142–47.
26. Laura Henry, *Red to Green: Environmental Activism in Post-Soviet Russia*, Ithaca, NY, and London, 2010, p. 34.
27. N. Rogozhina, 'The Caspian: Oil Transit and Problems of Ecology', *Problems of Economic Transition*, vol. 53, no. 5, 2010, p. 90.
28. Friends of the Earth in Russia website: www.foeeurope.org/russia
29. V. Pope, 'Poisoning Russia's River of Plenty', *US News and World Report*, no. 112, 1992, p. 49.
30. Henry, *Red to Green*, pp. 234–36.
31. Quoted in Jo Crotty, 'Managing Civil Society: Democratisation and the Environmental Movement in a Russian Region', *Communist and Post-Communist Studies*, vol. 36, 2003, p. 498.
32. *Vesti*, 8 June 2019.

BIBLIOGRAPHY

ARCHIVAL SOURCES

Astrakhan', Russian Federation: Gosudarstvennyi arkhiv Astrakhanskoi oblasti (GAAO)
Fond 1, delo 743, Astrakhan' Provincial Soviet.

Iaroslavl', Russian Federation: Gosudarstvennyi arkhiv Iaroslavskoi oblasti (GAIO)
Fond 72, opis' 1, delo 122; opis' 2, delo 837, delo 1970, Chancellery of the Iaroslavl' Governor-General.
Fond 73, opis' 1, tom 2, delo 4399; tom 3, delo 6372, Chancellery of the Iaroslavl' Governor.

Kazan', Russian Federation: Natsional'nyi arkhiv Respubliki Tatarstan (NART)
Fond 1, opis' 1, delo 112, delo 1107; opis' 2, delo 294, delo 393, delo 399, delo 1107, delo 1231; opis' 3, delo 218, delo 856, delo 1632, delo 1834, delo 3068, delo 5196, delo 9400; opis' 6, delo 91, delo 612, Chancellery of the Kazan' Governor.
Fond 2, opis' 1, delo 1920, delo 2264; opis' 2, delo 63, delo 100, delo 755, delo 1966, delo 4432, Kazan' Provincial Office.
Fond 22, opis' 2; opis' 3, Kazan' Tatar ratusha.

Kew, United Kingdom: National Archives, United Kingdom (NA)
FO 371/71659, Foreign Office Northern/Soviet Union: 'The trip by G.M. Warr and P.J. Kelly to Stalingrad and Astrakhan in 1948'.
PC1/2673 Privy Council Papers, report on an epidemic in Vetlanka.

London, United Kingdom: British Library (BL)
Add. MS 47431, Baki estate papers, 1819–25.

St Petersburg, Russian Federation: Sankt Peterburgskii Institut istorii RAN (StPII).
Fond 36, delo 477, Report from Iaroslavl' 1778.

Samara, Russian Federation: Gosudarstvennyi arkhiv Samarskoi oblasti (GASO)
Fond 1, opis' 1, tom 1, delo 198, delo 528, delo 2060, delo 2201; opis' 1, tom 2, delo 5502; opis' 8, tom 1, delo 108, Samara Provincial Chancellery.

PRIMARY PUBLISHED SOURCES

1905 god v Tsaritsyne (vospominaniia i dokumenty), compiled by V.I. Tomarev and E.N. Shkodina, Stalingrad, 1960.

BIBLIOGRAPHY

Abbott, James. *Narrative of a Journey from Heraut to Khiva, Moscow and St Petersburgh, during the late Russian Invasion of Khiva*, London, vol. 2, 1884.

Aksakov, Sergei. *A Russian Schoolboy*, translated by J.D. Duff, Oxford, 1983.

Akunin, Boris. *Pelagia and the Red Rooster*, translated by Andrew Bromfield, London, 2008.

Andreev, N. *Illiustrirovannyi putevoditel' po Volge i eia pritokam Oke i Kame*, 2nd edition, Moscow, 1914.

Anthology of Chuvash Poetry, An, compiled by Gennady Aygi, translated by Peter France, London and Boston, 1991.

Armstrong, Terence, ed. *Yermak's Campaign in Siberia*, London, Haklyut Society, 1975.

Babine, Alexis. *A Russian Civil War Diary: Alexis Babine in Saratov, 1917–1922*, edited by Donald J. Raleigh, Durham, NC, and London, 1989.

Baedeker, Karl. *Karl Baedeker's Russia 1914*, London, 1971.

Bechhofer, C.E. *Through Starving Russia. Being the Record of a Journey to Moscow and the Volga Provinces in August and September 1921*, Westport, CT, 1921.

Bell, John. *Travels from St Petersburg in Russia, to Diverse Parts of Asia*, Glasgow, 1763.

Bolotov, A.B. *Zhizn' i prikliucheniia Andreia Bolotova, opisannye samim im dlia svoikh potomkov*, Moscow–Leningrad, 1931, reprint Cambridge, MA, vol. 3, 1973.

Bremner, Robert. *Excursions in the Interior of Russia*, London, vol. 2, 1839.

Bunin, Ivan. *Sunstroke: Selected Stories of Ivan Bunin*, translated by Graham Hettlinger, Chicago, IL, 2002.

Burton, Reginald G. *Tropics and Snow: A Record of Travel and Adventure*, London, 1898.

Butyrkin, V. 'Razskazy iz sluzhby na Volge', *Morskoi sbornik*, vol. 69, 1863, pp. 405–18, vol. 71, 1864, pp. 97–124.

Cairns, Andrew. *The Soviet Famine 1932–33: An Eye Witness Account of Conditions in the Spring and Summer of 1932*, Edmonton, Canadian Institute of Ukrainian Studies, 1989.

Chamberlain, Lesley. *Volga, Volga: A Voyage down the Great River*, London and Basingstoke, 1995.

Chekhov, Anton. 'The Grasshopper', from *The Wife and Other Stories*, translated by Constance Garnett, New York, 1918.

Chekhov, Anton. *Letters of Anton Chekhov*, selected and edited by Avrahm Yarmolinsky, New York, 1973.

Chirikov, Evgenii. *Marka of the Pits,* translated by L. Zarine, London, n.d. [1930?].

Chirikov, Evgenii. *Otchii dom: semeinaia khronika*, Moscow, 2010.

Clarke, Edward Daniel. *Travels in Various Countries of Europe, Asia and Africa*, vol. 1, *Russia Tartary and Turkey*, London, 1810.

Collins, Samuel. *The Present State of Russia*, London, 1671.

Cook, John. *Voyages and Travels through the Russian Empire, Tartary, and Part of the Kingdom of Persia*, Edinburgh, vol. 1, 1770.

Coxe, William. *Travels into Poland, Russia, Sweden and Denmark*, Dublin, vol. 1, 1784.

Cross, A.G., ed. *An English Lady at the Court of Catherine the Great*, Cambridge, 1989.

Custine, A.-L.-L. de, *The Empire of the Czar*, 2nd edition, London, vol. 3, 1844.

Davis, John P. 'The Struggle with Cholera in Tsarist Russia and the Soviet Union, 1892–1927', PhD dissertation, University of Kentucky, 2012.

Dem'ianov, G.P. *Putevoditel' po Volge ot Tveri do Astrakhani*, 6th edition, Nizhnii Novgorod, 1900.

Fenomen 19-oi. Vospominaniia i razmyshleniia, edited by V.V. Rozen and O.V. Shimel'fenig, Saratov, 2010.

Forster, George. *A Journey from Bengal to England . . . and into Russia by the Caspian Sea*, London, vol. 2, 1808.

Forsyth, William. *The Great Fair of Nijni Novgorod and How We Got There*, London, 1865.

Georgi, Johann Gottlieb. *Russia or a Compleat Historical Account of all the Nations which Compose that Empire*, translated by William Tooke, London, 4 vols, 1780–83.

Gerth, Nathan M. 'A Model Town: Tver', the Classical Imperial Order, and the Rise of Civic Society in the Russian Provinces, 1763–1861', PhD dissertation, University of Notre Dame, 2014.

Gheith, Jehanne M. and Katherine R. Jolluck, eds. *Gulag Voices: Oral Histories of Soviet Incarceration and Exile*, New York, 2011.

Ginzburg, Eugenia Semyonovna. *Journey into the Whirlwind*, translated by Paul Stevenson and Max Hayward, New York and London, 1967.

BIBLIOGRAPHY

Ginzburg, Eugenia Semyonovna. *Within the Whirlwind*, translated by Ian Boland, San Diego, London and New York, 1981.

Glen, William. *Journal of a Tour from Astrachan to Karass*, Edinburgh, 1823.

Golod v SSSR 1930–1934 gg., Moscow, 2009.

Goncharov, Ivan. *Oblomov*, translated by David Magarshack, Harmondsworth, 1954.

Gorky, Maksim. *My Childhood*, translated by Ronald Wilks, London, 1966.

Gorky, Maksim. *Selected Short Stories*, New Delhi, 2000.

Gornostaev, Andrey. 'Eighteenth-Century Chichikovs and Purchasing Runaway Souls', paper given at the Study Group on Eighteenth-Century Russia 10th International Conference in Strasbourg, July 2018.

Grossman, Vasily. *Life and Fate*, translated by Robert Chandler, London, 2006.

Grossman, Vasily. *Stalingrad*, translated by Robert and Elizabeth Chandler, London, 2019.

Guber (a Lutheran pastor). 'Dnevnik Pastora Gubera', *Russkaia starina*, vol. 22, no. 8, 1878, pp. 581–90.

Guthrie, Katherine Blanche. *Through Russia: From St Petersburg to Astrakhan and the Crimea*, London, 1874.

Hakluyt, Richard. *The Principal Navigations, Voyages, Traffiques and Discoveries of the English Nation*, edited by Jack Beeching, London, 1972.

Hanway, Jonas, *An Historical Account of the British Trade over the Caspian Sea*, Dublin, vols 1–2, 1754.

Haxthausen, August von. *The Russian Empire: Its People, Institutions, and Resources*, translated by Robert Fairie, London, vol. 1, 1856.

Herberstein, Sigismund von. *Notes upon Russia*, translated by R.H. Major, London, vol. 2, 1852.

Historical Anthology of Kazan Tatar Verse: Voices of Eternity, compiled and translated by David J. Matthews and Ravil Bukharev, Richmond, 2000.

Holderness, Mary. *New Russia: Journey from Riga to the Crimea, by the Way of Kiev*, London, 1823.

Ibn Fadlān and the Land of Darkness: Arab Travellers in the Far North, translated by Paul Lunde and Caroline Stone, London, 2012.

Ilf and Petrov (Il'ia Fainzil'berg and Evgenii Kataev). *The Twelve Chairs*, translated by John Richardson, London, 1965.

Ilic, Melanie. *Life Stories of Soviet Women: The Interwar Generation*, London and New York, 2013.

Istoriia gubernskogo goroda Iaroslavlia, compiled by A.M. Rutman, Iaroslavl', 2006.

Istoriia Kazani v dokumentakh i materialakh: XX vek, Kazan', 2004.

Istoriia Kazani v dokumentakh i materialakh: XIX vek, Kazan', 2005.

Istoriia Saratovskogo kraia 1590–1917, Saratov, 1964.

Istoriia Saratovskogo kraia 1917–1965, Saratov, 1967.

Kak eto bylo: dokumental'nyi sbornik, compiled by V.A. Ugriumov, Nizhnii Novgorod, 2011.

Khudozhniki brat'ia Grigorii i Nikanor Chernetsovy: grecheskii mir v russkom iskusstve, St Petersburg, 2000.

Kissel', F.Kh. *Istoriia goroda Uglicha*, reprint of 1844 edition, Uglich, 1994.

Kollektivizatsiia sel'skogo khoziaistva v srednem Povolzh'e (1927–1937), edited by N.N. Panov and F.A. Karevskii, Kuibyshev, 1970.

Konovalov, S. 'Ludvig Fabritius's Account of the Razin Rebellion', *Oxford Slavonic Papers*, vol. 6, 1955, pp. 72–101.

Kooperativno-kolkhoznoe stroitel'stvo v Nizhegorodskoi gubernii (1917–1927): dokumenty i materialy, Gor'kii, 1980.

Koroloff, Rachel. 'Seeds of Exchange: Collecting for Russia's Apothecary and Botanical Gardens in the Seventeenth and Eighteenth Centuries', PhD dissertation, University of Illinois at Urbana-Champaign, 2014.

Kreshchenie Tatar (sbornik dokumentov), Kazan', 2002.

Krest'ianskaia voina 1773–1775 gg. v Rossii, Moscow, 1973.

Krest'ianskoe dvizhenie v Simbirskoi gubernii v period revoliutsii 1905–1907 godov: dokumenty i materialy, Ul'ianovsk, 1955.

Kul'tura Simbirskogo-Ul'ianovskogo kraia: sbornik dokumentov i materialov, Ul'ianovsk, 2014.

Livanov, Konstantin. *Zapiski doktora (1926–1929)*, Rybinsk, 2017.

Margeret, Jacques. *The Russian Empire and Grand Duchy of Muscovy: A 17th-Century French Account*, translated and edited by Chester S.L. Dunning, Pittsburgh, PA, 1983.

BIBLIOGRAPHY

Munro-Butler-Johnstone, H.A. *A Trip up the Volga to the Fair of Nijni-Novgorod*, Oxford and London, 1875.

Narodnoe sudostroenie v Rossii: entsiklopedicheskii slovar' sudov narodnoi postroiki, edited by P.A. Filin and S.P. Kurnoskin, St Petersburg, 2016.

Naumov, Aleksandr Nikolaevich. *Iz utselevshikh vospominanii, 1868–1917*, 2 vols, New York, 1954.

Nikitin, Afanasii, *Khozhenie za tri moria Afanasiia Nikitina*, Tver', 2003.

Oktiabr' v Povolzh'e, Saratov, 1967.

Olearius, Adam. *The Travels of Olearius in Seventeenth-Century Russia*, translated and edited by Samuel H. Baron, Stanford, CA, 1967.

Oliphant, Laurence. *The Russian Shores of the Black Sea in the Autumn of 1852, with a Voyage down the Volga and a Tour through the Country of the Don Cossacks*, Edinburgh and London, 1854.

'Opis' 1000 del kazach'iago otdela', *Trudy Orenburgskoi uchenoi arkhivnoi komissii*, Orenburg, vol. 24, 1913.

Ostrovsky, Aleksandr. *Four Plays*, translated by Stephen Mulrine, London, 1997.

Pallas, P.S. *Travels through the Southern Provinces of the Russian Empire in the Years 1793 and 1794*, London, vol. 1, 1802.

Parkinson, John. *A Tour of Russia, Siberia and the Crimea 1792–1794*, London, 1971.

Perrie, Maureen. *The Image of Ivan the Terrible in Russian Folklore*, Cambridge, 1987.

Perry, John. *The State of Russia under the Present Czar*, reprint of the 1716 edition, London, 1967.

Petrov-Vodkin, K.S. *Moia povest' (Khlynovsk)*, St Petersburg, 2013.

Pilnyak, Boris. *The Volga Flows to the Caspian Sea*, London, 1932.

Poety 1790–1810-kh godov, Leningrad, 1971.

Polnoe sobranie zakonov rossiiskoi imperii, first and second series, 1649–1881, St Petersburg, 1830, 1881.

Povolzh'e: spravochnik-putevoditel' po Volge, Kame, Oke . . . na 1933, compiled by A.S. Insarov, G.G. Sitnikov and I.I. Fedenko, 2nd edition, Moscow, 1933.

Provintsial v Velikoi Rossiiskoi Revoliutsii. Sbornik dokumentov. Simbirskaia guberniia v ianvare 1917 – marte 1918 gg., edited by N.V. Lipatova, Ul'ianovsk, 2017.

Pushkin, Alexander. *The Complete Prose Tales of Alexander Sergeyevitch Pushkin*, translated by G.R. Aitken, London, 2008.

Raikov, A.F. *Voina glazami ochevidtsa: vospominaniia o Velikoi Otechestvennoi Voine*, Tver', 2010.

Repin, Il'ia. *Dalekoe blizkoe i vospominaniia*, Moscow, 2002.

Revoliutsionnoe dvizhenie v Astrakhani i Astrakhanskoi gubernii v 1905–1907 godakh: sbornik dokumentov i materialov, Astrakhan', 1957.

Revoliutsionnoe dvizhenie v Chuvashii v pervoi russkoi revoliutsii 1905–1907; dokumenty i materialy, Cheboksary, 1965.

Romaniello, Matthew P. 'Absolutism and Empire. Governance of Russia's Early Modern Frontier', PhD dissertation, Ohio State University, 2003.

Roth, Henry Ling. *A Sketch of the Agriculture and Peasantry of Eastern Russia*, London, 1878.

Rybushkin, M. *Zapiski ob Astrakhani*, 3rd edition, Astrakhan', 2008.

Samara-Kuibyshev. Khronika sobytii 1886–1986, Kuibyshev, 1985.

Sbornik Imperatorskogo Russkogo istoricheskogo obshchestva, St Petersburg, vol. 10, 1872, vol. 19, 1876, vol. 115, 1903.

Scott, Charles. *The Baltic, the Black Sea and the Crimea*, London, 1854.

Sebba, Mark. *Ideology and Alphabets in the Former USSR*, Lancaster University Working papers, 2003.

Semenov, Nikolai. *O chem ne uspel*, Saratov, 2016.

Sheehy, Ann. *The Crimean Tatars, Volga Germans and Meskhetians: Soviet Treatment of Some National Minorities*, Minority Rights Group, report no. 6, 1971.

Simbirskaia guberniia v gody grazhdanskoi voiny (mai 1918 g. – mart 1919 g.): sbornik dokumentov, Ul'ianovsk, vol. 1, 1958.

Spilman, James. *A Journey through Russia by Two Gentlemen who Went in the Year 1739*, London, 1742.

Spottiswoode, William. *Tarantasse Journey through Eastern Russia in the Autumn of 1856*, London, 1857.

[T.C.] *The New Atlas: or, Travels and Voyages in Europe, Asia, Africa and America*, London, 1698.

BIBLIOGRAPHY

Tagirova, N.F. 'Khlebnyi rynok Povolzh'ia vo vtoroi polovine XIX – nachale XX vv.', dissertation, Samara State University, 1999.

Tolmachev, N.A. *Putevye zametki N.A. Tolmacheva o zhizni i byte krest'ian Kazanskoi gubernii v seredine XIX v. Sbornik dokumentov i materialov*, compiled by Kh.Z. Bagautdinova, Kazan', 2019.

Tolstoy, Alexey. *Selected Stories*, Moscow, 1949.

Tolstoy, Alexey. *Collected Works*, vol. 6, *Ordeal*, Moscow, 1982.

Travels of the Ambassadors from the Duke of Holstein, The, London, 1669.

Treasures of Catherine the Great, Catalogue of an Exhibition at Somerset House, London, 2000.

Turnerelli, Edward Tracy. *Russia on the Borders of Asia: Kazan, Ancient Capital of the Tatar Khans*, 2 vols, London, 1854.

Volneniia urzhumskikh Mariitsev v 1889 godu, compiled by L. Shemier et al., Ioshkar-Ola, 2017.

Watrous, Stephen D., ed. *John Ledyard's Journey through Russia and Siberia 1787–1788*, Madison, WI, and London, 1966.

Weber, Friedrich Christian. *The Present State of Russia*, reprint, London, vol. 1, 1968.

Weber, Jacob [Veber, Iakov]. *Iakov Iakovlevich Veber; Katalog k 125-letiu so dnia rozhdeniia. Zhivopis'*, Saratov, 1995.

Wilmot, M. and C. Wilmot. *The Russian Journals of Martha and Catherine Wilmot 1803–1808*, London, 1934.

Yakhina, Guzel, *Zuleikha: A Novel*, translated by Lisa C. Hayden, London, 2015.

Yakhina, Guzel. *Deti moi: roman*, printed by Amazon, 2018.

Zemtsova, Oxana. 'Russification and Educational Policies in the Middle Volga Region (1860–1914)', PhD dissertation, European University of Florence, 2014.

'Zhaloba Saratovskikh krest'ian na zemskii sud', *Russkii arkhiv*, vol. 46, 1908, pp. 215–16.

'Zolotoi vek' Iaroslavlia: Opyt kul'turografii russkogo goroda XVII – pervoi treti XVIII veka, compiled by V.V. Gorshkova et al., Iaroslavl', 2004.

SECONDARY SOURCES

Abdullin, Yahya G. 'Islam in the History of the Volga Kama Bulgars and Tatars', *Central Asian Survey*, vol. 9, no. 2, 1990, pp. 1–11.

Alabin, P.V. *Samara: 1586–1886 gody*, compiled by P.S. Kabytov, Samara, 1992.

Aleksandrova, S.A. and T.I. Vedernikova. *Sel'skie dvorianskie usad'by Samarskogo zavolzh'ia v XIX – XX vv.*, Samara, 2015.

Alekseev, V.V. *Zemlia Borskaia: vekhi istorii*, Samara, 2016.

Alexander, John T. *Autocratic Politics in a National Crisis: The Imperial Russian Government and Pugachev's Revolt, 1773–1775*, Bloomington, IN, 1969.

Alexander, John T. *Emperor of the Cossacks: Pugachev and the Frontier Jacquerie of 1773–1775*, Lawrence, KS, 1973.

Alexander, John T. *Bubonic Plague in Early Modern Russia: Public Health and Urban Disorder*, Oxford, 2003.

Alishev, S.Kh. *Istoricheskie sud'by narodov Srednego Povolzh'ia XVI – nachalo XIX v.* Moscow, 1990.

Alishev, S.Kh. *Bolgaro-kazanskie i zolotoordynskie otnosheniia v XII–XVI vv.*, Kazan', 2009.

Ananian, Zh.A. 'Armianskoe kupechestvo v Rossii', in *Kupechestvo v Rossii XV – pervaia polovina XIX veka*, Moscow, 1997, pp. 232–61.

Andrusz, Gregory D. *Housing and Urban Development in the USSR*, London, 1984.

Appelbaum, Anne. *Gulag: A History*, London, 2003.

Arik, Durmuş. 'Islam among the Chuvashes and its Role in the Change of Chuvash Ethnicity', *Journal of Muslim Minority Affairs*, vol. 27, no. 1, 2010, pp. 37–54.

Artamonov, M.I. *Istoriia khazar*, Leningrad, 1962.

Ascher, Abraham. *The Revolution of 1905*, vol. 1, *Russia in Disarray*, Stanford, CA, 2004.

Attwood, Lynne. *Gender and Housing Space in Soviet Russia: Private Life in a Public Space*, Manchester, 2010.

Avrich, Paul. *Russian Rebels 1600–1800*, New York, 1972.

Badcock, Sarah. 'From Saviour to Pariah: A Study of the Role of Karl Ianovich Grasis in Cheboksary during 1917', *Revolutionary Russia*, vol. 15, no. 1, 2002, pp. 69–96.

BIBLIOGRAPHY

Badcock, Sarah. *Politics and the People in Revolutionary Russia: A Provincial History*, Cambridge, 2007.

Bariev, R.Kh. *Volzhskie Bulgary: Istoriia i kul'tura*, St Petersburg, 2005.

Bartlett, Roger. *Human Capital: The Settlement of Foreigners in Russia 1762–1804*, Cambridge, 1979.

Bashkirov, A.S. *Pamiatniki Bulgaro-Tatarskoi kul'tury na Volge*, Kazan', 1928.

Baumann, R.F. 'Subject Nationalities in the Military Service of Imperial Russia: The Case of the Bashkirs', *Slavic Review*, vol. 46, 1987, pp. 489–502.

Bazhanov, E.A. *Sviashchennye reki Rossii*, Samara, 2008.

Becker, Seymour. *Nobility and Privilege in Late Imperial Russia*, DeKalb, IL, 1985.

Beevor, Antony. *Stalingrad*, London, 1999.

Beliakov, A.V. 'Sluzhilye Tatary XV – XVI vv.', in *Bitva na Vozhe – predtecha vozrozhdeniia sredneve-kovoi Rusi; sbornik nauchnykh statei*, Riazan', 2004, pp. 81–86.

Beliakov, Aleksandr and Stanislav Smirnov. *Lishentsy Nizhegorodskogo kraia (1918–1936 gg.)*, Nizhnii Novgorod, 2018.

Bellamy, Chris. *Absolute War: Soviet Russia in the Second World War*, London, 2009.

Bennigsen, Alexandre and S. Enders Wimbush. *Muslims of the Soviet Empire: A Guide*, C. Hurst & Company, Bloomington, IN, 1986.

Berk, Stephen M. 'The Democratic Counterrevolution: Komuch and the Civil War on the Volga', *Russian History*, vol. 7, 1980, pp. 176–90.

Beskrovnyi, L.G. *Russkaia armiia i flot v XVIII veke (ocherki)*, Moscow, 1958.

Beskrovnyi, L.G. *The Russian Army and Fleet in the Nineteenth Century*, Gulf Breeze, FL, 1996.

Bikeikin, E.N. *Agrarnaia modernizatsiia i razvitie sel'skogo khoziaistva Srednego Povolzh'ia: 1953–1991 gg. (na materialakh Mariiskoi, Mordovskoi i Chuvashskoi ASSR)*, Saransk, 2017.

Biktasheva, A.N. *Kazanskie gubernatory v dialogakh vlastei (pervaia polovina XIX veka)*, Kazan', 2008.

Biran, Michel. 'The Mongol Empire and Inter-Civilization Exchange', in Benjamin Z. Kedar, ed., *The Cambridge World History*, vol. 5, *Expanding Webs of Exchange and Conflict, 500 CE to 1500 CE*, Cambridge, 2015, pp. 534–58.

Bohac, Rodney D. 'The Mir and the Military Draft', *Slavic Review*, vol. 47, no. 4, 1988, pp. 652–66.

Boiko, Ivan, Iuri Markov and Valentina Kharitonova. 'The Chuvash Republic', *Anthropology & Archaeology of Eurasia*, vol. 44, no. 2, 2005, pp. 41–60.

Bol'shaia Volga: iz istorii stroitel'stva Verkhnevolzhskikh GES, Rybinsk, 2015.

Borisov, P.G. *Kalmyki v nizov'iakh Volgi*, Moscow, 1917.

Bowlt, John E. 'Russian Painting in the Nineteenth Century', in Theofanis G. Stavrou, ed., *Art and Culture in Nineteenth-Century Russia*, Bloomington, IN, 1983, pp. 113–39.

Braslavskii, L.Iu. 'Raskol Russkoi pravoslavnoi tserkvi i ego posledstviia v istorii narodov Srednego Povolzh'ia', in *Istoriia khristianizatsii narodov Srednego Povolzh'ia: kriticheskie suzhdenie i otsenka: mezhvuzovskii sbornik nauchnykh trudov*, Cheboksary, 1988, pp. 34–42.

Broxup, Marie Benningsen. 'Volga Tatars', in Graham Smith, ed., *The Nationalities Question in the Soviet Union*, London and New York, 1990, pp. 277–90.

Brunson, Molly. 'Wandering Greeks: How Repin Discovers the People', *Ab Imperio*, vol. 2, 2012, pp. 83–105.

Burdin, Evgenii. *Volzhskii kaskad GES: Triumf i tragediia Rossii*, Moscow, 2011.

Burdina, L.V. 'Staroobriadchestvo v Kostromskom krae', in *Staroobriadchestvo: istoriia, kul'tura, sovremennost'. Materialy*, Moscow, 2000, pp. 216–22.

Burmistrova, L.P. 'Publichnye lektsii v Kazanskom Universitete (30 – 60-e gody XIX v.)', in *Stranitsy istorii goroda Kazani*, Kazan', 1981, pp. 29–42.

Bushkanets, E.G. *Iunost' L'va Tolstogo: kazanskie gody*, Kazan', 2008.

Bushuev, I.A. 'Pogrom v provintsii: o sobytiiakh 19 oktiabria 1905 g. v Kostrome', in *Gosudarstvo, obshchestvo, tserkov' v istorii Rossii XX veka*, part 2, Ivanovo, 2014, pp. 398–402.

Busygin, E.P. *Russkoe naselenie Srednego Povolzh'ia*, Kazan', 1966.

Busygin, E.P., N.V. Zorin and E.V. Mikhailichenko. *Obshchestvennyi i semeinyi byt russkogo sel'skogo naseleniia Srednego Povolzh'ia*, Kazan', 1973.

Cavender, Mary W. *Nests of Gentry: Family, Estate, and Local Loyalties in Provincial Russia*, Newark, DE, 2007.

Cherniak, E.V. and A.B. Madiiarov. *Gorodskoe samoupravlenie v Kazani 1870–1892 gg.*, Kazan', 2003.

BIBLIOGRAPHY

Cholakhian,V.A. *Sotsial'no-demograficheskie posledstviia industrial'nogo razvitiia Nizhnego Povolzh'ia (konets XIX v. – 1930-e gg.)*, Saratov, 2008.

Cholakhian, V.A. *Armiane na Saratovskoi zemle*, Saratov, 2018.

Cioc, Marc. *The Rhine: An Eco-Biography, 1815–2000*, Seattle, WA, and London, 2002.

Conquest, Robert. *The Harvest of Sorrow: Soviet Collectivisation and the Terror-Famine*, London, 1986.

Cross, S.H. 'The Scandinavian Infiltration into Early Russia', *Speculum*, vol. 21, no. 4, 1946, pp. 505–14.

Crotty, Jo. 'Managing Civil Society: Democratisation and the Environmental Movement in a Russian Region', *Communist and Post-Communist Studies*, vol. 36, 2003, pp. 489–508.

Crummey, Robert O. *The Formation of Muscovy 1304–1613*, London and New York, 1987.

Cunliffe, Barry. *By Steppe, Desert and Ocean: The Birth of Eurasia*, Oxford, 2015.

Curta, Florin. 'Markets in Tenth-Century al-Andalus and Volga Bulghāria: Contrasting Views of Trade in Muslim Europe', *Al-Masaq*, vol. 25, no. 3, 2013, pp. 305–30.

Dale, Robert. 'Divided We Stand: Cities, Social Unity and Post-War Reconstruction in Soviet Russia, 1945–1953', *Contemporary European History*, vol. 24, no. 4, 2015, pp. 493–516.

Davletshin, G.M. *Volzhskaia Bulgariia: dukhovnaia kul'tura*, Kazan', 1990.

Dehaan, Heather D. *Stalinist City Planning: Professionals, Performance, and Power*, Toronto, Buffalo, NY, and London, 2013.

Demchenko, A.A. *Literaturnaia i obshchestvennaia zhizn' Saratova (iz arkhivnykh razyskanii)*, Saratov, 2008.

Denisov, V.V. *Istoriia monastyrei Verkhnego Povolzh'ia vtoraia polovina XVIII – nachalo XX vv.*, Iaroslavl', 2012.

Denningkhaus, V. 'Russkie nemtsy i obshchestvennye nastroeniia v Povolzh'e v period Pervoi Mirovoi Voiny', in *Voenno-istoricheskie issledovaniia v Povolzh'e*, Saratov University, vol. 7, 2006, pp. 171–80.

Dennison, Tracy K. 'Serfdom and Household Structure in Central Russia: Voshchazhnikovo, 1816–1858', *Continuity and Change*, vol. 18, no. 3, 2003, pp. 395–429.

Derechinskaia, E.L. 'Istoriia Nizhegorodskoi evreiskoi obshchiny v kontekste istorii evreev Rossii', in *Proshloe i nastoiashchee evreiskikh obshchin Povolzh'ia i Tsentral'noi Rossii*, Nizhnii Novgorod, 2011, pp. 59–69.

Dickinson, Sara. *Breaking Ground: Travel and National Culture in Russia from Peter I to the Era of Revolution*, New York, 2003.

Dimitriev, V.D. *Chuvashskii narod v sostave Kazanskogo khanstva: predystoriia i istoriia*, Cheboksary, 2014.

Dimitriev, V.D. and S.A. Selivanova. *Cheboksary: ocherki istorii goroda XVIII veka*, Cheboksary, 2011.

Dimnik, Martin. 'Kievan Rus', the Bulgars and the Southern Slavs', in D. Luscombe and J. Riley-Smith, eds, *The New Cambridge Medieval History*, vol. 4, *c. 1024 – c. 1198*, Cambridge, 2004, pp. 254–76.

Dixon, Simon. 'The "Mad Monk" Iliador in Tsaritsyn', *The Slavonic and East European Review*, vol. 88, nos 1–2, 2010, pp. 377–415.

Dubman, Eduard. *Khoziaistvennoe osvoenie Srednego Povolzh'ia v XVI veka. Po materialam tserkov-no-monastyrskikh vladenii*, Kuibyshev, 1991.

Dubman, Eduard. *Promyslovoe predprinimatel'stvo i osvoenie Ponizovogo Povolzh'ia v kontse XVI – XVII vv.*, Samara, 1999.

Dubman, E.L. and P.S. Kabytov, eds, *Povolzh'e – 'vnutrenniaia okraina' Rossii: gosudarstvo i obshchestvo v osvoenii novykh territorii (konets XVI – nachalo XX vv.)*, Samara, 2007.

Dubman, E.L., P.S. Kabytov and O.B. Leont'ev. *Istoriia Samary (1586–1917 gg.)*, Samara, 2015.

Dubov, I.V. *Velikii Volzhskii put'*, Leningrad, 1989.

Dubov, I.V. 'Velikii Volzhskii put'' v istorii drevnei Rusi', in *Mezhdunarodnye sviazi, torgovye puti i goroda Srednego Povolzh'ia IX – XII vekov: materialy mezhdunarodnogo simpoziuma Kazan', 8–19 sentiabria 1998 g.*, Kazan', 1999, pp. 88–93.

Dunlop, Douglas. *The History of the Jewish Khazars*, Princeton, NJ, 1954.

Dutov, N.V. *Iaroslavl': istoriia i toponimika ulits i ploshchadei goroda*, 2nd edition, Iaroslavl', 2015.

Dzhundzhuzov, S.V. *Kalmyki v Srednem Povolzh'e i na iuzhnom Urale*, Orenburg, 2014.

El'chaninova, O.Iu. *Sel'skoe naselenie Srednego Povolzh'ia v period reform 1953–1964 gg.*, Samara, 2006.

BIBLIOGRAPHY

Ely, Christopher. *This Meager Nature: Landscape and National Identity in Imperial Russia*, DeKalb, IL, 2002.

Ely, Christopher. 'The Origins of Russian Scenery: Volga River Tourism and Russian Landscape Aesthetics', *Slavic Review*, vol. 62, no. 4, 2003, pp. 666–82.

d'Encausse, Hélène Carrière. 'Party and Federation in the USSR: The Problem of the Nationalities and Power in the USSR', *Government and Opposition*, vol. 13, no. 2, 1978, pp. 133–50.

Engelstein, Laura. *Russia in Flux: War, Revolution, Civil War 1914–21*, Oxford, 2018.

Ermolaeva, L.K. 'Krupnoe kupechestvo Rossii v XVII – pervoi chetverti XVIII v. (po materialam astrakhanskoi torgovli)', *Istoricheskie zapiski*, vol. 114, 1986, pp. 302–25.

Evdokimov, Ivan. *Levitan: povest'*, Moscow, 1958.

Evtuhov, Catherine. *Portrait of a Province: Economy, Society and Civilization in Nineteenth-Century Nizhnii Novgorod*, Pittsburgh, PA, 2011.

Fakhrutdinov, R.G., *Ocherki po istorii Volzhskoi Bulgarii*, Moscow, 1984.

Fakhrutdinov, R.G. 'Zolotaia Orda i ee rol' v istorii Tatarskogo naroda', in *Iz istorii Zolotoi Ordy*, Kazan', 1993, pp. 6–17.

Fedorov-Davydov, G.A. 'Money and Currency', in *Great Bolgar*, Kazan', 2015, pp. 114–23.

Fennell, John. *The Crisis of Medieval Russia 1200–1304*, London and New York, 1983.

Figes, Orlando. *Peasant Russia, Civil War: The Volga Countryside in Revolution 1917–1921*, London, 1989.

Filatov, N.F. *Goroda i posady Nizhegorodskogo Povolzh'ia v XVIII veke*, Gor'kii, 1989.

Filtzer, Donald. *The Hazards of Urban Life in Late Stalinist Russia: Health, Hygiene and Living Conditions 1943–53*, Cambridge, 2010.

Fitzpatrick, Ann Lincoln. *The Great Russian Fair: Nizhnii Novgorod 1846–90*, London and New York, 1990.

Fitzpatrick, Sheila. *Stalin's Peasants: Resistance and Survival in the Russian Village after Collectivization*, Oxford and New York, 1994.

Fowkes, Ben. *The Disintegration of the Soviet Union: A Study in the Rise and Triumph of Nationalism*, London and Basingstoke, 1997.

Frame, Murray. *School for Citizens: Theatre and Civil Society in Imperial Russia*, New Haven, CT, and London, 2006.

Frank, Allen J. 'Russia and the Peoples of the Volga-Ural Region: 1600–1850', in Nicola Di Cosmo, Allen J. Frank and Peter B. Golden, eds, *The Cambridge History of Inner Asia: The Chinggisid Age*, Cambridge, 2009, pp. 380–91.

Frank, Allen J. 'The Western Steppe in Volga-Ural Region, Siberia and the Crimea', in Nicola Di Cosmo, Allen J. Frank and Peter B. Golden, eds, *The Cambridge History of Inner Asia: The Chinggisid Age*, Cambridge, 2009, pp. 237–59.

Frank, Allen and Ronald Wixman. 'The Middle Volga: Exploring the Limits of Sovereignty', in Ian Bremmer and Ray Taras, eds, *New States, New Politics: Building the Post-Soviet Nations*, Cambridge, 1996, pp. 140–82.

Franklin, Simon. 'Rus'', in David Abulafia, ed., *The New Cambridge Medieval History*, vol. 5, *c. 1198–1300*, Cambridge, 1999, pp. 796–808.

Franklin, Simon. 'Kievan Rus' (1015–1125)', in Maureen Perrie, ed., *The Cambridge History of Russia*, vol. 1, *From Early Rus' to 1698*, Cambridge, 2006, pp. 73–97.

Franklin, Simon and Jonathan Shepard. *The Emergence of Rus 750–1200*, London and New York, 1996.

French, R.A. 'Canals in Pre-Revolutionary Russia', in J.H. Bater and R.A. French, eds, *Studies in Russian Historical Geography*, vol. 2, London, 1983, pp. 451–81.

Fyodorov-Davydov [Fedorov-Davydov], G.A. *The Culture of the Golden Horde Cities*, translated by H. Bartlett Wells, Oxford, 1984.

Gabdrafikova, L.R. and Kh.M. Abdulin. *Tatary v gody Pervoi Mirovoi Voiny (1914–1918 gg.)*, Kazan', 2015.

Galanin, I.F., A.N. Ananin, V.A. Kuznetsov and A.S. Sergeev. 'Changes in the Species Composition and Abundance of Young-of-the-Year Fishes in the Upper Volga Stretch of the Kuibyshev Reservoir during the Period of 1991 to 2009', *Russian Journal of Ecology*, vol. 45, no. 5, 2014, pp. 407–13.

Garipova, Z.G. *Kazan': obshchestvo, politika, kul'tura*, Kazan', 2004.

BIBLIOGRAPHY

German, A.A. *Bol'shevistskaia vlast' i nemetskaia avtonomiia na Volge (1918–1941)*, Saratov, 2004.

Gibadullina, E.M. 'Osobennosti mechetestroitel'stva v Samarskoi gubernii vo vtoroi polovine XIX – nachale XX vv.', in *Islamskaia tsivilizatsiia v Volgo-Ural'skom regione: doklady*, Ufa, 2008, pp. 67–73.

Golden, Peter B. *Nomads and their Neighbours in the Russian Steppe: Turks, Khazars and Qipchaqs*, Aldershot, 2003.

Golden, Peter B. 'Aspects of the Nomadic Factor in the Economic Development of Kievan Rus'', in Peter B. Golden, *Nomads and their Neighbours in the Russian Steppe: Turks, Khazars and Qipchaqs*, Aldershot, 2003, pp. 58–101.

Golden, Peter B. 'The Conversion of the Khazars to Judaism', in Peter B. Golden, Haggai Ben-Shammai and András Róna-Tas, eds, *The World of the Khazars: New Perspectives*, Leiden and Boston, MA, 2007, pp. 123–62.

Golden, Peter B., Haggai Ben-Shammai and András Róna-Tas, eds. *The World of the Khazars: New Perspectives*, Leiden and Boston, MA, 2007.

Goldman, Wendy Z. *Women, the State and the Revolution: Soviet Family Policy and Social Life 1917–36*, Cambridge, 1993.

Golikova, N.B. 'Iz istorii formirovaniia kadrov naemnykh rabotnikov v pervoi chetverti XVIII v.', *Istoriia SSSR*, 1965, no. 1, pp. 75–98.

Golikova, N.B. *Naemnyi trud v gorodakh Povolzh'ia v pervoi chetverti XVIII veka*, Moscow, 1965.

Golikova, N.B. 'Torgovlia krepostnymi bez zemli v 20-kh godakh XVIII v. (po materialam krepost-nykh knig gorodov Povolzh'ia)', *Istoricheskie zapiski*, vol. 90, 1972, pp. 303–31.

Golikova, N.B. *Astrakhanskoe vosstanie 1705–1706*, Moscow, 1975.

Golovkin, Anatolii. *Zhernova. Kniga pamiati Tverskikh Karel*, Tver', 2017.

Gorevoi, G., T. Kirillova and A. Shitokov. 'Byvshie dvorianskie usad'by po beregam rek Tudovki i Volgi', in *Po Volge pod flagom 'Tverskoi zhizni'. Sbornik statei*, Staritsa, 2008.

Gorod Tol'iatti, Kuibyshev, 1967.

Goroda nashei oblasti: geografiia, istoriia, ekonomika, naselenie, kul'tura, Gor'kii, 1974.

Gorokhov, V.S. *Po zakonam kolkhoznoi zhizni*, Saratov, 1979.

Grachev, S.V. *Geopolitika i prosveshchennie nerusskikh narodov Povolzh'ia (60 gg. XIX – nachalo XX v.)*, Saransk, 2000.

Gray, Rosalind P. *Russian Genre Painting in the Nineteenth Century*, Oxford, 2000.

Great Bolgar, Kazan', 2015.

Gregory, Serge. *Antosha and Levitasha: The Shared Lives and Art of Anton Chekhov and Isaac Levitan*, DeKalb, IL, 2015.

Grekov, B.D. and Iakubovskii, A.Iu. *Zolotaia Orda i ee padenie*, Moscow and Leningrad, 1950.

Gubaidullin, A.M. *Fortifikatsiia gorodishch Volzhskoi Bulgarii*, Kazan', 2002.

Gur'ianov, E. *Drevnie vekhi Samary*, Kuibyshev, 1986.

Halperin, Charles J. *Russia and the Golden Horde: The Mongol Impact on Medieval Russian History*, Bloomington, IN, 1987.

Halperin, Charles J. 'Omissions of National Memory: Russian Historiography on the Golden Horde as Politics of Inclusion and Exclusion', *Ab Imperio*, vol. 3, 2004, pp. 131–44.

Hamamoto [Khamamoto], Mami. 'Sviazuiushchaia rol' Tatarskikh kuptsov Volgo-Ural'skogo regiona v tsentral'noi evrazii: zveno "shelkovogo puti novogo vremeni"', in N. Norikhiro et al., eds, *Volgo-Ural'skii region v imperskom prostranstve XVIII – XX vv.*, Moscow, 2011, pp. 39–57.

Hart, James G. 'From Frontier Outpost to Provincial Capital: Saratov, 1590–1860', in Rex A. Wade and Scott J. Seregny, *Politics and Society in Provincial Russia: Saratov, 1590–1917*, Columbus, OH, 1989, pp. 10–27.

Hartley, Janet M. 'Russia in 1812, Part 1: The French Presence in the *Gubernii* of Smolensk and Mogilev', *Jahrbücher für Geschichte Osteuropas*, vol. 38, 1990, pp. 178–98.

Hartley, Janet M. *Alexander I*, London and New York, 1994.

Hartley, Janet M. *A Social History of the Russian Empire 1650–1825*, London and New York, 1999.

Hartley, Janet M. 'Bribery and Justice in the Provinces in the Reign of Catherine II', in Stephen Lovell, Alena Ledeneva and Andrei Rogachevskii, eds, *Bribery and Blat in Russia: Negotiating Reciprocity from the Middle Ages to the 1990s*, Basingstoke, 2000, pp. 48–64.

BIBLIOGRAPHY

Hartley, Janet M. *Russia 1762–1825: Military Power, the State and the People*, Westport, CT, and London, 2008.

Hartley, Janet M. *Siberia: A History of the People*, London and New York, 2014.

Hartley Janet M. 'Education and the East: The Omsk Asiatic School', in Maria Di Salvo, Daniel H. Kaiser and Valerie A. Kivelson, eds, *Word and Image in Russian History: Essays in Honor of Gary Marker*, Boston, MA, 2015, pp. 253–68.

Hausmann, Guido. 'The Volga Source: Sacralization of a Place of Memory', in *Velikii volzhskii put'*, Kazan', 2001, pp. 340–46.

Hausmann, Guido. *Mütterchen Wolga. Ein Fluss als Erinnerungsort vom 16. bis ins frühe 20. Jahrhundert*, Frankfurt and New York, 2009.

Hays, Jeffrey. 'Water Pollution in Russia', http://factsanddetails.com/russia/Nature_Science_Animals/sub9_8c/

Haywood, Richard M. 'The Development of Steamboats on the Volga and its Tributaries, 1817–1865', *Research in Economic History*, vol. 6, 1981, pp. 127–92.

Heilbronner, Hans. 'The Russian Plague of 1878–79', *Slavic Review*, vol. 21, no. 1, 1962, pp. 89–112.

Henry, Laura, *Red to Green: Environmental Activism in Post-Soviet Russia*, Ithaca, NY, and London, 2010.

Henze, Charlotte. *Disease, Health Care and Government in Late Imperial Russia: Life and Death on the Volga, 1823–1914*, London and New York, 2011.

History of the Tatars since Ancient Times, The, Academy of Sciences of the Republic of Tatarstan, 7 vols, Kazan', 2017.

Hosking, Geoffrey. *Russia: People and Empire, 1552–1917*, London, 1998.

Huttenbach, Henry R. 'Muscovy's Conquest of Muslim Kazan and Astrakhan', 1552–56', in Michael Rywkin, ed., *Russian Colonial Expansion to 1917*, London, 1988, pp. 45–69.

Ialtaev, I.F. *Derevnia Mariiskoi avtonomnoi oblasti v gody kollektivizatsiia (1929–1936)*, Ioshkar-Ola, 2015.

Iaroslavskii krai: istoriia, traditsii, liudi, Iaroslavl', 2017.

Ibneeva, G.B. 'Puteshestvie Ekateriny II po Volge v 1767 godu: uznavanie imperiii', *Ab Imperio*, vol. 2, 2000, pp. 87–104.

Ibneeva, G.B. *Imperskaia politika Ekateriny II v zerkale ventsenosnykh puteshestvii*, Moscow, 2009.

Ibragimov, Ruslan R. *Vlast' i religiia v Tatarstane v 1940–1980-e gg.*, Kazan', 2005.

Ibragimov, Ruslan R., 'Islam among the Tatars in the 1940s–1980s', in *The History of the Tatars since Ancient Times*, vol. 7, Kazan', 2017, pp. 455–64.

Il'in, B. *Saratov: istoricheskii ocherk*, Saratov, 1952.

Iskhakov, D.M. *Problemy stanovleniia i transformatsii Tatarskoi natsii*, Kazan', 1997.

Iskhakov, D.M. 'Kriasheny (istoriko-etnograficheskii ocherk)', in *Tatarskaia natsiia: istoriia i sovremennost'*, Kazan', 2002, pp. 108–24.

Istomina, E.G. *Vodnye puti Rossii: vo vtoroi polovine XVIII – nachale XIX veka*, Moscow, 1982.

Istomina. E.G. *Vodnyi transport Rossii v doreformennyi period*, Moscow, 1991.

Istoriia Chuvashskoi ASSR, Cheboksary, vol. 1, 1966.

Istoriia Tatarskoi ASSR, Kazan', vol. 1, 1955, vol. 2, 1960.

Iukht, A.I. 'Torgovye sviazi Astrakhani v 20-kh godakh XVIII v.', in *Istoricheskaia geografiia Rossii XVII – nachalo XX v.*, Moscow, 1975, pp. 177–92.

Iukht, A.I. *Torgovlia s vostochnymi stranami i vnutrennii rynok Rossii (20–60-e gody XVIII veka)*, Moscow, 1994.

Ivanov, A.G. *Mariitsy Povolzh'ia i Priural'ia*, Ioshkar-Ola, 1993.

Ivanov, A.G. *Istochniki po istorii i kul'ture narodov Volgo-Viatskogo regiona (XVIII – nachalo XX vv.)*, Ioshkar-Ola, 2013.

Ivanov, V.P. 'Vliianie migratsii iz sela na demograficheskie i etnicheskie kharakteristiki gorodskikh semei', in *Sel'skoe khoziaistvo i krest'ianstvo Srednego Povolzh'ia v usloviiakh razvitogo sotsializma*, Cheboksary, 1982, pp. 100–24.

Izmaylov, Iskander and Murat Kaveyev. 'Bulgar City in the 13–15th Centuries', in *The History of the Tatars since Ancient Times*, vol. 3, Kazan', 2017, pp. 207–13.

Izmaylov, Iskander and Yuri Zeleneev. 'The Lower Volga Region and Capitals', in *The History of the Tatars since Ancient Times*, vol. 3, Kazan', 2017, pp. 228–38.

BIBLIOGRAPHY

Jansson, Ingmar. '"Oriental Import" into Scandinavia in the 8th–12th Centuries and the Role of Volga Bulgaria', in *Mezhdunarodnye sviazi, torgovye puti i goroda Srednego Povolzh'ia IX – XII vekov: materialy mezhdunarodnogo simpoziuma Kazan', 8–19 sentiabria 1998 g.*, Kazan', 1999, pp. 116–22.

Jones, Robert E. 'Catherine II and the Provincial Reform of 1775: A Question of Motivation', *Canadian Slavic Studies*, vol. 4, no. 3, 1970, pp. 497–512.

Jones, Robert E. 'Urban Planning and the Development of Provincial Towns in Russia during the Reign of Catherine II', in J.G. Garrard, ed., *The Eighteenth Century in Russia*, Oxford, 1973, pp. 321–44.

Jones, Robert E. *Provincial Development in Russia: Catherine II and Jacob Sievers*, New Brunswick, NJ, 1984.

Jones, Robert E. *Bread upon the Waters: The St Petersburg Grain Trade and the Russian Economy, 1703–1811*, Pittsburgh, PA, 2013.

Josephson, P. et al. *An Environmental History of Russia*, Cambridge, 2013.

Kabirova, Ayslu. 'Political Repression in the TASSR in the 1930s', in *The History of the Tatars since Ancient Times*, vol. 7, Kazan', 2017, pp. 366–70.

Kabuzan, V.M. *Narody Rossii v XVIII veke. Chislennost' i etnicheskii sostav*, Moscow, 1990.

Kahn, Jeffrey. *Federalism, Democratization, and the Rule of Law in Russia*, Oxford, 2002.

Kalinina, E.E. 'Evrei Udmurtii', in *Material'naia i dukhovnaia kul'tura narodov Urala i Povolzh'ia: istoriia i sovremennost'*, Glazov, 2016, pp. 288–90.

Kaplunovskii, Aleksandr. 'Tatary, musul'mane i russkie v meshchanskikh obshchinakh Srednego Povolzh'ia v kontse XIX – nachale XX veka', *Ab Imperio*, vol. 1, 2000, pp. 101–22.

Kappeler, A. *The Russian Empire: A Multiethnic History*, translated by Alfred Clayton, Harlow, 2001.

Karabushchenko, P.L. *Astrakhanskaia guberniia i ee gubernatory v svete kul'turno-istoricheskikh traditsii XVIII – XIX stoletii*, Astrakhan', 2011.

Kataev, I.M. *Na beregakh Volgi i istorii usol'skoi votchiny grafov Orlovykh*, Cheliabinsk, 1948.

Kazanskii Universitet 1804–1979: Ocherk istorii, Kazan', 1975.

Kazarin, V. *Volzhskie stranitsy Samarskoi istorii*, Samara, 2011.

Kefeli, A.N. *Becoming a Muslim in Imperial Russia: Conversion, Apostasy, and Literacy*, Ithaca, NY, 2014.

Kenez, Peter. *Civil War in South Russia, 1918*, Berkeley, CA, and London, vol. 1, 1971, vol. 2, 1977.

Khakimzyanov, F.S. and I.I. Izmailov. 'Language and Writing in Bolgar Town', in *Great Bolgar*, Kazan', 2015, pp. 300–11.

Khalin, A.A. *Istoriia volzhskogo rechnogo parokhodstva (seredina XIX – nachalo XX v.)*, Nizhnii Novgorod, 2017.

Khalin, A.A. et al. *Ocherki istorii volzhskogo rechnogo parokhodstva v XX – nachale XXI v.*, Nizhnii Novgorod, 2018.

Khasanov, Kh.Kh. 'Iz istorii formirovaniia Tatarskoi natsii', in *Tatariia v proshlom i nastoiashchem: sbornik statei*, Kazan', 1975, pp. 186–96.

Khayrutdinov, Ramil R. 'The System of the State Village Government of the Kazan Governorate in the Early 18th – the First Third of the 19th Centuries', *Journal of Sustainable Development*, vol. 8, no. 5, 2015, pp. 1–11.

Khazanov, Anatoly M. *After the USSR: Ethnicity, Nationalism, and Politics in the Commonwealth of Independent States*, Madison, WI, 1995.

Khlebnikov, A.V. *Razvitie sovetskoi avtonomii Mariiskogo naroda 1929–1936*, Ioshkar-Ola, 1976.

Khlebnikova, T.A. 'Tanning', in *Great Bolgar*, Kazan', 2015, pp. 168–71.

Khlopina, L.F. *Memorial nad Volgoi*, Ul'ianovsk, 2010.

Khodarkovsky, Michael. *Where Two Worlds Met: The Russian State and the Kalmyk Nomads, 1660–1771*, Ithaca, NY, and London, 1992.

Khodarkovsky, Michael. 'The Stepan Razin Uprising: Was it a "Peasant War"?', *Jahrbücher für Geschichte Osteuropas*, vol. 42, no.1, 1994, pp. 1–19.

Khodarkovsky, Michael. 'Taming the "Wild Steppe": Muscovy's Southern Frontier', *Russian History*, vol. 26, no. 3, 1999, pp. 241–97.

Khodarkovsky, Michael. 'The Conversion of Non-Christians in Early Modern Russia', in R.P. Geraci and M. Khodarkovsky, eds, *Of Religion and Empire: Missions, Conversion, and Tolerance in Tsarist Russia*, Ithaca, NY, and London, 2001, pp. 115–43.

BIBLIOGRAPHY

Khodarkovsky, Michael. *Russia's Steppe Frontier: The Making of a Colonial Empire, 1500–1800*, Bloomington, IN, 2002.

Khramkov, A.V. *Trudiashchiesia Kuibyshevskoi oblasti v gody Velikoi Otechestvennoi Voiny 1941–1945*, Kuibyshev, 1985.

Khuzin, Fayaz. 'Great City on the Cheremshan and the Town of Bolgar on the Volga', in *The History of the Tatars since Ancient Times*, vol. 2, Kazan', 2017, pp. 163–79.

Kirokos'ian, M.A. *Piraty Kaspiiskogo Moria*, Astrakhan', 2007.

Kirokos'ian, M.A. *Astrakhanskii kupets G.V. Tetiushinov*, Astrakhan', 2014.

Kirpichnikov, Anatoly. 'The Great Volga Route: Trade Relations with Northern Europe and the East', in *The History of the Tatars since Ancient Times*, vol. 2, Kazan', 2017, pp. 298–315.

Kobzov, V.S. *Voenno-administrativnaia struktura Orenburgskogo kazach'ego voiska v XVII – pervaia polovina XIX veka*, Cheliabinsk, 1996.

Koch, F.C. *The Volga Germans in Russia and the Americas from 1763 to the Present*, University Park, PA, and London, 1977.

Kochkina, Anna. 'Prichernomorsko-sredizemnomorskie sviazi Volzhskoi Bulgarii v X – nachale XIII vv. (arkheologicheskie dannye o torgovykh putiakh)', in *Mezhdunarodnye sviazi, torgovye puti i goroda Srednego Povolzh'ia IX – XII vekov: materialy mezhdunarodnogo simpoziuma Kazan', 8–19 sentiabria 1998 g.*, Kazan', 1999, pp. 132–38.

Kogan, E.S. *Ocherki istorii krepostnogo khoziaistva po materialam votchin Kurakinykh 2 –i poloviny XVIII veka*, Moscow, 1960.

Kollman, Nancy Shields. *The Russian Empire 1450–1801*, Oxford, 2017.

Kondrashov, Sergei. *Nationalism and the Drive for Sovereignty in Tatarstan, 1988–92: Origins and Development*, London and Basingstoke, 2000.

Koroleva, L.E., A.I. Lomovtsev and A.A. Korolev. *Vlast' i musul'mane Srednego Povolzh'ia (vtoraia polovina 1940-kh – pervaia polovina 1980-kh gg.*, Penza, 2001.

Kotilaine, Jarmo T. *Russia's Foreign Trade and Economic Expansion in the Seventeenth Century: Windows on the World*, Leiden and Boston, MA, 2005.

Kotliarov, D.A. *Moskovskaia Rus' i narody Povolzh'ia v XV – XVI vekakh*, Izhevsk, 2005.

Koval'chenko, I.D. *Russkoe krepostnoe krest'ianstvo v pervoi polovine XIX v.* Moscow, 1967.

Koznova, Irina. *Stalinskaia epokha v pamiati krest'ianstva Rossii*, Moscow, 2016.

Kugrysheva, Evelina. *Istoriia armian v Astrakhani*, Astrakhan', 2007.

Kulik, V.N. 'Uchastie Tveritianok v kul'turnoi zhizni gubernii v pervoi polovine XIX veka', in V. Uspenskaia and V. Kulik, eds, *Zhenshchiny v sotsial'noi istorii Tveri*, Tver', 2006.

Kurmacheva, M.D. *Goroda Urala i Povolzh'ia v krest'ianskoi voine 1773–1775 gg.*, Moscow, 1991.

Kuz'min, E.P. 'Povsednevnyi byt Tsarevokokshaiskikh voevod XVIII veka', in *Goroda Srednego Povolzh'ia: istoriia i sovremennost': sbornik statei*, Ioshkar-Ola, 2014, pp. 41–46.

Kuznetsov, I.A. *Krest'ianstvo Chuvashii v period kapitalizma*, Cheboksary, 1963.

Kuznetsov, I.A. *Ocherki po istorii Chuvashskogo krest'ianstva*, Cheboksary, vol. 1, 1957, vol. 2, 1969.

Lankina, Tomila. *Governing the Locals: Local Self-Government and Ethnic Mobilization in Russia*, Lanham, MD, and Oxford, 2004.

Lazzarini, Edward J. 'Ethnicity and the Uses of History: The Case of the Volga Tatars and Jadidism', *Central Asian Survey*, vol. 1, nos 2–3, 1982, pp. 61–69.

LeDonne, John P. *The Grand Strategy of the Russian Empire 1650–1831*, Oxford, 2004.

Lehtinen, Ildikó. *From the Volga to Siberia: The Finno-Ugric Peoples in Today's Russia*, Helsinki, 2002.

Lepin, K.M. *Vodnyi transport SSSR za 15 let*, Moscow, 1932.

Lieven, Dominic. *The Aristocracy in Europe 1815–1914*, Basingstoke and London, 1992.

Lieven, Dominic. *Empire: The Russian Empire and its Rivals*, London, 2000.

Lisovskaia, S. 'Istoriia obrazovaniia, formirovaniia i razvitiia evreiskoi obshchiny Iaroslavlia', in *Proshloe i nastoiashchee evreiskikh obshchin Povolzh'ia i Tsentral'noi Rossii*, Nizhnii Novgorod, 2011, pp. 37–42.

Logan, F. Donald. *The Vikings in History*, 3rd edition, New York and London, 2005.

Long, James. 'Agricultural Conditions in the German Colonies of Novouzensk District, Samara Province, 1861–1914', *The Slavonic and East European Review*, vol. 57, no. 4, 1979, pp. 531–51.

Long, James. *From Privileged to Dispossessed: The Volga Germans 1860–1917*, Lincoln, NE, and London, 1988.

BIBLIOGRAPHY

Long, James. 'The Volga Germans and the Famine of 1921', *Russian Review*, vol. 51, no. 4, 1992, pp. 510–25.

Longworth, Philip. *The Cossacks*, London, 1971.

Longworth, Philip. 'The Pretender Phenomenon in Eighteenth-Century Russia', *Past and Present*, vol. 66, 1975, pp. 61–83.

Macey, David. 'Reflections on Peasant Adaptation in Rural Russia of the Beginning of the Twentieth Century: The Stolypin Agrarian Reforms', *Journal of Peasant Studies*, vol. 31, nos 3–4, 2004, pp. 400–26.

McGrew, Roderick E. *Russia and the Cholera 1823–1832*, Madison, WI, 1965.

Madariaga, Isabel de. *Russia in the Age of Catherine the Great*, London, 1981.

Madariaga, Isabel de. *Ivan the Terrible*, New Haven, CT, and London, 2005.

Makhmutova, A.Kh. 'Shkola L'iabiby Khusianovoi v Kazani', in *Stranitsy istorii goroda Kazani*, Kazan', 1981, pp. 55–65.

Mamaeva, S.V. *Promyshlennost' Nizhnego Povolzh'ia v period voennogo kommunizma (1918 – vesna 1921 g.)*, Astrakhan', 2007.

Manning, Roberta. *The Crisis of the Old Order in Russia: Gentry and Government*, Princeton, NJ, 1982.

Marasanova, V.M. *Mestnoe upravlenie v rossiiskoi imperii (na materialakh Verkhnego Povolzh'ia)*, Moscow, 2004.

Mariiskaia ASSR za gody sovetskoi vlasti, Ioshkar-Ola, 1957.

Marsden, Thomas. *The Crisis of Religious Toleration in Imperial Russia: Bibikov's System for the Old Believers, 1841–1855*, Oxford, 2015.

Martin, Janet. 'The Land of Darkness and the Golden Horde: The Fur Trade under the Mongols XIII – XIVth Centuries', *Cahiers du Monde Russe et Soviétique*, vol. 19, no. 4, 1978, pp. 401–21.

Martin, Janet. 'Trade on the Volga: The Commercial Relations of Bulgar with Central Asia and Iran in the 11th–12th Centuries', *International Journal of Turkish Studies*, vol. 1, no. 2, 1980, pp. 85–98.

Martin, Janet. 'Muscovite Travelling Merchants: The Trade with the Muslim East (15th and 16th Centuries)', *Central Asian Survey*, vol. 4, no. 3, 1985, pp. 21–38.

Martin, Janet. *Treasures of the Land of Darkness: The Fur Trade and its Significance for Medieval Russia*, Cambridge, 1986.

Martin, Janet. *Medieval Russia 980–1584*, Cambridge, 1995.

Martin, Janet. 'Multiethnicity in Muscovy: A Consideration of Christian and Muslim Tatars in the 1550s–1560s', *Journal of Early Modern History*, vol. 5, no. 1, 2001, pp. 1–23.

Martin, Janet. 'The Emergence of Moscow (1359–1462)', in Maureen Perrie, ed., *The Cambridge History of Russia*, vol. 1, *From Early Rus' to 1698*, Cambridge, 2006, pp. 158–87.

Martin, Janet. 'North-Eastern Russia and the Golden Horde', in Maureen Perrie, ed., *The Cambridge History of Russia*, vol. 1, *From Early Rus' to 1689*, Cambridge, 2006, pp. 127–57.

Matsuzato, Kimitaka. 'The Regional Context of Islam in Russia: Diversities along the Volga', *Eurasian Geography and Economics*, vol. 47, no. 4, 2006, pp. 452–54.

Medvedev, E.I. *Grazhdanskaia voina v Srednem Povolzh'e (1918–1919)*, Saratov, 1974.

Mel'nikova, Elena. 'Baltiisko-Volzhskii put' v rannei istorii Vostochnoi Evropy', in *Mezhdunarodnye sviazi, torgovye puti i goroda Srednego Povolzh'ia IX – XII vekov: materialy mezhdunarodnogo simpoziuma Kazan', 8–19 sentiabria 1998 g.*, Kazan', 1999, pp. 80–87.

Melton, E. 'Enlightened Seigniorialism and its Dilemmas in Serf Russia, 1750–1830', *Journal of Modern History*, vol. 62, no. 4, 1990, pp. 675–708.

Merridale, Catherine. *Ivan's War: The Red Army 1939–45*, London, 2005.

Micklin, Philip P. 'International Environmental Implications of Soviet Development of the Volga River', *Human Ecology*, vol. 5, no. 2, 1977, pp. 113–35.

Mikhailova, S.M. *Kazanskii Universitet i prosveshchenie narodov Povolzh'ia i Priural'ia (XIX vek)*, Kazan', 1979.

Mironov, Boris and Brian A'Hearn. 'Russian Living Standards under the Tsars: Anthropometric Evidence from the Volga', *Journal of Economic History*, vol. 68, no. 3, 2008, pp. 900–09.

Mixter, Timothy. 'Of Grandfather-Beaters and Fat-Heeled Pacifists: Perceptions of Agricultural Labor and Hiring Market Disturbances in Saratov, 1872–1905', *Russian History*, vol. 7, nos 1–2, 1980, pp. 139–68.

BIBLIOGRAPHY

Mixter, Timothy. 'Peasant Collective Action in Saratov Province, 1902–1906', in Rex A. Wade and Scott J. Seregny, *Politics and Society in Provincial Russia: Saratov, 1590–1917*, Columbus, OH, 1989, pp. 191–232.

Moon, David. 'The Russian Academy of Sciences Expeditions to the Steppes in the Late Eighteenth Century', in *Personality and Place in Russian Culture: Essays in Memory of Lindsey Hughes*, edited by Simon Dixon, London, 2010, pp. 204–25.

Moon, David. *The Plough that Broke the Steppes: Agriculture and Environment on Russia's Grasslands, 1700–1914*, Oxford, 2013.

Morokhin, A.V. 'Prikhodskoe dukhovenstvo i staroobriadchestvo v Nizhegorodskom Povolzh'e v pervoi polovine XVIII v.', in *Staroobriadchestvo: istoriia, kul'tura, sovremennost'. Materialy*, Moscow, 2000, pp. 67–72.

Mosse, W.E. 'Revolution in Saratov (October – November 1910)', *The Slavonic and East European Review*, vol. 49, 1971, pp. 586–602.

Muckeston, Keith W. 'The Volga in the Prerevolutionary Industrialization of Russia', in *Yearbook of the Association of Pacific Coast Geographers*, 1, January 1965, pp. 67–77.

Naganawa, Norihiro. 'Holidays in Kazan: The Public Sphere and the Politics of Religious Authority among Tatars in 1914', *Slavic Review*, vol. 71, no. 1, 2012, pp. 25–48.

Narody Povolzh'ia i Priural'ia: Komi-zyriane, Komi permiaki, Mariitsy, Mordva, Udmurty, edited by N.F. Mokshin, T.F. Fedianovich and L.S. Khristoliubova, Moscow, 2000.

Nedashkovskii, Leonid. 'Mezhdunarodnaia i vnutrenniaia torgovlia', in *Zolotaia Orda v mirovoi istorii*, Kazan', 2016, pp. 608–13.

Nersisian, M.G., ed. *Dekabristy ob Armenii i Zakavkaz'e*, Erevan, 1985.

Nikitin, A.S. 'Deiatel'nost' Cheboksarskogo gorodskogo soveta narodnykh deputatov po upravlenii sotsial'no-kul'turnym razvitiem goroda (1917–1980 gg.)', in *Voprosy istorii politicheskogo, ekonomicheskogo i sotsial'no-kul'turnogo razvitiia Chuvashskoi ASSR*, Cheboksary, 1983, pp. 87–108.

Nikolaev, Gennady. 'Ethnocultural Interaction of the Chuvash and Tatars', in *The History of the Tatars since Ancient Times*, vol. 6, Kazan', 2017, pp. 623–32.

Nikolaev, Gennady. 'The World of a Multiconfessional Village', in *The History of the Tatars since Ancient Times*, vol. 6, Kazan', 2017, pp. 608–16.

Noack, Christian. 'State Policy and the Impact on the Formation of a Muslim Identity in the Volga-Urals', in *Islam in Politics in Russia and Central Asia (Early Eighteenth to Late Twentieth Centuries)*, edited by S.A. Dudoignon and Komatsu Hisau, London, New York and Bahrain, 2001, pp. 3–24.

Noack, Christian. 'The Western Steppe: The Volga-Ural Region, Siberia and the Crimea under Russian Rule', in Nicola Di Cosmo, Allen J. Frank and Peter B. Golden, eds, *The Cambridge History of Inner Asia: The Chinggisid Age*, Cambridge, 2009, pp. 303–30.

Noonan, Thomas. 'Suzdalia's Eastern Trade in the Century before the Mongol Conquest', *Cahiers du Monde Russe et Soviétique*, vol. 19, no. 4, 1978, pp. 371–84.

Noonan, Thomas. 'Monetary Circulation in Early Medieval Rus': A Study of Volga Bulgar Dirham Finds', *Russian History*, vol. 7, no. 3, 1980, pp. 294–311.

Noonan, Thomas. 'Why Dirhams First Reached Russia: The Role of Arab-Khazar Relations in the Development of the Earliest Islamic Trade with Eastern Europe', *Archivum Eurasiae Medii Aevi*, vol. 4, 1984, pp. 151–282.

Noonan, Thomas. *The Islamic World, Russia and the Vikings, 750–900: The Numismatic Evidence*, Aldershot, 1998.

Noonan, Thomas. 'European Russia, c. 500 – c. 1050', in Timothy Reuter, ed., *The New Cambridge Medieval History*, vol. 3, c. 900–1024, Cambridge, 2000, pp. 487–513.

Noonan, Thomas. 'Some Observations on the Economy of the Khazar Khaganate', in Peter B. Golden, Haggai Ben-Shammai and András Róna-Tas, eds, *The World of the Khazars: New Perspectives*, Leiden and Boston, MA, 2007, pp. 207–44.

Normanov, Aider. *Tatary Srednego Povolzh'ia i Priural'ia v rossiiskom zakondatel'stve vtoroi poloviny XVI – XVIII vv.*, Kazan', 2002.

Novosel'tsev, A.P. *Khazarskoe gosudartsvo i ego rol' v istorii vostochnoi Evropy i Kavkaza*, Moscow, 1990.

Obshchestvenno-politicheskoe dvizhenie i klassovaia bor'ba na Srednei Volge (konets XIX – nachalo XX veka), Kazan', 1972.

Ocherki istorii Mariiskoi ASSR, Ioshkar-Ola, 1965.

BIBLIOGRAPHY

Okenfuss, Max J. 'The Jesuit Origins of Petrine Education', in J.G. Garrard, *The Eighteenth Century in Russia*, Oxford, 1973, pp. 106–30.

O'Meara, Patrick. *The Russian Nobility in the Age of Alexander I*, London, 2019.

Ostrowski, Donald. *Muscovy and the Mongols: Cross-Cultural Influences on the Steppe Frontier, 1304–1589*, Cambridge, 1998.

Ostrowski, Donald. 'The Growth of Muscovy (1462–1533)', in Maureen Perrie, ed., *The Cambridge History of Russia*, vol. 1, *From Early Rus' to 1698*, Cambridge, 2006, pp. 213–39.

Overy, Richard. *Russia's War*, London, 1997.

Pallot, Judith. 'Agrarian Modernisation on Peasant Farms in the Era of Capitalism', in J.H. Bater and R.A. French, eds, *Studies in Russian Historical Geography*, vol. 2, London, 1983, pp. 423–49.

Pallot, Judith. *Land Reform in Russia, 1906–17: Peasant Responses to Stolypin's Project of Rural Transformation*, Oxford, 1999.

Palmer, Scott W. 'How Memory was Made: The Construction of the Memorial to the Heroes of the Battle of Stalingrad', *Russian Review*, vol. 68, no. 3, 2009, pp. 373–407.

Papmehl, K.A. *Metropolitan Platon of Moscow (Peter Levshin), 1737–1812: The Enlightened Prelate, Scholar and Educator*, Newtonville, MA, 1983.

Pelenski, Jaroslaw. *Russia and Kazan: Conquest and Imperial Ideology (1438–1560s)*, The Hague and Paris, 1974.

Peretiatkovich, G. *Povolzh'e v XVII i nachale XVIII veka (ocherki iz istorii kolonizatsii kraia)*, Odessa, 1882.

Petrukhin, Vladimir. 'Khazaria and Rus': An Examination of their Historical Relations', in Peter B. Golden, Haggai Ben-Shammai and András Róna-Tas, eds, *The World of the Khazars: New Perspectives*, Leiden and Boston, MA, 2007, pp. 245–68.

Phillips, Hugh. '"A Bad Business" – The February Revolution in Tver", *The Soviet and Post-Soviet Review*, vol. 23, no. 2, 1996, pp. 123–41.

Phillips, Hugh. 'Riots, Strikes and Soviets: The City of Tver in 1905', *Revolutionary Russia*, vol. 17, no. 2, 2004, pp. 49–66.

Pleshakov, I.N. 'Gardkouty v Saratovskom Povolzh'e: iz istorii rechnoi strazhi XVIII – pervoi poloviny XIX vv.', in *Voenno-istoricheskie issledovaniia v Povolzh'e*, Saratov, vol. 7, n.d., pp. 20–27.

Pliutsinskii, S.S. 'Pereselennye nemtsy na territorii Astrakhanskoi gubernii v gody Pervoi Mirovoi Voiny (1914–1918 gg.)', in *Istoricheskaia i sovremennaia regionalistika Verkhovnogo Dona i Nizhnego Povolzh'ia: sbornik nauchnykh statei*, Volgograd, 2005, pp. 218–31.

Polianskii, F.Ia. *Gorodskoe remeslo i manufaktura v Rossii XVIII v.*, Moscow, 1960.

Poluboiarinova, M.D. *Rus' i Volzhskaia Bolgaria v X – XV vv.*, Moscow, 1993.

Poluboyarinova [Poluboiarinova], M.D. 'Bolgar Trade', in *Great Bolgar*, Kazan', 2015, pp. 100–13.

Poluboyarinova [Poluboiarinova], M.D. 'Glasswear', in *Great Bolgar*, Kazan', 2015, pp. 160–71.

Polyakova [Poliakova], G.F. 'Non-Ferrous and Precious Metal Articles', in *Great Bolgar*, Kazan', 2015, pp. 132–37.

Potapova, E.V. 'Vlast' i staroobriadtsy: iz istorii staroobriadcheskoi obshchiny goroda Rzheva v pervoi polovine XIX v.', in *Staroobriadchestvo v Tverskom krae: proshloe i nastoiashchee*, Tver' and Rzhev, 2007, pp. 45–50.

Poten'kin, P.M. *Krest'ianskie volneniia v Saratovskoi gubernii v 1861–1863 gg.*, Saratov, 1940.

Prawdin, Michael. *The Mongol Empire: Its Rise and Legacy*, translated by Eden and Cedar Paul, London, 1940.

Primako, D.D. 'Staroobriadcheskaia obshchina goroda Rzheva v XIX v.', in *Staroobriadchestvo v Tverskom krae: proshloe i nastoiashchee*, Tver' and Rzhev, 2007, pp. 51–64.

Prokhorov, M.F. 'Otkhodnichestvo krest'ian v gorodakh Verkhnego Povolzh'ia v seredine XVIII veka', *Russkii gorod*, Moscow, vol. 9, 1990, pp. 144–64.

Prokof'eva, L.S. *Krest'ianskaia obshchina v Rossii vo vtoroi polovine XVIII – pervoi polovine XIX veka (na materialakh votchin Sheremetevykh)*, Leningrad, 1981.

Raeff, M. 'Pugachev's Rebellion', in R. Forster and J.P. Greene, eds, *Preconditions of Revolution in Early Modern Europe*, Baltimore, MD, and London, 1970, pp. 161–201.

Raffensperger, Christian A. *Kingdom of Rus'*, Kalamazoo, MI, 2017.

Raleigh, Donald J. 'Revolutionary Politics in Provincial Russia: The Tsaritsyn "Republic" in 1917', *Slavic Review*, vol. 40, no. 2, 1981, pp. 194–209.

BIBLIOGRAPHY

Raleigh, Donald J. *Revolution on the Volga: 1917 in Saratov*, Ithaca, NY, and London, 1986.

Raleigh, Donald J. *Experiencing Russia's Civil War: Politics, Society, and Revolutionary Culture in Saratov, 1917–1922*, Princeton, NJ, and Oxford, 2002.

Ransel, David L. *A Russian Merchant's Tale: The Life and Adventures of Ivan Alekseevich Tolchenëv, based on his Diary*, Bloomington, IN, 2009.

Razmolodin, M.L. *Chernosotennoe dvizhenie v Iaroslavle i guberniiakh Verkhnego Povolzh'ia v 1905–1915 gg.*, Iaroslavl', 2001.

Rechnoi transport SSSR 1917–1957: sbornik statei o razvitii rechnogo transporta SSSR za 40 let, Moscow, 1957.

Repinetskii, A.I. and M.A. Rumiantseva. *Gorodskoe naselenie Srednego Povolzh'ia v poslevoennoe dvatsatiletie 1945–1965 gg.*, Samara, 2005.

Riabov, Oleg. ' "Let us Defend Mother Volga": The Material Symbol of the River in the Discourse of the Stalinist Battle', *Women in Russian Society*, no. 2, 2015, pp. 11–27.

Riabov, Oleg. ' "Mother Volga" and "Mother Russia": On the Role of the River in Gendering Russianness', in Jane Costlow and Arja Rosenholm, eds, *Meanings and Values of Water in Russian Culture*, Abingdon, 2017, pp. 81–97.

Riabtsev, A.L. *Gosudarstvennoe regulirovanie vostochnoi torgovli Rossii v XVII – XVIII vekov*, Astrakhan', 2012.

Robbins, Richard G. *Famine in Russia 1891–1892: An Imperial Government Responds to a Crisis*, New York and London, 1975.

Rodin, F.N. *Burlachestvo v Rossii*, Moscow, 1975.

Rogozhina, N. 'The Caspian: Oil Transit and Problems of Ecology', *Problems of Economic Transition*, vol. 53, no. 5, 2010, pp. 86–93.

Romaniello, Matthew P. 'Controlling the Frontier: Monasteries and Infrastructure in the Volga Region, 1552–1682', *Central Asian Survey*, vol. 19, nos 3–4, 2000, pp. 429–43.

Romaniello, Matthew P. *The Elusive Empire: Kazan and the Creation of Russia 1552–1671*, Madison, WI, and London, 2012.

Rorlich, Azade-Ayşe. 'History, Collective Memory and Identity: The Tatars of Sovereign Tatarstan', *Communist and Post-Communist Studies*, vol. 32, no. 4, 1999, pp. 379–96.

Rorlich, Azade-Ayşe. *The Volga Tatars: A Profile in National Resilience*, Stanford, CA, 2017.

Rossman, Vadim. 'Lev Gumilev, Eurasianism and Khazaria', *East European Jewish Affairs*, vol. 32, no. 1, 2002, pp. 30–51.

Rotermel', Boris. *Tverskie nemtsy. Die Russlanddeutschen von Twer*, Tver', 2011.

Ruder, Cynthia. 'Water and Power: The Moscow Canal and the "Port of Five Seas" ', in Jane Costlow and Arja Rosenholm, eds, *Meanings and Values of Water in Russian Culture*, Abingdon, 2017, pp. 175–88.

Russkie monastyri: Sredniaia i Nizhniaia Volga, Moscow, 2004.

Rywkin, Michael. 'The Prikaz of the Kazan Court: First Russian Colonial Office', *Canadian Slavonic Papers*, vol. 18, no. 3, 1976, pp. 293–300.

Samatova, Ch.Kh. *Imperskaia vlast' i Tatarskaia shkola vo vtoroi polovine XIX – nachale XX veka*, Kazan', 2013.

Samogorov, V., V. Pastushenko, A. Kapitonov and M. Kapitonov. *Iubileinyi Ul'ianovsk*, Ekaterinburg, 2013.

Samsonov, A.M. *Stalingradskaia bitva*, 2nd edition, Moscow, 1968.

Sanukov, K.N. 'Repressii 1930-kh i krest'ianstvo Mariiskoi ASSR', in V.A. Iurchenkov, ed., *Krest'ianstvo i vlast' Srednego Povolzh'ia*, Saransk, 2004, pp. 352–57.

Saunders, J.J. *The History of the Mongol Conquests*, London, 1971.

Semenova, E.Iu. *Mirovozzrenie gorodskogo naseleniia Povolzh'ia v gody Pervoi Mirovoi Voiny (1914 – nachalo 1918 gg.): sotsial'nyi, ekonomicheskii, politicheskii aspekty*, Samara, 2012.

Semenova, E.Iu. *Rossiiskii gorod v gody Pervoi Mirovoi Voiny (na materialakh Povolzh'ia)*, Samara, 2016.

Seniutkin, S.B. *Istoriia Tatar Nizhegorodskogo Povolzh'ia s poslednei treti XVI do nachala XX vv. (istoricheskaia sud'ba misharei Nizhegorodskogo kraia)*, Moscow and Nizhnii Novgorod, 2009.

Shabanov, Andrey. *Art and Commerce in Late Imperial Russia: The Peredvizhniki, a Partnership of Artists*, London, 2019.

BIBLIOGRAPHY

Sharifzhanov, Izmail. 'The Parliament of Tatarstan, 1990–2005: Vain Hopes, or the Russian Way towards Parliamentary Democracy in a Regional Dimension', *Parliaments, Estates and Representation*, vol. 27, no. 1, 2007, pp. 239–50.

Sharoshkin, N.A. *Promyshlennost' i rabochie Povolzh'ia v 1920-e gody*, Penza, 2008.

Shaw, D.J.B. 'Southern Frontiers of Muscovy, 1550–1700', in J.H. Bater and R.A. French, eds, *Studies in Russian Historical Geography*, vol. 1, London, 1983, pp. 118–43.

Shaydullin, Rafael. 'Peasantry and the State and the Tatar Autonomous Soviet Socialist Republic', in *The History of the Tatars since Ancient Times*, vol. 7, Kazan', 2017, pp. 279–92.

Shaykhudinov, Renat. 'The Terrorist Attacks on the Volga Region, 2012–13: Hegemonic Narratives and Everyday Understandings of (In)Security', *Central Asian Survey*, vol. 37, no. 1, 2018, pp. 50–67.

Shepard, Jonathan. 'The Origins of Rus' (c. 900–1015)', in Maureen Perrie, ed., *The Cambridge History of Russia*, vol. 1, *From Early Rus' to 1698*, Cambridge, 2006, pp. 47–72.

Simms, James. 'The Economic Impact of the Russian Famine of 1891–92', *The Slavonic and East European Review*, vol. 60, no. 1, 1982, pp. 63–74.

Slepnev, I.N. 'Vliianie sozdaniia zheleznodorozhnoi seti na tovarizatsiiu zernovogo proizvodstva Rossii (vtoraia polovina XIX v.)', in *Povolzh'e v sisteme vserossiiskogo rynka: istoriia i sovremennost'*, Cheboksary, 2000, pp. 53–68.

Smith, Alison K., 'Peasant Agriculture in Pre-Reform Kostroma and Kazan' Provinces', *Russian History*, vol. 26, no. 4, 1999, pp. 355–424.

Smith, Alison K. 'Provisioning Kazan': Feeding the Provincial Russian Town', *Russian History*, vol. 30, no. 4, 2003, pp. 373–401.

Smith, R.E.F. and David Christian. *Bread and Salt: A Social and Economic History of Food and Drink in Russia*, Cambridge, 1984.

Smith-Peter, Susan. *Imagining Russian Regions: Subnational Identity and Civil Society in Nineteenth-Century Russia*, Leiden and Boston, MA, 2018.

Sokolov, K.I. *Proletarii protiv 'Proletarskoi' vlasti: protestnoe dvizhenie rabochikh v Tverskoi gubernii v kontse 1917–1922 gg.*, Tver', 2017.

Sokolova, N.B. 'Khoziaistvenno-torgovaia deiatel'nost' Makar'evskogo zheltovodskogo monastyria', in *Verkhnee i Srednee Povolzh'e v period feodalizma: mezhvuzovskii sbornik*, Gor'kii, 1985, pp. 43–49.

Stampfer, Shaul. 'Did the Khazars Convert to Judaism?', *Jewish Social Studies*, vol. 19, no. 3, 2013, pp. 1–72.

Starikov, S.V. *Velikaia reka Rossii na rubezhe XIX–XX vekov: Volga ot Nizhnego Novgoroda do Kazani na starinnykh otkrytkakh*, Ioshkar-Ola, 2009.

Stites, Richard. *Russian Popular Culture: Entertainment and Society since 1900*, Cambridge, 1992.

Stites, Richard. *Serfdom, Society, and the Arts in Imperial Russia: The Pleasure and the Power*, New Haven, CT, and London, 2005.

Sultanov, F.M. *Islam i Tatarskoe natsional'noe dvizhenie v rossiiskom i mirovom musul'manskom kontekste: istoriia i sovremennost'*, Kazan', 1999.

Sunderland, Willard. *Taming the Wild Field: Colonization and Empire of the Russian Steppe*, Ithaca, NY, and London, 2014.

Sverdlova, L.M. *Kazanskoe kupechestvo: sotsial'no-ekonomicheskii portret (kon. XVIII – nach. XX v.)*, Kazan', 2011.

Swift, E. Anthony. *Popular Theater and Society in Tsarist Russia*, Berkeley, CA, and London, 2002.

Syzranov, A.V. *Islam v Astrakhanskom krae: istoriia i sovremennost'*, Astrakhan', 2007.

Tagirova, N.F. *Rynok Povolzh'ia (vtoraia polovina XIX – nachalo XX vv.)*, Moscow, 1999.

Taimasov, L. 'Etnokonfessional'naia situatsiia v Kazanskoi gubernii nakanune burzhuaznykh reform', in K. Matsuzato, ed., *Novaia Volga i izuchenii etnopoliticheskoi istorii Volgo-Ural'skogo regiona: Sbornik statei*, Sapporo, 2003, pp. 106–37.

Tarlovskaia, V.R. 'Torgovye krest'iane Povolzh'ia v kontse XVII – nachale XVIII veka', *Istoriia SSSR*, 1983, no. 2, pp. 149–58.

Tarlovskaia, V.R. *Torgovlia Rossii perioda pozdnego feodalizma*, Moscow, 1988.

Tatary Samarskogo kraia: istoriko-etnograficheskie i sotsial'no-ekonomicheskie ocherki, edited by Sh.Kh. Galimov, Samara, 2017.

BIBLIOGRAPHY

Toriyama, Yusuke. 'Images of the Volga River in Russian Poetry from the Reign of Catherine the Great to the End of the Napoleonic Wars', *Study Group on Eighteenth Century Russia*, 2013; www.sgecr.co.uk/newsletter2013/toriyama.html

Trubina, Elena. 'The Reconstructed City as Rhetorical Space: The Case of Volgograd', in Tovi Fenster and Haim Yacobi, *Remembering, Forgetting and City Builders*, Farnham, 2010, pp. 107–20.

Tsybin, V.M. and E.A. Ashanin. *Istoriia Volzhskogo kazachestva*, Saratov, 2002.

Tuna, Mustafa. *Imperial Russia's Muslims: Islam, Empire, and European Modernity 1788–1914*, Cambridge, 2015.

Tushkanov, L.V. *Chastnovladel'cheskoe khoziaistvo Saratovskoi gubernii v poreformennyi period (1861–1904 gg.)*, Volgograd, 2010.

Urazmanova, Raufa. 'Ceremonies and Festivals', in *The History of the Tatars since Ancient Times*, vol. 6, Kazan', 2017, pp. 686–94.

Urazmanova, Raufa. 'Festive Culture of Tatars in Soviet Times', in *The History of the Tatars since Ancient Times*, vol. 7, Kazan', 2017, pp. 464–72.

Urazmanova, Raufa. 'The Transformation of the Tatar Holiday Culture in the Post Soviet Period', *The History of the Tatars since Ancient Times*, vol. 7, Kazan', 2017, pp. 688–96.

Usova, T.V. 'Composition of Sturgeon Fry Migrating from Spawning Areas in the Lower Volga', *Russian Journal of Ecology*, vol. 36, no. 4, 2005, pp. 288–89.

Valeev, R.M. 'K voprosu o tovarno-denezhnykh otnosheniiakh rannikh Bulgar (VIII–X vv.)', in *Iz istorii rannikh Bulgar*, Kazan', 1981, pp. 83–96.

Valeeva-Suleimanova, G.F. 'Problemy izucheniia iskusstva Bulgar zolotoordskogo vremeni (vtoraia polovina XIII – nachalo XV vv.), in *Iz istorii Zolotoi Ordy*, edited by A.A. Arslanova and G.F. Valeeva-Suleimanova, Kazan', 1993, pp. 61–66.

Valkenier, Elizabeth. *Ilya Repin and the World of Russian Art*, New York, 1990.

Vásáry, István. 'The Jochid Realm: The Western Steppe and Eastern Europe', in Nicola Di Cosmo, Allen J. Frank and Peter B. Golden, eds, *The Cambridge History of Inner Asia: The Chinggisid Age*, Cambridge, 2009, pp. 67–86.

Vasil'ev, F.V., S.V. Dmitrievskii and M.Iu. Pukhov. *Russkoe sel'skoe naselenie iugo-vostochnykh uezdov Nizhegorodskoi gubernii v XIX – nachale XX v. (poseleniia, krest'ianskaia usad'ba, zhilishche)*, Nizhnii Novgorod, 2006.

Veshchev, P.V., G.I. Guteneva and R.S. Mukhanova. 'Efficiency of Natural Reproduction of Sturgeons in the Lower Volga under Current Conditions', *Russian Journal of Ecology*, vol. 43, no. 2, 2012, pp. 142–47.

Vinogradov, A.A. *Staroobriadtsy Simbirsko-Ul'ianovskogo Povolzh'ia serediny XIX – pervoi treti XX veka*, Ul'ianovsk, 2010.

Viola, Lynne. *Peasant Rebels under Stalin: Collectivization and the Culture of Peasant Resistance*, Oxford, 1996.

Vishlenkova, E.A., S.Iu. Malysheva and A.A. Sal'nikova. *Kul'tura povsednevnosti provintsial'nogo goroda: Kazan' i Kazantsy v XIX–XX vekakh*, Kazan', 2008.

Vodolagin, M.A. *Ocherki istorii Volgograda 1589–1967*, Moscow, 1968.

Voliani, A. *Elektrogigant na Volge*, Leningrad, 1934.

Volkov, I.V. 'Stolypinskaia agrarnaia reforma v Iaroslavskoi gubernii', in *Ocherki istorii Iaroslavskogo kraia*, Iaroslavl', 1974.

Volkov, M.Iu. *Goroda Verkhnego Povolzh'ia i severo-zapada Rossii. Pervaia chetvert' XVIII v.*, Moscow, 1994.

Vorob'ev, I.I. *Tatary Srednego Povolzh'ia i Priural'ia*, Moscow, 1967.

Vovina, Olessia P. 'Building the Road to the Temple: Religion and National Revival in the Chuvash Republic', *Nationalities Papers*, vol. 28, no. 4, 2000, pp. 695–706.

Wade, Rex A. and Seregny, Scott J., *Politics and Society in Provincial Russia: Saratov, 1590–1917*, Columbus, OH, 1989.

Waldron, Peter. *The End of Imperial Russia, 1855–1917*, Basingstoke and London, 1997.

Waldron, Peter. *Between Two Revolutions: Stolypin and the Politics of Renewal in Russia*, London, 1998.

Werth, Paul W. *At the Margins of Orthodoxy: Mission, Governance, and Confessional Politics in Russia's Volga-Kama Region, 1827–1905*, Ithaca, NY, and London, 2002.

BIBLIOGRAPHY

Werth, Paul W. 'Coercion and Conversion: Violence and the Mass Baptism of the Volga Peoples, 1740–55', *Kritika*, vol. 4, no. 8, 2003, pp. 543–69.

Werth, Paul W. and Radik Iskhakov. 'Christian Instruction and Movements of Christened Tatars for their Return to Islam in the Pre-Reform Period', *The History of the Tatars since Ancient Times*, vol. 6, Kazan', 2017, pp. 538–43.

Wigglesworth-Baker, Teresa. 'Language Policy and Post-Soviet Identities in Tatarstan', *Nationalities Papers*, vol. 44, no. 1, 2016, pp. 20–37.

Williams, Christopher. 'Tatar Nation Building since 1991: Ethnic Mobilisation in Historical Perspective', *Journal of Ethnopolitics and Minority Issues in Europe*, vol. 10, no. 1, 2011, pp. 94–123.

Yusupova, Guzal. 'The Islamic Representation of Tatarstan as an Answer to the Equalization of the Russian Regions', *Nationalities Papers*, vol. 44, no. 1, 2016, pp. 38–54.

Zagidullin, Ildus. 'The Movement of Converting Baptised Tatars to Islam in the Beginning of the 20th Century', in *The History of the Tatars since Ancient Times*, vol. 7, Kazan', 2017, pp. 130–36.

Zaitsev, I. *Astrakhanskoe khanstvo*, Moscow, 2004.

Zaitsev, I. 'The Astrakhan Khanate', in *The History of the Tatars since Ancient Times*, vol. 4, Kazan', 2017, pp. 197–202.

Zaitsev, I. 'The Astrakhan Tatars', in *The History of the Tatars since Ancient Times*, vol. 4, Kazan', 2017, pp. 784–86.

Zaitsev, I., 'The Astrakhan Yurt', in *The Golden Horde in World History*, Kazan', 2017, pp. 747–55.

Zaitsev, M.V. *Saratovskaia gorodskaia duma (1871–1917)*, Saratov, 2017.

Zakirova, A. 'Bone Carving', in *Great Bolgar*, Kazan', 2015, pp. 172–77.

Zeisler-Vralsted, Dorothy. 'The Aesthetics of the Volga and the National Narrative in Russia', *Environment and History*, vol. 20, 2014, pp. 93–122.

Zeisler-Vralsted, Dorothy. *Rivers, Memory and Nation-Building: A History of the Volga and the Mississippi Rivers*, New York, 2014.

Zetkina, I.A. *Natsional'noe prosvetitel'stvo Povolzh'ia: formirovanie i razvitie*, Saransk, 2003.

Zhivkov, Boris. *Khazaria in the Ninth and Tenth Centuries*, translated by Daria Manova, Leiden, 2015.

Zimonyi, István. *The Origins of the Volga Bulgars*, Szeged, 1990.

Zorin, A.N. *Goroda i posady dorevoliutsionnogo Povolzh'ia*, Kazan', 2001.

INDEX

INDEX

INDEX

Bulavin, Kondratii, 77
Buldyra, village, 251
Bulgaria, xiv, 19
Bunin, Ivan, 203
Bunting, Nikolai von, 232–3
Burnasheva, Zahida (Iffat Tutash), 304
Burton, Reginald, 158
Byzantine Empire, 14, 17, 19, 23, 27, 36

Cairns, Andrew, 257
canals, in imperial Russia, xvi, 3, 5, 125, 132, 175, 177, 179–80, 182, 188; in the Soviet Union, 132, 180, 264, 284, 306–10; see also Moscow-Volga canal; Volga-Don canal
Captain's Daughter, The, novel, 91
car industry, xvii, 4, 281–2
Caspian flotilla, 92
Caspian Sea, 1, 5–7, 18, 20, 26–9, 31–2, 37–39, 41, 47–8, 70, 72–3, 82, 104, 132, 137, 171, 174–5, 177, 181, 270, 287, 311–14
Catherine II, empress of Russia, 4, 6, 8, 77, 81, 86–7, 89, 93–5, 98–107, 110–13, 119, 132–4, 158–60, 165, 167–9, 180, 193, 199, 280
Catherine Pavlovna, sister of Alexander I, 166
Catholics, 7, 109, 111, 146–7, 160–1, 163, 224
Caucasus, mountain region, 14, 19–19, 22, 27–8, 32, 36, 38–9, 55, 66, 87–8, 90, 105, 132, 150, 187, 195, 220, 231, 240, 256, 270, 295, 297
caviar, 5, 37, 39, 174, 208, 312, 314; see also sturgeon
Cayley, Edward, 181
Central Asia, 6, 18–19, 21–3, 32, 37–9, 127, 151, 171–2, 177, 187, 189, 191, 224, 277
Chagatai khanate, 38
Charla, see Tsarevokokshaisk
Cheboksarov, Nikolai, 261
Cheboksary, district, 117, 119, 161, 257
Cheboksary, town, 3, 53, 68, 103, 113, 157, 163–4, 175, 186, 235, 281, 289, 292, 296, 300, 304
Cheboksary GES, 292, 308
Chechen Autonomous Soviet Socialist Republic, 293
Chechnya, 293
Chekhov, Anton, 203, 207–8
Chekhova, Mariia, 207
Cheremiss, people, see Mari
Cherepovets, 313
Cherkassk, 72
Chernetsov, Grigorii, 199–202
Chernetsov, Nikanor, 199–202
Chernigov regiment, 90

Chernobyl, nuclear plant, 313
Chernogubov, V.A., 259
Chernyi Iar, 57–8, 74–5, 80
Chernyshevsky, Nikolai, 168
Chesme, battle of, 111
chess, 154, 170
Chichagov, Dmitrii, 203
China, 6, 18, 21, 37–9, 41, 56, 100, 109, 178–9, 187, 189, 191
Chinese, people, 55, 190, 242
Chinggis Khan, khagan of the Mongol empire, 29, 32–3, 248
Chirikov, Evgenii, 198–9
Chistopol, district, 113–15, 144, 153, 161, 251
Chistopol, town, 159, 165, 168, 221
Chkalov, see Orenburg
Chkalov, Valerii, xv
Chokri, Gali, 47, 226
cholera, 122, 126–31, 213, 242
Chuikov, Vasilii, 271–2
Church of the Transfiguration (at Volga source), 205
Chuvash, people, 2, 20, 53–6, 61–2, 75, 83, 90, 94, 104, 107–10, 112–13, 117, 134, 137, 139, 141, 143–7, 150–5, 159, 186, 216, 223, 234–5 245, 252, 261, 281, 286, 288, 297, 299, 300–2, 304; see also identity, Chuvash; peasants, Chuvash
Chuvash Autonomous Soviet Socialist Republic, 289, 295, 299, 304
Chuvash Republic, 3, 53, 295, 300, 302
Chuvash Revolutionary Tribunal, 304
Circassians, 190
Civil War, Russian, xvi, xvii, 4, 8, 229–30, 236–48, 251–2, 268, 286, 288–9, 297
Clarke, Edward, 111–12
Cleaning Beetroots, painting, 200
Clergy, Russian Orthodox, see religion, Orthodox clergy
climate, 25, 126, 133, 150, 177; see also agriculture; famine
coinage, trade, 5, 13, 16, 18, 22, 23–6, 36–7
collectivization, see peasants, collectivization
Collins, Samuel, 116
Colorado, river, 311
Communist Party, Italian, xvii, 281
Communist Party, Soviet Union, xvi, 250–52, 260, 262, 290–2
Communist Party, Tatar ASSR, 290–1
Constantinople, 26, 27, 81, 118, 225; see also Istanbul
Constituent Assembly, 233–6
conversions, see religion, conversions
Cossacks, xv, 8, 58–9, 66, 68, 91, 93–4, 98–9, 105, 113–14, 158, 173, 190, 277; Astrakhan

366

INDEX

Cossack host, 91, 94, 178; in Civil War, 240; Don Cossack host, 72–3, 77, 82, 88; organization, 70–2, 88; origins, 70; Iaik Cossack host, 79, 87–8; in imperial army, 71, 88–9, 219; Orenburg Cossack host, 88; in revolt, 8, 66, 72–85, 96, 118, 140; in revolutions, 219, 235–6; in the Soviet Union, 71, 249, 277; Terek Cossack Host, 87–8, 124; Volga Cossack host, 72, 87, 94; Zaporozhian Cossack host, 88

Courland, 230

Crimea, 17, 32, 38, 53, 111–12, 132–3, 240, 270, 302

Crimean khanate, 42

Crimean Tatars, *see* Tatars, Crimean

Crimean War (1853–56), 143, 182, 210

Cuman, people, 32

Custine, Astolphe-Louis-Léonor, Marquis de, 152, 157

Czech Legion, 237, 239

Dagestan, 73, 301

Danube, river, 196–7

Davletkulovo, village, 277

Declaration of the Rights of the Peoples of Russia, 286

Declaration of the Sovereignty of the Tatar Republic, 293

Demianov, G.P., 207

Denikin, Anton, 240, 286

Denmark, 22

Derzhavin, Gavriil, 86

Desnitsky, S., 96

Dimitrov, Georgi, xiv

Dimitrovgrad, *see* Melekess

Dimsdale, Elizabeth, 139

Dmitrii (Donskoi), prince of Moscow and then grand prince of Vladimir, 38, 43

Dmitrii (Mikhailovich), prince of Tver, 43

Dnieper, river, 14, 26–7, 70, 88, 104, 230, 285

Dokuchaev, Vasilii, 134

Dolgoruky, G.F., 62

Dombrovsky, Viacheslav, 262

Don, region, 83, 183, 240

Don, river, 4, 26, 38, 70, 72–5, 84, 91, 99–100, 104, 132, 175, 180, 197, 240, 274–5

Don canal, *see* Volga-Don canal

Dorpat, 167

Down the Volga, guidebook, 206

Drankov, Aleksandr, 84

Dresden, 190

Dresden, battle of, 90

Drozhdov, Filaret, archbishop of Moscow, 113

Dubovka, Cossack *stanitsa*, 72, 80, 87, 158, 187

East India Company, 59

education, 19–20, 102, 111, 115–16, 134, 166–8, 221–5, 299; *see also* language policy; Russo-Tatar schools; Russo-Tatar Teachers' Seminary; Tatars, schools

Efrosin, monk, 195

Egorov, Platon, 90

Egypt, 36

Ehrenburg, Ilia, 266

Ekimov, Pavel, 185

Eletma, 306

Elizabeth, empress of Russia, 109–11, 118

Elton, lake, 65, 143, 178

Eltsin, Boris, *see* Yeltsin

Elvov, N., 260

emancipation of the serfs, *see* peasants, emancipation of the serfs

Enganaevo, village, 255

Engels, town, *see* Pokrovsk

Engels, Friedrich, xiv

Ermak, Cossack leader, 71–2, 81, 85

Ethnographic Museum, St Petersburg, 134, 328 n. 47

Evening Post, The, painting, 262

Evening on the Volga, painting, 202

'Evening on the Volga', poem, 194, 204

Evreinov, D.P. 181

Evsevev, M.E., 153

Evtushenko, Evgenii, 285

Extraordinary All-Russian Commission Investigating the Famine, 247

Fabritius, Ludvig, 73–5

factories, *see* industry

Fainzilberg, Ilia (Ilf), 283

fairs, in imperial Russia, 3, 39, 63–4, 131, 145, 153, 157, 169, 179, 185, 186–91, 207, 225

famine: in the Civil War, 242, 246–8; in imperial Russia, 128, 150, 199, 212–13, 306; in the Soviet Union, 134, 213, 249–50, 253, 256–7, 304, 306

Father's House, novel, 199

Filaret (Drozhdov), archbishop of Moscow, 113

Finland, 21, 22, 123, 262

First World War, xvi, 217, 220, 229–32, 248

fishing/fish trade, 5–6, 17, 21, 37, 39, 41, 63–5, 69, 71–2, 94, 133, 151, 163, 174, 177–9, 241, 307, 311–14, 316

Foreign Ministry, Russian, 165

Foreign Office (*posolskii prikaz*), Russian, 72

Forster, George, 92

Forsyth, William, 188–9

France, 108, 130, 134

Frankfurt, 190

367

INDEX

INDEX

Ibn Rusta (Ibn Rustah), 17, 20, 26
Ibragimova, Giuzel, 264–5
Ibrakhimov, Nikolai, 168
'Ice is Moving, The', story, 198
Iceland, 25
Idel-Ural state, 288
identity: Chuvash, 20, 302; Russian, 7, 9, 193,
    195–7, 204–9; Tatar, 9, 20, 47, 54, 146,
    209–10, 220–2, 224–6, 286–8, 299–305;
    see also language policy
Ievlev, Ivan, 120
Ikrainoe, village, 314
Ilf, Ilia (Fainzilberg), 283
Ilham Ghali, khan of Kazan (1479–84,
    1485–87), 44
Illustrated Guide to the Volga, An, 207
Ilmen, lake, 26
Ilminsky, Nikolai, 116, 222
Imperial Geographical Society, Russian, 134
'In the Year of Seventy-Three', poem, 83
India, 21–2, 37, 59, 68, 84, 109, 126, 171–2,
    174, 278; *see also* traders, Indian
industry: in imperial Russia, 6, 25, 61, 68, 77,
    79, 89–90, 119, 157, 161, 177–9, 191, 218,
    259; in revolutions and civil war, 219, 232,
    239–41, 248; in the Russian Federation,
    296, 314–15; in the Soviet Union, 6, 259,
    267–8, 280–2, 292, 296, 297, 299, 307,
    309, 312–13
Innopolis, 303
Instruction, by Catherine II, 94
Ioshkar-Ola, xv, *see* Tsarevokokshaisk
Iosif (Udalov), bishop of Chistopol, 263
Iran, *see* Persia, 151
Iraq, 151
Irbit, 264
Ireland, 24, 213
Irkutsk, 255
Isaev, Mukhamet, 160
Isfahan, 21
Isker, 71
Islam, *see* mosques; Muslims; religion,
    conversions
Ismail, Nogai leader, 46
Istomina, Enessa, 184
Italian peninsula/Italy, 193, 200, 202
Italians, 274
Itil, 2, 4, 15–19, 28, 36
Iurii, prince of Moscow, 42–3
Iurii II, prince of Vladimir-Suzdal, 29
Iusupov, Dmitrii, 121
Ivan, lake, 197
Ivan I, prince of Moscow and grand prince of
    Vladimir, 43

Ivan III, grand prince of Moscow and of all
    Russia, 43–4
Ivan IV, grand prince of Moscow, tsar of Russia,
    3, 5, 31, 40, 44–6, 48, 55–7, 59–60, 64, 71,
    86, 89, 104, 131, 172, 288, 301
Ivanovka, village, 121
Izhevsk, 289

Jani Beg, khan, 34
Japan, 261
Jenkinson, Anthony, 46–7
Jerusalem, 81
jewellery, 18, 23–5, 36, 190
Jews, 7, 17, 19, 28, 98, 101, 107, 110, 121,
    161–2, 164, 172–3, 182, 187, 190, 220,
    230–1, 234; *see also* Khazaria, religion
Jochi, Mongol khan, 32
Joseph (Joseph ben Aaron), ruler of Khazaria,
    28

Kaban, lake, 170
Kadet party, 225, 236–7
Kalach, 275
Kaliazin, 309
Kalinin, xvii, 269, 276; *see also* Tver
Kalinin, Mikhail, xvii
Kalka, battle of, 32
Kalmyks, people, xvii, 4, 6–7, 55–8, 65–6,
    69–71, 73, 90–1, 94, 98–101, 111, 133,
    154, 159, 190, 245, 281
Kama, river, 1, 39, 52, 72, 93, 109, 113, 178,
    181–2, 187, 207, 239, 311
Kamyshin, 178, 231, 241
Kamyshla, region, 277
Kanoat, Mumin, 303
Kappel, Vladimir, 239
Karamzin, Nikolai, 166, 204
Karelians, people, in Tver, 262
Kashin, 42
Kashirin, region, 255
Kassel, 275
Kassimov, 306
Kataev, Evgenii (Petrov), 283
Kazakh Autonomous Soviet Socialist Republic,
    xv
Kazakhs, people, 55–7, 62, 65, 297
Kazakhstan, xv, 6, 53, 65, 256, 287, 291–2,
    297, 311, 314
Kazan, galley, 103
Kazan, khanate, xv, 31, 33, 159, 287–8;
    conquest of, 5, 7, 31, 45–8, 51–2, 55, 292;
    relations with Moscow, 7, 31, 41, 44–8;
    religion, 7, 39; trade, 5, 39
Kazan, province, 52, 54, 62, 83, 90, 94, 98,
    100, 107, 109–10, 112–14, 120–1, 126–7,

INDEX

Morkvash, village, 200
Morozov textile works, 219
Moscow (Muscovy), principality, relations with Golden Horde, 7, 30, 32, 34–5, 37–8, 42–3; relations with Kazan khanate, 7, 31, 41, 44–8; relations with Rus principalities, 42–3; rise of, 27, 41–4; *see also* army, Muscovite
Moscow, province, 91
Moscow, town, 1–3, 35, 59–60, 67, 75, 79–85, 89, 92, 102–3, 118–19, 124–7, 129, 131–2, 139, 158, 162, 165, 169–70, 172, 176, 178–80, 182–3, 189, 196, 202, 218, 237, 240, 260, 266–70, 279, 283–4, 293, 308
Moscow University, 96, 167
Moscow-Volga canal, 284, 308–10
Moskva, river, 37
mosques, 17, 19–20, 22–3, 36–7, 48, 56, 102, 104, 107, 109–13, 145, 147–8, 165, 171, 189, 221–2, 262–3, 283, 299–301, 303; *see also* Bolgar state, religion; Muslims; Tatars, religion
Mother Volga, statue, 3, 309
Motherland Calls, The, statue, 278–79
Muhammad Amin, khan of Kazan (1487–95), 44
Muhammad Uzbeg, khan of the Golden Horde (1313–41), 42–3
Munro-Butler-Johnstone, H.A., 131, 166, 177
Muqtadir (al-Muqtadir), caliph, 16, 20
Murasa, village, 120
Murom, 306
Muscovy Company, 46
Muslim Military Congress, Second, 288
Muslim Spiritual Assembly, xv, 111
Muslims, 7, 17–20, 28, 47–8, 51, 53, 55, 101, 105, 106–16, 119, 144–6, 161, 221–4, 255, 262, 293, 298, 300–1, 303; *see also* mosques; Tatars, religion
musketeers, *see* army, musketeers
Mutiashin, peasant, 253
My Children, novel, 137

Nachalova, village, 254
Nagorno-Karabakh, 297
Napoleon Bonaparte, 83, 102, 196
Napoleonic Wars, 88, 112, 143, 169, 268, 282
Nasiri, Kayyum, 225
National Art Gallery, Ioshkar-Ola, 296
National Assembly of the Muslims of Inner Russia and Siberia, 287
Naumov, Aleksandr, 196, 213, 268
Naval Ministry, Russian, 195, 199; *see also* admiralty (ministry)
navigation, on the Volga, *see* steamships; transportation, shipping

Navy, Russian, on the Volga, 88, 92–3
Neidgart, P.P., 206
Nekrasov, Nikolai, 192–93, 197–8, 304
Nektarius (Trezvinsky), bishop of Iaransk, 263
Nemirovich-Danchenko, Vasilii, 206
Nesmelov, V.I., 263
Neufchâtel, 190
Neva, river, 103, 179
Nevsky, Alexander, prince of Novgorod, 34, 41
New Economic Policy (NEP), 250–1, 283
New Sarai, 36, 38; *see also* Sarai
Nicholas I, tsar, 93, 112, 119–20, 161, 199
Nicholas II, tsar, 231–2
Nikitin, Afanasii, 68, 84
Nikon, Patriarch of Moscow, 75, 117–18
Nikulina, Liubov, 170
Nile, river, 1, 31, 195
Nizhnii Novgorod, province, 52, 63, 107, 120, 145, 152, 160–1, 186, 231, 244, 259
Nizhnii Novgorod, town, xv, 1, 3, 6, 21, 29, 31–2, 35, 38, 41, 63–4, 68, 75, 79, 91, 93, 100, 103–4, 124–8, 131, 156–7, 159, 160–1, 167–70, 176, 179, 182, 184–91, 195–6, 198, 207–8, 213–15, 225, 232, 234, 237, 239, 259, 283, 295; *see also* Gorkii
Nizhnii Novgorod GES, 308
nobles: in revolutions and civil war, 239, 243–4, 246; Russian, 42–3, 60–2, 65, 67, 74, 76–7, 79–80, 94, 96–7, 99, 107–8, 113–15, 119, 139–40, 142–3, 149, 163, 168–9, 180, 182, 185, 193, 200, 209–14, 217, 236, 248; Tatar, 42, 51, 54, 59, 62, 107, 108, 110
Nogai horde, 42, 45–6, 55
Nogais, people, 6, 45, 55, 57, 62, 65–6, 70–1, 91, 99, 190
Norilsk, 264
Normanist theory, 24
Norway, 22, 24
Novgorod, principality, 27, 29, 37, 41–3
Novgorod, town, 22, 42
Novinka, village, 254
Novocherkassk, 72
Novoe Ermakovo, village, 277
Novoe Nikitino, 255
Novorossiisk, 240
Nyenskans, xvi

Oblomov, novel, 194–5
Obolensky-Nogotkov, Ivan, 296
Obraztsov, Vasilii, 119
Odoevsky, Iakov, 76
Office of Converts, 109–10
officials: in imperial Russia, 59–60, 64, 73–7, 79–80, 86–8, 91, 93–7, 99–100, 105, 113,

372

INDEX

INDEX

INDEX